Springer Optimization and Its Applications

Volume 22

Aims and Scope

Optimization has continued to expand in all directions at an astonishing rate. New algorithmic and theoretical techniques are continually developing and the diffusion into other disciplines is proceeding at a rapid pace, with a spot light on machine learning, artificial intelligence, and quantum computing. Our knowledge of all aspects of the field has grown even more profound. At the same time, one of the most striking trends in optimization is the constantly increasing emphasis on the interdisciplinary nature of the field. Optimization has been a basic tool in areas not limited to applied mathematics, engineering, medicine, economics, computer science, operations research, and other sciences.

The series **Springer Optimization and Its Applications** (SOIA) aims to publish state-of-the-art expository works (monographs, contributed volumes, textbooks, handbooks) that focus on theory, methods, and applications of optimization. Topics covered include, but are not limited to, nonlinear optimization, combinatorial optimization, continuous optimization, stochastic optimization, Bayesian optimization, optimal control, discrete optimization, multi-objective optimization, and more. New to the series portfolio include Works at the intersection of optimization and machine learning, artificial intelligence, and quantum computing.

Volumes from this series are indexed by Web of Science, zbMATH, Mathematical Reviews, and SCOPUS.

More information about this series at http://www.springer.com/series/7393

Urmila M. Diwekar

Introduction to Applied Optimization

Third Edition

 Springer

Urmila M. Diwekar
Vishwamitra Research Institute
Crystal Lake, IL, USA

ISSN 1931-6828   ISSN 1931-6836   (electronic)
Springer Optimization and Its Applications
ISBN 978-3-030-55406-4   ISBN 978-3-030-55404-0   (eBook)
https://doi.org/10.1007/978-3-030-55404-0

Mathematics Subject Classification: 80M50, 90C27, 49JXX, 35B50

This Springer imprint is published by the registered company Springer Nature Switzerland AG
The registered company address is: Gewerbestrasse 11, 6330 Cham, Switzerland

To my parents Leela and Murlidhar Diwekar for teaching me to be optimistic and to dream.
To my husband Sanjay Joag for supporting my dreams and making them a reality
and
To my niece Ananya whose innocence and charm provide optimism for the future.

Foreword

Optimization has pervaded all spheres of human endeavor. Although optimization has been practiced in some form or other from the early prehistoric era, this area has seen progressive growth during the last five decades. Modern society lives not only in an environment of intense competition but is also constrained to plan its growth in a sustainable manner with due concern for conservation of resources. Thus, it has become imperative to plan, design, operate, and manage resources and assets in an optimal manner. Early approaches have been to optimize individual activities in a standalone manner; however, the current trend is towards an integrated approach: integrating synthesis and design, design and control, production planning, scheduling, and control. The functioning of a system may be governed by multiple performance objectives. Optimization of such systems will call for special strategies for handling the multiple objectives to provide solutions closer to the systems requirement. Uncertainty and variability are two issues which render optimal decision-making difficult. Optimization under uncertainty would become increasingly important if one is to get the best out of a system plagued by uncertain components. These issues have thrown up a large number of challenging optimization problems which need to be resolved with a set of existing and newly evolving optimization tools.

Optimization theory had evolved initially to provide generic solutions to optimization problems in linear, nonlinear, unconstrained, and constrained domains. These optimization problems were often called mathematical programming problems with two distinctive classifications, namely linear and nonlinear programming problems. Although the early generation of programming problems was based on continuous variables, various classes of assignment and design problems required handling of both integer and continuous variables leading to mixed integer linear and nonlinear programming problems (MILP and MINLP). The quest to seek global optima has prompted researchers to develop new optimization approaches which do not get stuck at a local optimum, a failing of many of the mathematical programming methods. Genetic algorithms derived from biology and simulated annealing inspired by optimality of the annealing process are two such potent methods which have emerged in recent years. The developments in computing technology have placed at the disposal of the user a wide array of optimization codes with varying degrees of rigor and sophistication. The challenges to the user are manyfold. How to set up an optimization problem? What is the most suitable optimization method to use? How to perform a sensitivity analysis? An intrepid user may also want to extend the capabilities of an existing optimization method or integrate the features of two or more optimization methods to come up with more efficient optimization methodologies.

This book, appropriately titled *Introduction to Applied Optimization*, has addressed all the issues stated above in an elegant manner. The book has been structured to cover all key areas of optimization, namely deterministic and stochastic optimization and single and multiobjective optimization. In keeping with the application focus of the book, the reader is provided with deep insights into key aspects of an optimization problem: problem formulation, basic principles and structure of various optimization techniques, and computational aspects.

The book begins with a historical perspective on the evolution of optimization followed by the identification of key components of an optimization problem and its mathematical formulation. Types of optimization problems that can occur and the software codes available to solve these problems are presented. The book then moves on to treat in the next two chapters two major optimization methods, namely linear programming and nonlinear programming. Simple introductory examples are used to illustrate graphically the characteristics of the feasible region and location of optima. The simplex method used for the solution of the LP problem is described in great detail. The author has used an innovative example to develop the Karush–Kuhn–Tucker conditions for NLP problems. The Lagrangian formulation has been used to develop the relationships between primal-dual problems. The transition from the continuous to discrete optimization problem is made in Chapter 4. The distinctive character of the solution to the discrete optimization problem is demonstrated graphically with a suitable example. The efficacy of the branch-and-bound method for the solution of MILP and MINLP problems is brought out very clearly. Decomposition methods based on generalized Bender's decomposition (GBD) and outer approximation (OA) are projected as efficient approaches for the solution of MILP and MINLP problems. Developing optimal solutions using simulated annealing and genetic algorithms are also explained in great detail. The potential of combining simulated annealing and nonlinear programming (SA-NLP) to generate more efficient solutions for MINLP problems is stressed with suitable examples. Chapter 5 deals with strategies for optimization under uncertainty. The strategy of using the mean value of a random variable for optimization is shown to be suboptimal. Using probabilistic information on the uncertain variable, various measures such as the value of stochastic solution (VSS) and the expected value of perfect information (EVPI) are developed. The optimization problem with recourse is analyzed. Two policies are considered, namely "here and now" and "wait and see." The development of chance-constrained programming and L-shaped decomposition methods using probability information is shown. For simplification of optimization under uncertainty, the use of sampling techniques for scanning the uncertain parameter space is advocated. Among the various sampling methods analyzed, the Hammersley sequence sampling is shown to be the most efficient. The stochastic annealing algorithm with an adaptive choice of sample size is shown as an efficient method for handling stochastic optimization problems.

Multiobjective optimization is treated in the next chapter. The process of identification of a nondominated set from the set of feasible solutions is presented. Three methods, namely weighting method, constraint method, and goal programming method are discussed. STA-NLP framework is proposed as an alternate approach to handle multiobjective optimization problems.

The book ends with a treatment of optimal control in Chapter 7. The first part deals with well-known methods like calculus of variations, maximum principle, and dynamic programming. The next part deals with stochastic dynamic optimization. Stochastic formulation of dynamic programming is done using Ito's lemma. The book concludes with a detailed study of the dynamic optimization of batch distillation. The thorough treatment of the stochastic distillation case should provide a revealing study for the reader interested in solving dynamic optimization problems under uncertainty.

The material in the book has been carefully prepared to keep the theoretical development to a minimal level while focusing on the principles and implementation aspects of various algorithms. Numerous examples have been given to lend clarity to the presentations. Dr. Diwekar's own vast research experience in nonlinear optimization, optimization under uncertainty, process synthesis, and dynamic optimization has helped in focusing the reader's attention to critical issues associated with various classes of optimization problems. She has used the hazardous waste blending problem on which she has done considerable research as a complex enough process for testing the efficacy of various optimization methods. This example is used very skillfully to demonstrate the strengths and weaknesses of various optimization methods.

The book with its wide coverage of most of the well-established and emerging optimization methods will be a valuable addition to the optimization literature. The book will be a valuable guide and reference material to a wide cross-section of the user community comprising students, faculty, researchers, practitioners, designers, and planners.

Bombay, India   K. P. Madhavan
20 November, 2002

Preface: Second Edition

I am happy to present the second edition of this book. In this second edition, I have updated all the chapters and additional material has been added in Chapters 3 and 7. New examples have also been added in various chapters. The solution manual and case studies for this book are available online on the Springer website with the book link.

This book would not have been possible without the constant support from my husband Dr. Sanjay Joag and my sisters Dr. Anjali Diwekar and Dr. Prajakta Sambarey. Thanks are due to my graduate students Francesco Baratto, Saadet Ulas, Karthik Subramanyan, Weiyu Xu, and Yogendra Shastri for providing feedback on the first edition. Thanks are also due to the many readers around the world who sent valuable feedback.

Clarendon Hills, IL, USA

Urmila M. Diwekar

February 2007

Preface: Third Edition

I am happy to present the third edition of this book. In this edition, I have added more clarifications in each chapter. Also, several new sections are added. In Chapter 2, a section on LP degeneracy is added with an example. In Chapter 3, a new figure is added to show how, when the function is nonconvex can result in a saddle. Also, a description with example is added about the existence of Lagrange multipliers. In Chapter 4, a separate section on the cutting plane method with an example and a simple example for GBD and OA algorithms is added. In Chapter 5, we corrected the chance-constrained example and added a detailed example to generate LHS and Monte Carlo methods. Chapter 6 is unchanged as far as theory is concerned. In Chapter 7, a section on the derivation of the stochastic maximum principle is added. A new real-world case study related to sustainability is added in Chapters 2, 3, 4, 5, 6, and 7. I got feedback from a number of graduate students over the years, which resulted in this new edition. My special thanks to my ex-graduate student Professor Kirti Yenkie of Rowan University, who provided valuable feedback and notes. Thanks are also due to the many readers around the world who sent valuable feedback.

Crystal Lake, IL, USA   Urmila M. Diwekar
May 2020

Acknowledgments for the First Edition

आशा नाम मनुष्याणां काचिदाश्चर्य शृंह्रला ।

यया बध्दा प्रधावन्ति मुक्तां तिष्ठन्ति पङ्गुवत् ॥

–Sanskrit Subhashita

What an amazing chain the word hope is for humans, tied by it they surge forward, and unbound from it, they sit like the lame.

This book was written during some difficult times. My parents have taught me to be hopeful and optimistic. That and the constant support from my husband, Dr. Sanjay Joag, kept me going. This book would not exist without the personal support and encouragement I received from him, my parents, and my mother-in-law, Leela Joag.

It meant a lot to me that this book received a foreword from my advisor and guru Professor K. P. Madhavan. He taught me not only the first course on optimization but also how to be a disciplined and rigorous researcher. My heartfelt thanks to him. Thanks also go to the many who shared this writing experience. I appreciate the efforts of Dr. Keith Tomazi, Mallinckrodt Chemicals/Tyco-Health Care, Professor Panos Pardalos, University of Florida, Professor Antonis Kokossis, University of Surrey, UK, and Dr. Karl Schnelle, Dow Agro Sciences, for their many valuable suggestions. I am grateful to Dr. Kemal Sahin, my postdoctoral fellow, and my graduate students Dr. Ki-joo Kim, Dominic Boccelli, Renyou Wang, and Amit Goyal for their careful review of the initial drafts of manuscript and for their comments. Thanks also to Aaron Pavkov for meticulously correcting all the different versions of the manuscript.

To my parents for their untiring love and their lifetime of hard work behind the greatest gift to me—education—to my husband for his constant love, encouragement, and his unconditional sacrifices to support my career, and to my innocent and charming niece Ananya for being what she is, I dedicate this work.

Chicago, IL, USA   Urmila M. Diwekar
December 2002

Contents

List of Figures

List of Tables

Author Biography

Dr. Urmila M. Diwekar is the president of the Vishwamitra Research, a non-profit research institute that she founded to pursue multidisciplinary research in the areas of Optimization under Uncertainty and Computer-Aided Design applied to Energy, Environment, and Sustainability. She was a Professor in the Departments of Chemical Engineering, Bio-Engineering, and Industrial Engineering; in the Institute for Environmental Science and Policy; and at the University of Illinois at Chicago (UIC). She has a special formal arrangement with UIC where she remains as the main advisor for her Ph.D. and M.S. students and teaches a course on optimization. She was on the faculty of the Carnegie Mellon University, with early promotions to both the Associate and the Full Professor level.

1

Introduction

Since the fabric of the universe is most perfect, and is the work of a most wise Creator, nothing whatsoever takes place in the universe in which some form of maximum and minimum does not appear.

–Leonard Euler

Optimization is a part of life. The evolution process in nature reveals that it follows optimization. For example, animals that live in colder climates have smaller limbs than the animals living in hotter climates, to provide a minimum surface- area-to-volume ratio. In our day-to-day lives we make decisions that we believe can maximize or minimize our set of objectives, such as taking a shortcut to minimize the time required to reach a particular destination, finding the best possible house that can satisfy maximum conditions within cost constraints, or finding a lowest-priced item in the store. Most of these decisions are based on our years of knowledge of the system without resorting to any systematic mathematical theory. However, as the system becomes more complicated involving more and more decisions to be made simultaneously and becoming constrained by various factors, some of which are new to the system, it is difficult to take optimal decisions based on a heuristic and previous knowledge. Furthermore, many times the stakes are high and there are multiple stakeholders to be satisfied. Mathematical optimization theory provides a better alternative for decision making in these situations provided one can represent the decisions and the system mathematically.

With the advent of computers it is possible to exploit these theories to their maximum extent. Beightler et al. (1967) described optimization as a three-step decision-making process, where the first step is the knowledge of the system, which in mathematical terms can be considered as modeling the process. The second step involves finding a measure of system effectiveness, or the objective function. The third step is based on the theory of optimization, which is the main focus of this book.

© Springer Nature Switzerland AG 2020
U. M. Diwekar, *Introduction to Applied Optimization*, Springer
Optimization and Its Applications 22,
https://doi.org/10.1007/978-3-030-55404-0_1

The theory of optimization has its roots in the isoperimetric problem faced by Queen Dido in 1000 BC. She procured for the founding of Carthage the largest area of land that could be surrounded by the hide of a bull. From the hide she made a rope, which she arranged in a semicircle with the ends against the sea. Queen Dido's intuitive solution was correct. But it was many centuries before a formal proof was presented, and the mathematical and systematic solution to this problem proved to be a very difficult problem in the calculus of variations. The calculus of variations essentially handles problems where the decision variable is a vector. Figure 1.1 shows a rectangle, a square, and a circle having the same perimeter but different areas. It can be seen that the maximum area is covered by the circle.

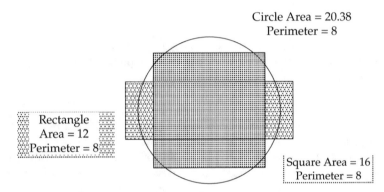

Fig. 1.1. Isoperimetric objects

The calculus of variations, or the first systematic theory of optimization, was born on June 4, 1694, when John Bernoulli posed the *Brachistochrone* (Greek for "shortest time") problem, and publicly challenged the mathematical world to solve it. The problem posed was, "What is a slide path down which a frictionless object would slip in the least possible time?" Earlier attempts to solve this problem were made by many well-known scientists including Galileo who proposed the solution to be a circular arc, an incorrect solution, and Leibnitz who presented ordinary differential equations without solving them. Then John Bernoulli proved the optimal path to be a cycloid. From that point, efforts continued in the area of the calculus of variations, leading to the study of multiple integrals, differential equations, control theory, problem transformation, and so on. Although this research mainly involved theory and analytical solutions, it formed the basis for numerical optimization developed during and after World War II. World War II made scientists aware of numerical optimization and solutions to physics and engineering problems. In 1947 Dantzig proposed the simplex algorithm for linear programming problems. Necessary conditions were presented by Kuhn and Tucker in the early 1950s, which formed a focal point for nonlinear programming research. Now, numerical optimization techniques constitute a fundamental part of theoretical and practical science and engineering (Diwekar, 1995).

1.1 Problem Formulation: A Cautionary Note

Consider the following optimization problem.

Example 1.1: A chemical manufacturer produces a chemical from two raw materials X_1 and X_2. Although X_1 can be purchased at \$5 per ton, X_2 is less expensive and can be obtained at \$1 per ton. The manufacturer wants to determine the amount of each raw material required to reduce the cost per ton of product to a minimum. Formulate the problem as an optimization problem.

Solution: Let x_1 be the amount of X_1 required to produce a ton of the product and x_2 be the amount of X_2 consumed in the process. Then the problem can be formulated as given below.

$$\text{Minimize} \quad Z = 5x_1 + x_2 \qquad (1.1)$$
$$x_1, \ x_2$$

subject to

$$x_1 \geq 0; x_2 \geq 0$$

From the objective function given in Equation (1.1), it is apparent that to minimize Z, the cost per ton of product produced, the manufacturer has to purchase zero amounts of X_1 and X_2.

Example 1.1 involves manufacturing a chemical from reactants X_1 and X_2 and we found that the optimal solution involves zero amounts of X_1 and X_2. This is a mathematically correct solution. However, from the realistic (thermodynamic) point of view, if reactant X_1 is not present at all, then there is no possibility of product formation. What did we do wrong? We forgot to provide information about the minimum amount of reactants required to form the product. As the problem formulation did not have insight about the reaction chemistry requirement, our solution is mathematically right but practically useless. This shows that correct problem formulation is the key step in optimization.

1.2 Degrees of Freedom Analysis

The degrees of freedom analysis provides the number of decision variables one can change to obtain the optimum design, and is crucial in optimization.

Consider Case 1 in Table 1.1 given. One can see that there are two variables and two equations, and hence there is no optimization problem because the two sets of Equations $D-2$ and $D-3$ can be solved directly to obtain the values of the decision variables x_1 and x_2. Equation $D-1$ does not play any role in finding the values of these variables. On the other hand, Case 2 has two variables and one equality, and hence one degree of freedom. Case 3 has two variables and two inequalities, leading to two degrees of freedom. In Cases 2 and 3, optimization can be used because the number corresponding to the degrees of freedom is greater than zero.

Table 1.1. Degrees of freedom (DOF) analysis

Case 1	Case 2	Case 3	Equation
Min $x_1^2 + x_2^2$	Min $x_1^2 + x_2^2$	Min $x_1^2 + x_2^2$	$D-1$
$x_1 - x_2 = 4.0$	$x_1 + x_2 = 2.0$	$x_1 - x_2 \le 4.0$	$D-2$
$x_1 + x_2 = 2.0$	$-$	$x_1 + x_2 \le 2.0$	$D-3$
0	1	2	DOF
No	Yes	Yes	Optimization
$x_1 = 3;\ x_2 = -1$	$x_1 = 1;\ x_2 = 1$	$x_1 = 0;\ x_2 = 0$	Solution

Is it possible to use optimization for a system when there are fewer variables than equations?

This is a typical situation in the parameter estimation problem presented below.

$$Y = f(x, \theta) \tag{1.2}$$

Here, the quantity Y is a function of variables x and θ, where θ represents m unknown parameters. In general, n experiments are performed to obtain various values of Y by changing values of x. Each dataset needs to satisfy the equation above, resulting in the n equations and $n > m$. In this case, one still can use optimization techniques by combining all n equations into an objective function containing the square of the errors for each equation. The least square problem ensures that there are degrees of freedom available to solve the optimization problem.

1.3 Objective Function, Constraints, and Feasible Region

Figure 1.2 plots the graph of the objective function Z versus a decision variable x. Figure 1.2a shows a linear programming problem where the linear objective function as well as its constraints (lines AB, BC, and CA) are linear. The constraints shown in Figure 1.2a are inequality constraints indicating that the solution should be above or on line AB, and below or on lines BC and CA. ABC

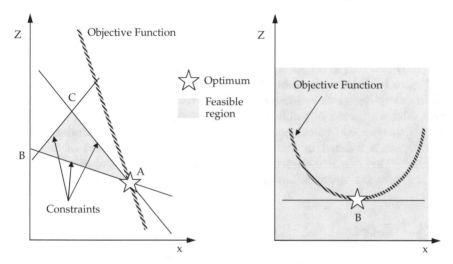

Fig. 1.2. Linear and nonlinear programming problems

represents the feasible region of operation within which the solution should lie. The constraints are binding the objective space, and hence the linear objective is lying at the edge of the feasible region (constraint). Figure 1.2b shows an unconstrained problem, hence the feasible region extends to infinity. The minimum of the objective function lies at point B, where the tangent to the curve is parallel to the x-axis, having a zero slope (the derivative of the objective function with respect to the decision variable is zero). This is one of the necessary conditions of optimality for a nonlinear programming problem where the objective function and/or constraints are nonlinear. The earlier theories involving calculus of variations use this condition of optimality to reach preferably an analytical solution to the problem. However, for many real-world problems, it is difficult to obtain an analytical solution and one has to follow an iterative scheme. This is numerical optimization.

1.4 Numerical Optimization

A general optimization problem can be stated as follows.

$$\text{Optimize}_{x} \quad Z = z(x) \tag{1.3}$$

subject to

$$h(x) = 0 \tag{1.4}$$

$$g(x) \leq 0 \tag{1.5}$$

The goal of an optimization problem is to determine the decision variables x that optimize the objective function Z, while ensuring that the model operates within established limits enforced by the equality constraints h (Equation(1.4)) and inequality constraints g (Equation (1.5)).

Figure 1.3 illustrates schematically the iterative procedure employed in a numerical optimization technique. As seen in the figure, the optimizer invokes the model with a set of values of decision variables x. The model simulates the phenomena and calculates the objective function and constraints. This information is utilized by the optimizer to calculate a new set of decision variables. This iterative sequence is continued until the optimization criteria pertaining to the optimization algorithm are satisfied.

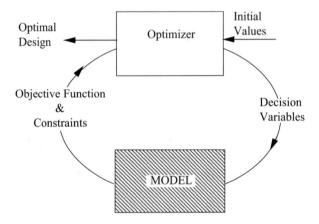

Fig. 1.3. Pictorial representation of the numerical optimization framework

There are a large number of software codes available for numerical optimization. Examples of these include solvers such as MINOS, CPLEX, CONOPT, and NPSOL. Also, many mathematical libraries, such as NAG, OSL, IMSL, and HARWELL have different optimization codes embedded in them. Popular software packages such as EXCEL, MATLAB, and SAS also have some optimization capabilities. There are algebraic modeling languages like AMPL, LINGO, AIMMS, GAMS, and ISIGHT specifically designed for solving optimization problems and software products such as Omega and Evolver have spreadsheet interfaces. However, a discussion of all the different accessible software is beyond the scope of this book. SIAM publications provides a comprehensive software guide by Moré and Wright (1993). Furthermore, the Internet provides a great source of information. A group of researchers at Argonne National Laboratory and Northwestern University launched a project known as the Network-Enabled Optimization System (NEOS). Its associated Optimization Technology Center maintains a Web site at: http://www.mcs.anl.gov/otc/

which includes a library of freely available optimization software, a guide to software selection, educational material, and a server that allows online execution (Carter and Price, 2001). Also, the site: http://OpsResearch.com/ OR-Objects
includes data structures and algorithms for developing optimization applications.

Optimization algorithms mainly depend upon the type of optimization problems described in the next section.

1.5 Types of Optimization Problems

Optimization problems can be divided into the following broad categories depending on the type of decision variables, objective function(s), and constraints.

- Linear programming (LP): The objective function and constraints are linear. The decision variables involved are scalar and continuous.
- Nonlinear programming (NLP): The objective function and/or constraints are nonlinear. The decision variables are scalar and continuous.
- Integer programming (IP): The decision variables are scalars and integers.
- Mixed integer linear programming (MILP): The objective function and constraints are linear. The decision variables are scalar; some of them are integers while others are continuous variables.
- Mixed integer nonlinear programming (MINLP): A nonlinear programming problem involving integer as well as continuous decision variables.
- Discrete optimization: Problems involving discrete (integer) decision variables. This includes IP, MILP, and MINLPs.
- Optimal control: The decision variables are vectors.
- Stochastic programming or stochastic optimization: Also termed optimization under uncertainty. In these problems, the objective function and/or the constraints have uncertain (random) variables. Often involves the above categories as subcategories.
- Multiobjective optimization: Problems involving more than one objective. Often involves the above categories as subcategories.

1.6 Summary

Optimization involves several steps: (1) understanding the system, (2) finding a measure of system effectiveness, and (3) degrees of freedom analysis and applying a proper optimization algorithm to find the solution. Optimization problems can be divided into various categories such as LP, NLP, MINLP, and stochastic programming depending on the type of objective function, constraints, and/or decision variables.

In short, optimization is a systematic decision-making process. Consider the following problem faced by Noah described in the Bible:

> and God said unto Noah, make thee an ark of gopher wood; rooms shalt thou make in this ark. The length of the ark shall be 300 cubits, the breadth of it 50 cubits, and the height of it 30 cubits. With lower, second, and third stories shalt thou make it. And of every living thing of all flesh two of every sort shalt thou bring in the ark, to keep them alive with thee, they shall be male and female. Thus did Noah, according to all that God commanded him, so did he.
>
> –Chapter 6, Genesis, Old Testament

Noah's problem: Determine the minimum number of rooms that allows a compatible assignment of the animals. What Noah faced is a mixed integer nonlinear programming problem. Consider adding the following lines to Noah's problem.

> ...and include the uncertainties associated with forecasting the consumption of food. Also consider the variabilities in weights and nature of the different animals in the assignment.

This is a mixed integer nonlinear programming problem under uncertainty and represents a challenge even today.

Exercises

1.1 For the problems below, indicate the degrees of freedom and the problem type (LP, NLP, IP, etc.)

(a)

$$\max f(x, y) = 3x + 4y$$
$$\text{s.t.} \quad x + 4y - z \leq 10$$
$$y + z \geq 6$$
$$x - y \leq 3$$

(b)

$$\min f(x, y) = 3 \cdot x^2 + 4 \cdot \sin(y \cdot z)$$
$$\text{s.t.} \quad x + 4y \leq 10$$
$$y + z = 6 + \pi$$
$$x - y \leq 3$$
$$z \in \{0, \pi/2, \pi\}$$

(c)

$$\min \; 4.35 \cdot x^2 \cdot y_1 + 1.74 \cdot x \cdot z \cdot y_2 - 2.5 \cdot k \cdot y_3$$
$$\text{s.t. } x - z + k \leq 10$$
$$y_1 + y_2 \leq 1$$
$$y_2 \leq y_3$$
$$x \leq 8$$
$$k \leq 7$$
$$x, k \geq 0$$
$$y_1, y_2, y_3 \in \{0, 1\}$$

(d)

$$\min \; \sigma_{R_B}^2 = \int_0^1 (R_B - \overline{R_B})^2 dF$$
$$\text{s.t. } \overline{R_B} = \int_0^1 R_B(\theta, x, u) dF$$

$$C_A = \frac{C_{A_i}}{1 + k_A^0 \cdot e^{-E_A/RT} \cdot \tau}$$
$$C_B = \frac{C_{B_i} + k_A^0 \cdot e^{-E_A/RT} \cdot \tau \cdot C_A}{1 + k_B^0 \cdot e^{-E_B/RT} \cdot \tau}$$
$$-r_A = k_A^0 \cdot e^{-E_A/RT}$$
$$-r_B = k_B^0 \cdot e^{-E_B/RT} - k_A^0 \cdot e^{-E_A/RT}$$
$$Q = F\rho C_p \cdot (T - T_i) + V \cdot (r_A H_{RA} + r_B H_{RB})$$
$$\tau = V/F$$
$$R_B = r_b \cdot V$$

where θ denotes the control variables corresponding to the degrees of free-dom, x are the state variables equal to the number of equality constraints, and u represents associated uncertainties.

1.2 Indicate whether the problem below is an NLP or an LP. What methods do you expect to be most effective for solving this problem?

$$\max \; f(x, y, z, m) = x - 3y + 1.25z - 2 \cdot \log(m)$$
$$\text{s.t. } m \cdot \exp(y) \geq 10$$
$$\log(m) - x + 4z \geq 6$$
$$x - 3y \leq 9$$

Bibliography

Beale E. M. (1977), *Integer Programming: The State of the Art in Numerical Analysis*, Academic Press, London.

Beightler C. S., D. T. Phillips, and D. J. Wilde (1967), *Foundations of Optimization*, Prentice-Hall I, Englewood Cliffs, NJ.

Biegler L., I. E. Grossmann, and A. W. Westerberg (1997), *Systematic Methods of Chemical Process Design*, Prentice-Hall, Upper Saddle River, NJ.

Birge J. R. (1997), Stochastic programming computation and application, *Informs Journal on Computing* **9**, 111.

Carter M. W. and C. C. Price (2001), *Operations Research: A Practical Introduction*, CRC Press, New York.

Diwekar U. M. (1995), *Batch Distillation: Simulation, Optimal Design and Control*, Taylor & Francis, Washington DC.

Moré J. J. and S. J. Wright (1993), *Optimization Software Guide*, SIAM Publications, Philadelphia.

Nocedal J. and S. Wright (1999), *Numerical Optimization*, Springer Series in Operations Research, Springer-Verlag, New York.

Reklaitis R., A. Ravindran, and K. M. Ragsdell (1983), *Engineering Optimization*, John Wiley & Sons, New York.

Taha H. A. (1997), *Operations Research: An Introduction*, Sixth Edition, Prentice Hall, Upper Saddle River, NJ.

Winston W. L. (1991), *Operations Research: Applications and Algorithms*, Second Edition, PWS-KENT, Boston.

2

Linear Programming

Linear programming (LP) problems involve linear objective function and linear constraints, as shown below in Example 2.1.

Example 2.1: Solvents are extensively used as process materials (e.g. extractive agents) or process fluids (e.g., CFC) in chemical process industries. Cost is a main consideration in selecting solvents. A chemical manufacturer is accustomed to a raw material X_1 as the solvent in his plant. Suddenly, he found out that he can effectively use a blend of X_1 and X_2 for the same purpose. X_1 can be purchased at \$4 per ton, however X_2 is an environmentally toxic material which can be obtained from other manufacturers. With the current environmental policy, this results in a credit of \$1 per ton of X_2 consumed. He buys the material a day in advance and stores it. The daily availability of these two materials is restricted by two constraints: (1) The combined storage (intermediate) capacity for X_1 and X_2 is 8 tons per day. The daily availability for X_1 is twice the required amount. X_2 is generally purchased as needed. (2) The maximum availability of X_2 is 5 tons per day. Safety conditions demand that the amount of X_1 cannot exceed the amount of X_2 by more than 4 tons. The manufacturer wants to determine the amount of each raw material required to reduce the cost of solvents to a minimum. Formulate the problem as an optimization problem.

Solution: Let x_1 be the amount of X_1 and x_2 be the amount of X_2 required per day in the plant. Then, the problem can be formulated as a linear programming problem as given below.

$$\text{Minimize} \quad Z = 4x_1 - x_2 \tag{2.1}$$

Electronic Supplementary Material: The online version of this chapter (https://doi.org/10.1007/978-3-030-55404-0_2) contains supplementary material, which is available to authorized users.

U. M. Diwekar, *Introduction to Applied Optimization*, Springer Optimization and Its Applications 22, https://doi.org/10.1007/978-3-030-55404-0_2

$$x_1, \ x_2$$

subject to

$$2x_1 + x_2 \leq 8 \qquad \text{Storage Constraint} \tag{2.2}$$
$$x_2 \leq 5 \qquad \text{Availability Constraint} \tag{2.3}$$
$$x_1 - x_2 \leq 4 \qquad \text{Safety Constraint} \tag{2.4}$$

$$x_1 \geq 0; x_2 \geq 0$$

As shown above, the problem is a two-variable LP problem, which can be easily represented in a graphical form. Figure 2.1 shows constraints (2.2) through (2.4), plotted as three lines by considering the three constraints as equality constraints. Therefore, these lines represent the boundaries of the inequality constraints. In the figure, the inequality is represented by the points on the other side of the hatched lines. The objective function lines are represented as dashed lines (isocost lines). It can be seen that the optimal solution is at the point $x_1 = 0$; $x_2 = 5$, a point at the intersection of constraint (2.3) and one of the isocost lines. All isocost lines intersect constraints either once or twice. The LP optimum lies at a vertex of the feasible region, which forms the basis of the simplex method. The simplex method is a numerical optimization method for solving linear programming problems developed by George Dantzig in 1947.

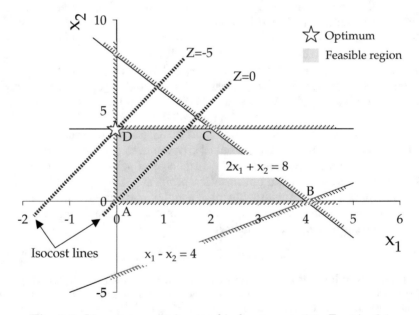

Fig. 2.1. Linear programming graphical representation, Exercise 2.1

2.1 The Simplex Method

The graphical method shown earlier can be used for two-dimensional problems; however, real-life LPs consist of many variables, and to solve these linear programming problems, one has to resort to a numerical optimization method such as the simplex method.

The generalized form of an LP can be written as follows.

$$\underset{x_i}{\text{Optimize}} \quad Z = \sum_{i=1}^{n} C_i x_i \tag{2.5}$$

subject to

$$\sum_{i=1}^{n} a_{ji} x_i \leq b_j \tag{2.6}$$

$$j = 1, 2, \ldots, m$$

$$x_j \in R$$

As stated in Chapter 1, a numerical optimization method involves an iterative procedure. The simplex method involves moving from one extreme point on the boundary (vertex) of the feasible region to another along the edges of the boundary iteratively. This involves identifying the constraints (lines) on which the solution will lie. In simplex, a slack variable is incorporated in every constraint to make the constraint an equality. Now, the aim is to solve the linear equations (equalities) for the decision variables x, and the slack variables s. The active constraints are then identified based on the fact that, for these constraints, the corresponding slack variables are zero.

The simplex method is based on the Gauss elimination procedure of solving linear equations. However, some complicating factors enter in this procedure: (1) all variables are required to be nonnegative because this ensures that the feasible solution can be obtained easily by a simple ratio test (Step 4 of the iterative procedure described below) and (2) we are optimizing the linear objective function, so at each step we want to ensure that there is an improvement in the value of the objective function (Step 3 of the iterative procedure given below).

The simplex method uses the following steps iteratively.

1. *Convert the LP into the standard LP form.*
   Standard LP
   - All the constraints are equations with a nonnegative right-hand side.
   - All variables are nonnegative.
     - Convert all negative variables x to nonnegative variables using two variables (e.g., $x = x^+ - x^-$); this is equivalent to saying if $x = -5$, then $-5 = 5 - 10$, $x^+ = 5$, and $x^- = 10$.

   - Convert all inequalities into equalities by adding slack variables (nonnegative) for less than or equal to constraints ($\leq$) and by subtracting surplus variables for greater than or equal to constraints ($\geq$).
- The objective function must be minimization or maximization.
- The standard LP involving m equations and n unknowns has m basic variables and $n - m$ nonbasic or zero variables. This is explained below using Example 2.1.

Consider Example 2.1 in the standard LP form with slack variables, as given below.

Standard LP:

$$\text{Maximize} \quad - Z \tag{2.7}$$

$$-Z + 4x_1 - x_2 = 0 \tag{2.8}$$
$$2x_1 + x_2 + s_1 = 8 \tag{2.9}$$
$$x_2 + s_2 = 5 \tag{2.10}$$
$$x_1 - x_2 + s_3 = 4 \tag{2.11}$$

$$x_1 \geq 0; x_2 \geq 0$$
$$s_1 \geq 0; s_2 \geq 0; s_3 \geq 0$$

The feasible region for this problem is represented by the region ABCD in Figure 2.1. Table 2.1 shows all the vertices of this region and the corresponding slack variables calculated using the constraints given by Equations (2.9)–(2.11) (note that the nonnegativity constraint on the variables is not included).

Table 2.1. Feasible region in Figure 2.1 and slack variables

Point	x_1	x_2	s_1	s_2	s_3
A	0.0	0.0	8.0	5.0	4.0
B	4.0	0.0	0.0	5.0	0.0
C	1.5	5.0	0.0	0.0	7.5
D	0.0	5.0	3.0	0.0	9.0

It can be seen from Table 2.1 that at each extreme point of the feasible region, there are $n - m = 2$ variables that are zero and $m = 3$ variables that are nonnegative. An extreme point of the linear program is characterized by these m basic variables.

In simplex the feasible region shown in Table 2.1 gets transformed into a tableau (Table 2.2).

Table 2.2. Simplex tableau from Table 2.1

Row	$-Z$	x_1	x_2	s_1	s_2	s_3	RHS	Basic
0	1	4	-1	0	0	0	0	$-Z = 0$
1	0	2	1	1	0	0	8	$s_1 = 8$
2	0	0	$\underline{1}$	0	1	0	5	$s_2 = 5$
3	0	1	-1	0	0	1	4	$s_3 = 4$

2. *Determine the starting feasible solution.* A basic solution is obtained by setting $n - m$ variables equal to zero and solving for the values of the remaining m variables.

3. *Select an entering variable* (in the list of nonbasic variables) using the optimality (defined as better than the current solution) condition; that is, choose the next operation so that it will improve the objective function. Stop if there is no entering variable.

   Optimality Condition:

   • Entering variable: The nonbasic variable that would increase the objective function (for maximization). This corresponds to the nonbasic variable having the most negative coefficient in the objective function equation or the row zero of the simplex tableau.

   In many implementations of simplex, instead of wasting the computation time in finding the most negative coefficient, any negative coefficient in the objective function equation is used.

4. *Select a leaving variable using the feasibility condition.*

   Feasibility Condition:

   • Leaving variable: The basic variable that is leaving the list of basic variables and becoming nonbasic. The variable corresponding to the smallest nonnegative ratio (the right-hand side of the constraint divided by the constraint coefficient of the entering variable).

5. *Determine the new basic solution by using the appropriate Gauss–Jordan Row Operation.*

   Gauss–Jordan Row Operation:

   • Pivot Column: Associated with the row operation.
   • Pivot Row: Associated with the leaving variable.
   • Pivot Element: Intersection of Pivot row and Pivot Column.
   ROW OPERATION
   • Pivot Row = Current Pivot Row ÷ Pivot Element.
   • All other rows: New Row = Current Row—(its Pivot Column Coefficients × New Pivot Row).

6. *Go to Step 2.*

The following example illustrates the simplex method.

Example 2.2: Solve Example 2.1 using the simplex method.

Solution:

- Convert the LP into the standard LP form. For simplicity, we are converting this minimization problem to a maximization problem with $-Z$ as the objective function. Furthermore, nonnegative slack variables s_1, s_2, and s_3 are added to each constraint.
Standard LP:

$$\text{Maximize} \quad -Z \qquad\qquad (2.12)$$

$$
\begin{aligned}
-Z + 4x_1 - x_2 &= 0 & & (2.13)\\
2x_1 + x_2 + s_1 &= 8 & \text{Storage Constraint} & \quad (2.14)\\
x_2 + s_2 &= 5 & \text{Availability Constraint} & \quad (2.15)\\
x_1 - x_2 + s_3 &= 4 & \text{Safety Constraint} & \quad (2.16)
\end{aligned}
$$

$$x_1 \geq 0; x_2 \geq 0$$

The standard LP is shown in Table 2.3 where x_1 and x_2 are nonbasic or zero variables and s_1, s_2, and s_3 are the basic variables. The starting solution is $x_1 = 0$; $x_2 = 0$; $s_1 = 8$; $s_2 = 5$; $s_3 = 4$ obtained from the RHS column.

Table 2.3. Initial tableau for Example 2.2

Row	$-Z$	x_1	x_2	s_1	s_2	s_3	RHS	Basic	Ratio
0	1	4	-1	0	0	0	0	$-Z=0$	$-$
1	0	2	1	1	0	0	8	$s_1=8$	8
2	0	0	1	0	1	0	5	$s_2=5$	5
3	0	1	-1	0	0	1	4	$s_3=4$	$-$

- Determine the entering and leaving variables.
Is the starting solution optimum? No, because Row 0 representing the objective function equation contains nonbasic variables with negative coefficients. This can also be seen from Figure 2.2. In this figure, the current basic solution is shown to be increasing in the direction of the arrow.
Entering Variable: The most negative coefficient in Row 0 is x_2. Therefore, the entering variable is x_2. This variable must now increase in the direction of the arrow. How far can this increase the objective function? Remember that the solution has to be in the feasible region. Figure 2.2 shows that the maximum increase in x_2 in the feasible region is given by point D, which is on constraint (2.3). This is also the intercept of this constraint with the y-axis, representing x_2. Algebraically, these intercepts are the ratios of the right-hand side of the equations to the corresponding constraint coefficient

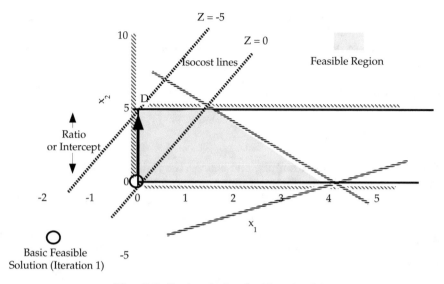

Fig. 2.2. Basic solution for Exercise 2.2

of x_2. We are interested only in the nonnegative ratios, as they represent the direction of increase in x_2. This concept is used to decide the leaving variable.

Leaving Variable: The variable corresponding to the smallest nonnegative ratio (5 here) is s_2. Hence, the leaving variable is s_2.

So, the Pivot Row is Row 2 and Pivot Column is x_2.

• The two steps of the Gauss–Jordan Row Operation are given below.

The pivot element is underlined in Table 2.3 and is 1.

Row Operation:

Pivot: $(0, 0, 1, 0, 1, 0, 5)$

Row 0: $(1, 4, -1, 0, 0, 0, 0) - (-1) (0, 0, 1, 0, 1, 0, 5) = (1, 4, 0, 0, 1, 0, 5)$
Row 1: $(0, 2, 1, 1, 0, 0, 8) - (1) (0, 0, 1, 0, 1, 0, 5) = (0, 2, 0, 1, -1, 0, 3)$
Row 3: $(0, 1, -1, 0, 0, 1, 4) - (-1) (0, 0, 1, 0, 1, 0, 5) = (0, 1, 0, 0, 1, 1, 9)$

These steps result in the following table (Table 2.4).

Table 2.4. The simplex tableau, Example 2.2, iteration 2

Row	$-Z$	x_1	x_2	s_1	s_2	s_3	RHS	Basic	Ratio
0	1	4	0	0	1	0	5	$-Z = 5$	–
1	0	2	0	1	-1	0	3	$s_1 = 3$	–
2	0	0	1	0	1	0	5	$x_2 = 5$	–
3	0	1	0	0	1	1	9	$s_3 = 9$	–

There is no new entering variable because there are no nonbasic variables with a negative coefficient in row 0. Therefore, we can assume that the solution is reached, which is given by (from the RHS of each row) $x_1 = 0$; $x_2 = 5$; $s_1 = 3$; $s_2 = 0$; $s_3 = 9$; $Z = -5$.

Note that at an optimum, all basic variables (x_2, s_1, s_3) have a zero coefficient in Row 0.

2.2 Infeasible Solution

Now consider the same example, and change the right-hand side of Equation (2.2) to -8 instead of 8. We know that constraint (2.2) represents the storage capacity and physics tells us that the storage capacity cannot be negative. However, let us see what we get mathematically.

Example 2.3: Constraint (2.2) is changed to reflect a negative storage capacity.

Solution: This results in the following LP.

$$\text{Maximize} \quad -Z \tag{2.17}$$
$$x_1,\ x_2$$

subject to

$$-Z + 4x_1 - x_2 = 0 \tag{2.18}$$
$$2x_1 + x_2 \leq -8 \quad \text{Storage Constraint} \tag{2.19}$$
$$x_2 \leq 5 \quad \text{Availability Constraint} \tag{2.20}$$
$$x_1 - x_2 \leq 4 \quad \text{Safety Constraint} \tag{2.21}$$

$$x_1 \geq 0; x_2 \geq 0$$

From Figure 2.3, it is seen that the solution is infeasible for this problem. Writing in standard LP form for simplex results in the following.
Standard LP: :

$$\text{Maximize} \quad -Z \tag{2.22}$$

$$-Z + 4x_1 - x_2 = 0 \tag{2.23}$$
$$-2x_1 - x_2 + s_1 = 8 \quad \text{Storage Constraint} \tag{2.24}$$
$$x_2 + s_2 = 6 \quad \text{Availability Constraint} \tag{2.25}$$
$$x_1 - x_2 + s_3 = 4 \quad \text{Safety Constraint} \tag{2.26}$$

$$x_1 \geq 0; x_2 \geq 0$$

Applying the simplex method results in Table 2.5 for the first step.

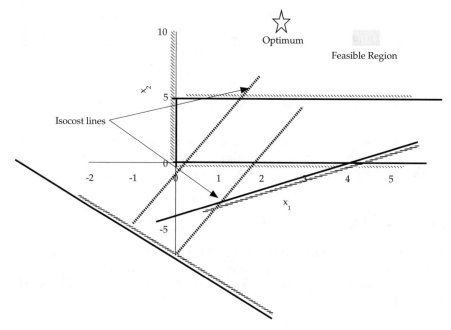

Fig. 2.3. Infeasible LP

Table 2.5. Initial simplex tableau, Example 2.3

Row	$-Z$	x_1	x_2	s_1	s_2	s_3	RHS	Basic	Ratio
0	1	4	-1	0	0	0	0	$-Z = 0$	$-$
1	0	-2	-1	1	0	0	8	$s_1 = 8$	$-$
2	0	0	$\underline{1}$	0	1	0	5	$s_2 = 5$	$\underline{5}$
3	0	1	-1	0	0	1	4	$s_3 = 4$	None

Standard LP:

$$\text{Maximize} \quad -Z + 4x_1 - x_2 \tag{2.27}$$

$$-2x_1 - x_2 - s_1 = 8 \tag{2.28}$$
$$x_2 + s_2 = 5 \tag{2.29}$$
$$x_1 - x_2 + s_3 = 4 \tag{2.30}$$

As can be seen, the entering variable with the most negative coefficient is x_2 and the leaving variable corresponding to the smallest nonnegative ratio is s_2.

Applying the Gauss–Jordan row operation results in Table 2.6.

Table 2.6. The simplex tableau, Example 2.3, iteration 2

Row	$-Z$	x_1	x_2	s_1	s_2	s_3	RHS	Basic	Ratio
0	1	4	0	0	1	0	5	$-Z = 0$	–
1	0	−2	0	−1	1	0	13	$s_1 = -13$	–
2	0	0	1	0	1	0	5	$x_2 = 5$	–
3	0	1	0	0	1	1	9	$s_3 = 9$	–

The solution to this problem is the same as before: $x_1 = 0$; $x_2 = 5$. However, this solution is not a feasible solution because the slack variable (artificial variable defined to be always positive), s_1, is negative.

2.3 Unbounded Solution

If constraints (2.19) and (2.20) are removed in the above example, the solution is unbounded, as can be seen in Figure 2.4. This means there are points in the feasible region with arbitrarily large objective function values (for maximization).

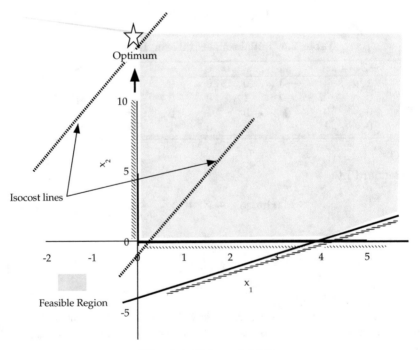

Fig. 2.4. Unbounded LP

Example 2.4: Constraints (2.19) and (2.20) removed.

Solution:

$$\text{Minimize} \quad Z = 4x_1 - x_2 \qquad (2.31)$$
$$x_1, \ x_2$$

subject to

$$x_1 - x_2 \leq 4 \quad \text{Safety Constraint} \qquad (2.32)$$

$$x_1 \geq 0; x_2 \geq 0$$

The simplex tableau for this problem is shown in Table 2.7.

Table 2.7. The simplex tableau, Example 2.4

Row	$-Z$	x_1	x_2	s_3	RHS	Basic	Ratio
0	1	4	-1	0	0	$-Z = 0$	$-$
1	0	1	-1	1	4	$s_3 = 4$	None

The entering variable is x_2 as it has the most negative coefficient in row 0. However, there is no leaving variable corresponding to the binding constraint (the smallest nonnegative ratio or intercept). That means x_2 can take as high a value as possible. This is also apparent in the graphical solution shown in Figure 2.4.

The LP is unbounded when (for a maximization problem) a nonbasic variable with a negative coefficient in row 0 has a nonpositive coefficient in each constraint, as shown in the above table.

2.4 Multiple Solutions

In the following example, the cost of X_1 is assumed to be negligible as compared to the credit of X_2. This LP has infinite solutions given by the isocost line ($x_2 = 5$) in Figure 2.5. The simplex method generally finds one solution at a time. Special methods such as goal programming or multiobjective optimization can be used to find these solutions. These methods are described in Chapter 6.

Example 2.5: Assume that in Example 2.1, the cost of X_1 is negligible. Find the optimal solution.

$$\text{Minimize} \quad Z = - x_2 \qquad (2.33)$$

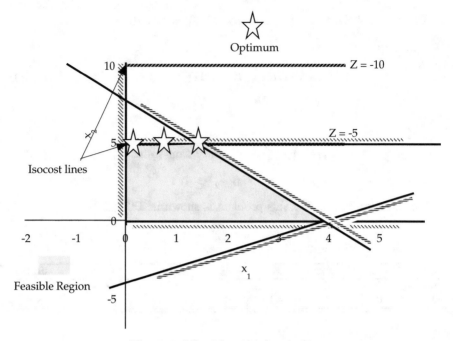

Fig. 2.5. LP with multiple solutions

$$x_1, \ x_2$$

subject to

$$2x_1 + x_2 \leq 8 \qquad \text{Storage Constraint} \tag{2.34}$$
$$x_2 \leq 5 \qquad \text{Availability Constraint} \tag{2.35}$$
$$x_1 - x_2 \leq 4 \qquad \text{Safety Constraint} \tag{2.36}$$

$$x_1 \geq 0; \ x_2 \geq 0$$

Solution: The graphical solution to this problem is shown in Figure 2.5. The simplex solution iteration summary is presented in Tables 2.8 and 2.9. The simplex method found the first solution to the problem; that is, $x_1 = 0$, $x_2 = 5$. Can simplex recognize that there are multiple solutions? Note that in Example 2.2, we stated that in the final simplex tableau solution, all basic variables have a zero coefficient in row 0. However, in the optimal tableau, there is a nonbasic variable x_1, which also has a zero coefficient.

Let us see if we make x_1 as an entering variable from the list of basic variables (Table 2.10). From the ratio test, one can see that s_1 would be the leaving variable. This results in the simplex tableau presented in Table 2.11.

Table 2.8. Initial tableau for Example 2.5

Row	$-Z$	x_1	x_2	s_1	s_2	s_3	RHS	Basic	Ratio
0	1	0	-1	0	0	0	0	$-Z = 0$	$-$
1	0	2	1	1	0	0	8	$s_1 = 8$	8
2	0	0	$\underline{1}$	0	1	0	5	$s_2 = 5$	$\underline{5}$
3	0	1	-1	0	0	1	4	$s_3 = 4$	$-$

Table 2.9. The Simplex tableau, Example 2.5, iteration 2

Row	$-Z$	x_1	x_2	s_1	s_2	s_3	RHS	Basic	Ratio
0	1	0	0	0	1	0	5	$-Z = 5$	$-$
1	0	2	0	1	-1	0	3	$s_1 = 3$	$-$
2	0	0	1	0	1	0	5	$x_2 = 5$	$-$
3	0	1	0	0	1	1	9	$s_3 = 9$	$-$

Table 2.10. The simplex tableau, Example 2.5, iteration 3

Row	$-Z$	x_1	x_2	s_1	s_2	s_3	RHS	Basic	Ratio
0	1	0	0	0	1	0	5	$-Z = 5$	$-$
1	0	$\underline{2}$	0	1	-1	0	3	$s_1 = 3$	1.5
2	0	0	1	0	1	0	5	$x_2 = 5$	$-$
3	0	1	0	0	1	1	9	$s_3 = 9$	9

Table 2.11. The simplex tableau, Example 2.5, iteration 4

Row	$-Z$	x_1	x_2	s_1	s_2	s_3	RHS	Basic	Ratio
0	1	0	0	0	1	0	5	$-Z = 5$	$-$
1	0	1	0	0.5	-0.5	0	1.5	$x_1 = 1.5$	$-$
2	0	0	1	0	1	0	5	$x_2 = 5$	$-$
3	0	0	0	-0.5	1.5	1	7.5	$s_3 = 7.5$	$-$

An alternate solution to the simplex is $x = (1.5, 5.0)$. Remember that this is also an optimum solution because there are only nonnegative coefficients left in row 0.

2.5 Degeneracy in LP

An LP is degenerate if, in a basic feasible solution, one of the basic variables takes on a zero value. Degeneracy could result in the cycling of a solution for the basic simplex method. If a sequence of pivots starting from some basic feasible solution ends up at the exact same basic feasible solution, then we refer to this as "cycling." If the simplex method cycles, it can cycle forever. The following example illustrates the degeneracy in LP.

Example 2.6: Consider the same example (Example 2.1), we have been using in this chapter. If instead of starting the feasible solution at x=(0,0) with x1 and x2 as nonbasic variables, if we start at point B, as shown in Figure 2.6. with x2 and s1 as nonbasic variables. This results in the following iteration summary presented in Tables 2.12, 2.13, and 2.14

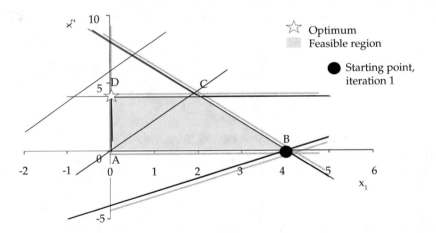

Fig. 2.6. Starting point Example 2.6

Table 2.12. Simplex tableau for Example 2.6, Iteration 1

Row	$-Z$	x_1	x_2	s_1	s_2	s_3	RHS	Basic
0	1	4	-1	0	0	0	0	$-Z = -6$
1	0	2	1	1	0	0	8	$x_1 = 4$
2	0	0	$\underline{1}$	0	1	0	5	$s_2 = 5$
3	0	1	-1	0	0	1	4	$s_3 = 0$

In iteration 1, the basic variable s_3 has zero value, this is degenerate LP. After iteration 2, the simplex tableau (Table 2.13) has no nonbasic variable with -negative coefficient. However, one of the nonbasic variable s_1 has zero coefficient, so like the last example, we can choose this as an entering variable. We can see that the leaving variable corresponds to constraint 3. However, constraint 3 has two basic variables x_1 and s_3. If we choose s_3 as the leaving variable, this results in Tableau given in Table 2.14. We can see that we have reached the same solution as iteration 1. This is cycling.

Table 2.13. The simplex tableau, Example 2.6, iteration 2

Row	$-Z$	x_1	x_2	s_1	s_2	s_3	RHS	Basic	Ratio
0	1	4	0	0	1	0	5	$-Z = -1$	–
1	0	2	0	1	−1	0	3	$x_1 = 1.5$	–
2	0	0	1	0	1	0	5	$x_2 = 5$	–
3	0	1	0	0	1	1	9	$s_3 = 7.5$	9

Table 2.14. Simplex tableau for Example 2.6, Iteration 3

Row	$-Z$	x_1	x_2	s_1	s_2	s_3	RHS	Basic
0	1	4	−1	0	0	0	0	$-Z = -6$
1	0	2	1	1	0	0	8	$x_1 = 4$
2	0	0	1	0	1	0	5	$s_2 = 5$
3	0	1	−1	0	0	1	4	$s_3 = 0$

If we chose the leaving variable as x_1, then we reach the optimum.

2.6 Sensitivity Analysis

The sensitivity of the linear programming solution is expressed in terms of shadow prices and opportunity (reduced) cost.

- Shadow Prices/Dual Prices/Simplex Multipliers: A shadow price is the rate of change (increase in the case of maximization and decrease in the case of minimization) of the optimal value of the objective function with respect to a particular constraint. Shadow prices are also called dual prices from the dual representations of LP problems used in the dual simplex method described in the next section.

  Figure 2.7 shows the shadow prices for various constraints in Example 2.1. As shown in the figure, if one changes the right-hand side of constraints (2.2) and (2.4) and uses the same basis, the optimal value is unchanged, so the shadow prices for these constraints are zero. This shows that if the management of the manufacturing company wants to increase their storage capacity, this decision will not have any implications as far as the solvent optimal cost is concerned. Similarly, if the company decides to relax the constraint on excess component volume (constraint (2.3)), that will also

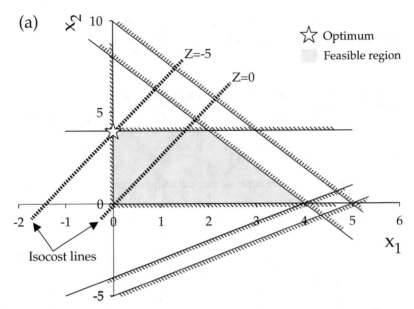

No change in optimum with unit change in constraint
Show price for this constraint = 0

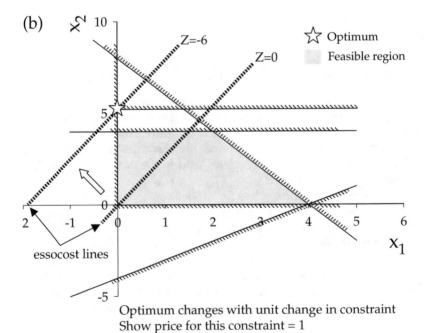

Optimum changes with unit change in constraint
Show price for this constraint = 1

Fig. 2.7. Shadow prices

not affect their solvent costs. However, if they can have access to more chemical X_2 per day (please see the LP formulation and corresponding simplex iteration summary, (Table 2.15 and 2.16 given) then that reduces the cost (objective function), as the shadow price for this constraint is 1.

Table 2.15. Initial tableau for the new LP

Row	$-Z$	x_1	x_2	s_1	s_2	s_3	RHS	Basic	Ratio
0	1	4	-1	0	0	0	0	$-Z = 0$	$-$
1	0	2	1	1	0	0	8	$s_1 = 8$	8
2	0	0	$\underline{1}$	0	1	0	6	$s_2 = 6$	$\underline{6}$
3	0	1	-1	0	0	1	4	$s_3 = 4$	$-$

Table 2.16. The simplex tableau, iteration 2

Row	$-Z$	x_1	x_2	s_1	s_2	s_3	RHS	Basic	Ratio
0	1	4	0	0	1	0	6	$-Z = 6$	$-$
1	0	2	0	1	-1	0	2	$s_1 = 2$	$-$
2	0	0	1	0	1	0	6	$x_2 = 6$	$-$
3	0	1	0	0	1	1	10	$s_3 = 10$	$-$

Standard LP:

$$\text{Maximize} \quad -Z \tag{2.37}$$

$$-Z + 4x_1 - x_2 = 0 \tag{2.38}$$
$$2x_1 + x_2 + s_1 = 8 \quad \text{Storage Constraint} \tag{2.39}$$
$$x_2 + s_2 = 6 \quad \text{Availability Constraint} \tag{2.40}$$
$$x_1 - x_2 + s_3 = 4 \quad \text{Safety Constraint} \tag{2.41}$$

$$x_1 \geq 0; x_2 \geq 0$$

Table 2.16 demonstrates that the slack variables for the two constraints with shadow prices of zero are positive (row 1 and 3). A less than or equal to ($\leq$) constraint will always have a nonnegative shadow price; a less than or equal to ($\leq$) constraint with positive slack variable (constraints 1 and 3) will have a zero shadow price; a greater than or equal to ($\geq$) constraint will always have a nonpositive shadow price; and an equality constraint may have a positive, a negative, or a zero shadow price.

The shadow prices are important for the following reasons.

–   To identify which constraints might be the most beneficially changed, and to initiate these changes as a fundamental means to improve the solution
–   To react appropriately when external circumstances create opportunities or threats to change the constraints

• Opportunity Cost/Reduced Cost: This is the rate of degradation of the optimum per unit use of a nonbasic (zero) variable in the solution.

Figure 2.8 shows that the opportunity cost for the nonbasic variable x_1 is 5. It can be seen that with the unit change in x_1, the solution lies on a different constraint (Equation (2.2)) changing the optimal objective function value from -5 to 0.

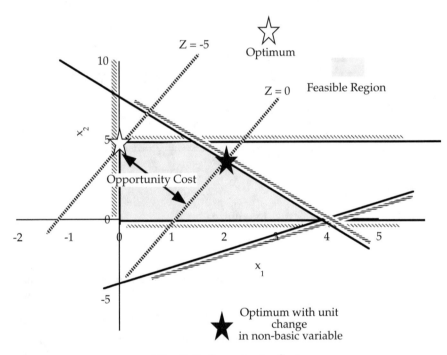

Fig. 2.8. Opportunity Cost

2.7 Other Methods

As a general rule, LP computational effort depends more on the number of constraints than the number of variables. The dual simplex method uses the dual representation of the original (primal) standard LP problem where the number of constraints is changed to the number of variables and vice versa. For large numbers of constraints, the dual simplex method is more efficient than the conventional simplex method. Table 2.17 shows the primal and dual representation of a standard LP. In the table, μ_j are the dual prices, or simplex multipliers. In nonlinear programming (Chapter 3) terminology, they are also known as the Lagrange multipliers. Using the NLP notations in Chapter 3, Example 3.8 shows the equivalence between the primal and dual representation shown in Table 2.17.

Table 2.17. The Primal and dual representation for an LP

Primal	Dual
Maximize $Z = \sum_{i=1}^{n} C_i x_i$	Minimize $Z_d = \sum_{j=1}^{m} b_j \mu_j$
$x_i, \ i = 1, 2, \ldots, n$	$\mu_j \ j = 1, 2, \ldots, m$
$\sum_{i=1}^{n} a_{ij} x_i \leq b_j$	$\sum_{j=1}^{m} a_{ij} \mu_j \geq C_i$
$j = 1, 2, \ldots, m$	$i = 1, 2, \ldots, n$
$x_i \geq 0$	$\mu_j \geq 0$

The simplex method requires the initial basic solution to be feasible. The Big M method and the two-phase simplex method circumvent the basic initial feasibility requirement of the simplex method. For details of these methods, please refer to Winston (1991).

Simplex methods move from boundary to boundary within the feasible region. On the other hand, interior-point methods visit points within the interior of the feasible region, which is more in line with the nonlinear programming techniques described in the next chapter. These methods are derived from the nonlinear programming techniques developed and popularized in the 1960s by Fiacco and McCormick, but their application to linear programming dates back only to Karmarkar's innovative analysis in 1984. The following example provides the basic concepts behind the interior point method.

Example 2.6: Take Example 2.5 and eliminate constraints (2.34) and (2.36). This converts the problem into a one-dimensional LP. Provide the conceptual steps for the interior point method using this LP.

$$\underset{x_2}{\text{Minimize}} \quad Z = -x_2 \tag{2.42}$$

subject to

$$x_2 \leq 5 \quad \text{Availability Constraint} \tag{2.43}$$

Solution: Just as we did in the simplex method earlier, let us add a variable s_2 to constraint (2.43). This results in the following two-dimensional problem.

$$\underset{x_2, s_2}{\text{Maximize}} \quad -Z = x_2 + 0s_2 \tag{2.44}$$

subject to

$$x_2 + s_2 = 5 \qquad (2.45)$$

This LP problem is shown in Figure 2.9. The constraint line represents the feasible region. Now consider a feasible point A on this constraint as a starting point. We need to take a step towards increasing the objective function (maximization) in the x_2 space, that is, the direction parallel to the x-axis. However, because this will be going out of the feasible region, this gradient is projected back to the feasible region at point B. As can be seen, this point is closer to the optimum than A. This gradient projection step is repeated until one reaches the optimum. Note that the step towards the gradient should not be too large to overshoot the optimum or too small to increase the number of iterations. It should also not get entrapped in the non-optimum solution. Karmarkar's interior point algorithm addresses these two concerns.

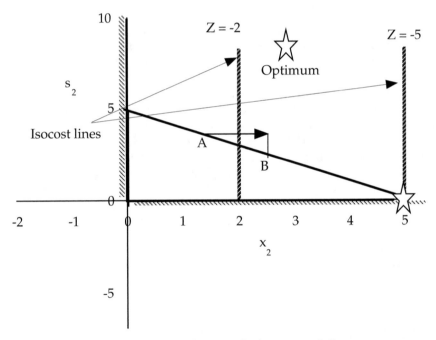

Fig. 2.9. The interior Point method conceptual diagram

Prior to 1987, all of the commercial codes for solving general linear programs made use of the simplex algorithm. This algorithm, invented in the late 1940s, has fascinated optimization researchers for many years because its performance on practical problems is usually far better than the theoretical worst case. During the period of 1984–1995, the interior point methods were the subject of intense theoretical and practical investigation, with practical code first appearing around 1989. These methods appear to be faster than the simplex method on large problems, but the advent of a serious rival spurred significant improvements in simplex codes. Today, the relative merits of the two approaches on any given problem depend strongly on the particular geometric or algebraic properties of the problem. In general, however, good interior point codes continue to perform as well or better than good simplex codes on larger problems when no prior information about the solution is available. When such "warm start" information is available, simplex methods are able to make much better use of it than the interior point methods (Wright, 1999).

2.8 Hazardous Waste Blending Problem as an LP

The Hanford site in southeastern Washington has produced nuclear materials using various processes for nearly 50 years. Radioactive hazardous waste was produced as byproducts of the processes. This waste will be retrieved and separated into high-level and low-level portions. The high-level and low-level wastes will be immobilized for future disposal.

The high-level waste will be converted into a glass form for disposal. The glass must meet both processability and durability restrictions. The processability conditions ensure that during processing, the glass melt has properties such as viscosity, electrical conductivity, and liquidus temperature, which lie within ranges known to be acceptable for the vitrification process. Durability restrictions ensure that the resultant glass meets the quantitative criteria for disposal in a repository. There are also bounds on the composition of the various components in the glass. In the simplest case, waste and appropriate glass forms (frit) are mixed and heated in a melter to form a glass that satisfies the constraints. It is desirable to keep the amount of frit added to a minimum for two reasons. First, this keeps the frit costs to a minimum. Second, the amount of waste per glass log formed is to be maximized, which keeps the waste disposal costs to a minimum. When there is only a single type of waste, the problem of finding the minimum amount of frit is relatively easy (Narayan et al., 1996).

Hanford has 177 tanks (50,000–1 million gallons) containing radioactive waste. Because these wastes result from a variety of processes, these wastes vary widely in composition, and the glasses produced from these wastes will be limited by a variety of components. The minimum amount of frit would be used if all the high-level wastes were combined to form a single feed to the vitrification process. Because of the volume of waste involved and the time span over which it will be processed, this is logistically impossible. However, much of the same benefit can be obtained by forming blends from sets of tanks. The problem is how to divide all the tanks into sets to be blended together so that a minimal amount of frit is required.

In this discrete blending problem, there are N different sources of waste that have to form a discrete number of blends B, with the number of blends being less than the number of sources or tanks. All the waste from any given tank is required to go to a single blend, and each blend contains waste from N/B sources. Blends of equal size (same number of wastes per blend) were specified; alternatively, blends could be formulated to have approximately the same waste masses. Figure 2.10 shows a set of four wastes that needs to be partitioned into two parts to form two blends. If neither of these were specified as constraints, all the waste would go to a single blend. In this chapter, we look at the single-blend problem. Table 2.18 shows the chemical composition of the high-level waste in three different tanks to be combined to form a single blend. The table shows the waste mass expressed as a total of the first ten chemicals, including the chemical termed as "other." Frit added to the blend consists of these ten chemicals. The waste mass is scaled down by dividing it by 1000 so as to numerically simplify the solution process. The rest of the chemicals are expressed as the fraction of the total.

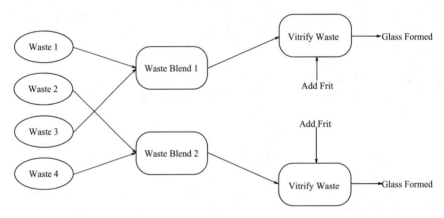

Fig. 2.10. Conversion of waste to glass

In order to form glass, the blend must satisfy certain constraints. These constraints are briefly described below.

1. *Individual Component Bounds:* There are upper $(p_{UL}^{(i)})$ and lower $(p_{LL}^{(i)})$ limits on the fraction of each component $p^{(i)}$ in glass. Therefore,

$$p_{LL}^{(i)} \le p^{(i)} \le p_{UL}^{(i)} \tag{2.46}$$

These bounds are shown in Table 2.19.

Table 2.18. Waste composition

Fractional composition of wastes				
Component	Comp. ID	AY-102	AZ-101	AZ-102
		Tank 1	Tank 2	Tank 3
SiO_2	1	0.072	0.092	0.022
B_2O_3	2	0.026	0.000	0.006
Na_2O	3	0.105	0.264	0.120
Li_2O	4	0.000	0.000	0.000
CaO	5	0.061	0.012	0.010
MgO	6	0.040	0.000	0.003
Fe_2O_3	7	0.328	0.323	0.392
Al_2O_3	8	0.148	0.157	0.212
ZrO_2	9	0.002	0.057	0.063
Other	10	0.217	0.096	0.173
Total	–	1.000	1.000	1.000
Cr_2O_3	11	0.016	0.007	0.005
F	12	0.006	0.001	0.001
P_2O_5	13	0.042	0.001	0.021
SO_3	14	0.001	0.018	0.009
Noble Metals	15	0.000	0.000	0.000
Waste Mass (kgs)		**59,772**	**40,409**	**143,747**

Table 2.19. Component bounds

Component	Lower bound, $p_{LL}^{(i)}$	Upper bound, $p_{UL}^{(i)}$
SiO_2	0.42	0.57
B_2O_3	0.05	0.20
Na_2O	0.05	0.20
Li_2O	0.01	0.07
CaO	0.00	0.10
MgO	0.00	0.08
Fe_2O_3	0.02	0.15
Al_2O_3	0.00	0.15
ZrO_2	0.00	0.13
Other	0.01	0.10

2. *Crystallinity Constraints:* The crystallinity constraints, or multiple component constraints, specify the limits on the combined fractions of different components. There are five such constraints.
   (a) The ratio of the mass fraction of SiO_2 to the mass fraction of Al_2O_3 should be greater than C_1 ($C_1 = 3.0$).
   (b) The sum of the mass fraction of MgO and the mass fraction of CaO should be less than C_2 ($C_2 = 0.08$).

(c) The combined sum of the mass fractions of Fe_2O_3, Al_2O_3, ZrO_2 and *Other* should be less than C_3 ($C_3 = 0.225$).

(d) The sum of the mass fraction of Al_2O_3 and the mass fraction of ZrO_2 should be less than C_4 ($C_4 = 0.18$).

(e) The combined sum of the mass fractions of MgO, CaO, and ZrO_2 should be less than C_5 ($C_5 = 0.18$).

3. *Solubility Constraints:* These constraints limit the maximum value for the mass fraction of one or a combination of components.

(a) The mass fraction of Cr_2O_3 should be less than 0.005.

(b) The mass fraction of F should be less than 0.017.

(c) The mass fraction of P_2O_5 should be less than 0.01.

(d) The mass fraction of SO_2 should be less than 0.005.

(e) The combined mass fraction of Rh_2O_3, PdO, and Ru_2O_2 should be less than 0.025.

4. *Glass Property Constraints:* Additional constraints govern the properties of viscosity, electrical conductivity, and durability but are not considered here.

Blending is most effective when the limiting constraint is one of the first three types, and for the LP formulation these three types of constraints are considered here.

Solution:

Hanford scientists have to decide the amount of each component to be added in the blend to obtain the minimum amount of glass satisfying the first three constraints. We define the decision variables first.

- w_{ij} = amount of component i (where i corresponds to the component ID) in the tank j.
- $W^{(i)}$ = amount of component i in the waste blend.
- $f^{(i)}$ = mass of ith component in the frit.
- $g^{(i)}$ = mass of ith component in the glass.
- G = total mass of glass
- $p^{(i)}$ = fraction of ith component in the glass.

Definition of the above decision variables implies that

$$W^{(i)} = \sum_{j=1}^{3} w_{ij} \tag{2.47}$$

$$g^{(i)} = W^{(i)} + f^{(i)} \tag{2.48}$$

$$G = \sum_{i=1}^{n} g^{(i)} \tag{2.49}$$

$$p^{(i)} = g^{(i)}/G \tag{2.50}$$

Note that G is composed of a known component $W^{(i)}$ and an unknown component $f^{(i)}$ representing degrees of freedom. Also, all these variables are nonnegative because frit can only be added to comply with the constraint. The objective is to minimize the total amount of waste to be vitrified. This can be formulated as:

$$MinG \equiv Min \sum_{i=1}^{n} f^{(i)} \tag{2.51}$$

Subject to the following constraints.

1. Component bounds:
   (a) $0.42 \leq p^{(SiO_2)} \leq 0.57$
   (b) $0.05 \leq p^{(B_2O_3)} \leq 0.20$
   (c) $0.05 \leq p^{(Na_2O)} \leq 0.20$
   (d) $0.01 \leq p^{(Li_2O)} \leq 0.07$
   (e) $0.0 \leq p^{(CaO)} \leq 0.10$
   (f) $0.0 \leq p^{(MgO)} \leq 0.08$
   (g) $0.02 \leq p^{(Fe_2O_3)} \leq 0.15$
   (h) $0.0 \leq p^{(Al_2O_3)} \leq 0.15$
   (i) $0.0 \leq p^{(ZrO_2)} \leq 0.13$
   (j) $0.01 \leq p^{(other)} \leq 0.10$
2. Five glass crystallinity constraints:
   (a) $p^{(SiO_2)} > p^{(Al_2O_3)} * C_1$
   (b) $p^{(MgO)} + p^{(CaO)} < C_2$
   (c) $p^{(Fe_2O_3)} + p^{(Al_2O_3)} + p^{(ZrO_2)} + p^{(Other)} < C_3$
   (d) $p^{(Al_2O_3)} + p^{(ZrO_2)} < C_4$
   (e) $p^{(MgO)} + p^{(CaO)} + p^{(ZrO_2)} < C_5$
3. Solubility Constraints:
   (a) $p^{(Cr_2O_3)} < 0.005$
   (b) $p^{(F)} < 0.017$
   (c) $p^{(P_2O_5)} < 0.01$
   (d) $p^{(SO_3)} < 0.005$
   (e) $p^{(Rh_2O_3)} + p^{(PdO)} + p^{(Ru_2O_3)} < 0.025$
4. Nonnegativity Constraint:
   (a) $f^{(i)} \geq 0$

Note that Equation (2.50) is a nonlinear equation, making the problem an NLP. We can eliminate this constraint if we can write all four types of constraint equations in terms of the mass of the component $g^{(i)}$ instead of the fraction $p^{(i)}$.

The LP Formulation

$$Min \sum_{i=1}^{n} f^{(i)} \tag{2.52}$$

$$W^{(i)} = \sum_{j=1}^{3} w_{ij} \tag{2.53}$$

$$g^{(i)} = W^{(i)} + f^{(i)} \tag{2.54}$$

$$G = \sum_{i=1}^{n} g^{(i)} \tag{2.55}$$

1. Component bounds:
   (a) $0.42G \le g^{(SiO_2)} \le 0.57G$
   (b) $0.05G \le g^{(B_2O_3)} \le 0.20G$
   (c) $0.05G \le g^{(Na_2O)} \le 0.20G$
   (d) $0.01G \le g^{(Li_2O)} \le 0.07G$
   (e) $0.0 \ \ \le g^{(CaO)} \le 0.10G$
   (f) $0.0 \ \ \le g^{(MgO)} \le 0.08G$
   (g) $0.02G \le g^{(Fe_2O_3)} \le 0.15G$
   (h) $0.0 \ \ \le g^{(Al_2O_3)} \le 0.15G$
   (i) $0.0 \ \ \le g^{(ZrO_2)} \le 0.13G$
   (j) $0.01G \le g^{(other)} \le 0.10G$
2. Five Glass crystallinity constraints:
   (a) $g^{(SiO_2)} > g^{(Al_2O_3)} * C_1$
   (b) $g^{(MgO)} + g^{(CaO)} < C_2 * G$
   (c) $g^{(Fe_2O_3)} + g^{(Al_2O_3)} + g^{(ZrO_2)} + g^{(Other)} < C_3 * G$
   (d) $g^{(Al_2O_3)} + g^{(ZrO_2)} < C_4 * G$
   (e) $g^{(MgO)} + g^{(CaO)} + g^{(ZrO_2)} < C_5 * G$
3. Solubility Constraints:
   (a) $g^{(Cr_2O_3)} < 0.005G$
   (b) $g^{(F)} < 0.017G$
   (c) $g^{(P_2O_5)} < 0.01G$
   (d) $g^{(SO_3)} < 0.005G$
   (e) $g^{(Rh_2O_3)} + g^{(PdO)} + g^{(Ru_2O_3)} < 0.025G$
4. Nonnegativity Constraint:
   (a) $f^{(i)} \ge 0$

This problem is then solved using iterative solution procedures using GAMS, and the solution to the LP is given in Table 2.20. The GAMS input files for this problem and the solution can be found online on Springer website with the book link. Thus, Hanford should add approximately 590 kgs of frit to the blend of these three tanks. Although this appears to be a small amount as compared to the total mass of the glass, when all the tanks are considered, blending and optimization can reduce the amount of total glass formed by more than half.

Table 2.20. Composition for the optimal solution

Component	Mass in the waste, $W^{(i)}$	Mass in frit $f^{(i)}$
SiO_2	11.2030	464.2909
B_2O_3	2.4111	110.1268
Na_2O	34.1980	7.5120
Li_2O	0.0000	8.3420
CaO	5.5436	
MgO	2.8776	
Fe_2O_3	89.0097	
Al_2O_3	45.5518	
ZrO_2	11.4111	
Other	41.7223	
Total	243.9281	590.2718

2.9 Sustainable Mercury Management: An LP

Mercury has been recognized as a global threat to our ecosystem, and it is fast becoming a major concern to environmentalists and policymakers. The adverse effects of mercury are increasingly acknowledged. For humans, the primary targets for the toxicity of mercury and mercury compounds are the nervous system, kidney, and developing fetus. Other systems in the human body that may be affected include respiratory, cardiovascular, gastrointestinal, hematological, immune, and reproductive. Mercury is also known to adversely affect mortality and reproduction rates in aquatic and terrestrial biota.

Mercury can cycle in the environment in all media as part of both natural and anthropogenic activities. Figure 2.11 gives a pictorial representation of the atmospheric mercury cycling. The majority of mercury is emitted in the air in elemental or inorganic form, mainly by coal fired power plants, waste incinerators, industrial and domestic utility boilers, and chlor-alkali plants. However, most of the mercury in the air is deposited into various water bodies such as lakes, rivers, and oceans through processes of dry and wet deposition. In addition, the water bodies are enriched in mercury due to direct additions, such as industrial wastewater discharge, stormwater runoffs, agricultural runoffs, and others. Once present in water, mercury is highly dangerous not only to the aquatic communities but also to humans through direct and indirect effects. Methylation of inorganic mercury leads to the formation of methyl mercury, which accumulates up the aquatic food chains so that organisms in higher trophic levels have higher mercury concentrations. The consumption of these aquatic animals by humans, and wild animals further aids bioaccumulation along the food chain leading to previously mentioned adverse effects. As a result, contaminated fish consumption is the most predominant path of human exposure to mercury. This has resulted in fish consumption advisories at various water bodies throughout the USA.

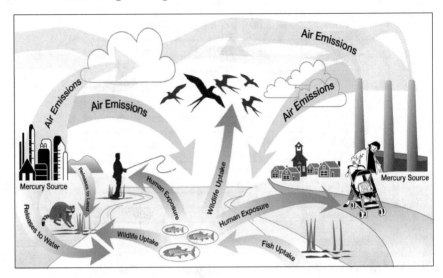

Fig. 2.11. Environmental cycling of mercury

2.9.1 Mercury Management Approach

Owing to the complex issues in mercury cycling, it might be argued that a sustainability based approach is essential. It points towards an approach with a broader perspective, accounting for interacting systems over disparate spatial and temporal scales. Sustainability also calls for a synergic approach to technological, regulatory, and social decision making. With this approach, successful management of mercury pollution should consider management strategies at various stages of the cycle. While doing so, it is essential to juxtapose the environmental, economic, and social objectives. The work presented by Diwekar and Shastri (2010) is a step in this direction. Since the idea of sustainability changes with the system being considered, different objectives, management options, and tools are applied for different applications. The work presents management strategies at two different stages of the mercury cycle. The case study details are derived from the papers by Shastri and Diwekar (2008; 2009), Diwekar and Shastri (2010) and Shastri et al. (2011). The input files and codes for GAMS, AIMMS, and MATLAB for this case study can be found online on the Springer website with the book link.

- Industrial sector (inter-industry) level management: Industries strive to reduce their regulatory compliance cost to ensure profitability and economic sustainability. This needs to be balanced with strict regulations for an environmental cause. Therefore, at the industrial sector level, this work analyzes the option of pollutant trading in water, and proposes a decision-making framework to optimize economic and ecological objectives. Through this case study, it also highlights the importance of correct modeling on optimal decisions. This problem is presented as an LP (this chapter), an NLP(Chapter 3), a MILP, and an MINLP (Chapter 4).

- Ecosystem level management: Effective control of mercury bioaccumulation in water bodies can further reduce the harmful effects and affect regulatory decisions. This work investigates the idea of lake pH control through time-dependent liming to reduce bioaccumulation. Optimal control theory has been used to derive the time-dependent liming decisions. This problem is posed as an optimal control problem in Chapter 7.

Uncertainties, static as well as dynamic, increase as one moves from individual industry level, where model and tools are well developed, to ecosystem level. The stochastic versions of this problem are presented in Chapters 5 and 7 and a multiobjective version in Chapter 6.

2.9.2 Watershed Based Trading

The primary reason for pollutant trading to be attractive is the flexibility it offers to all stakeholders. The main stakeholders are polluters (industries), and regulatory authority, which also represents the interests of the society. Although the flexibility offered by trading is desirable, it also adds new dimensions to decision making for these stakeholders. It is, therefore, important to understand individual perspective while designing a trading framework.

Following on the lines of air pollutant trading, trading principles have also been sanctioned by USEPA for water pollution problems on a limited basis since early 1980s. USEPA has since then formalized the concept into a framework to guide the effective implementation of water pollutant trading (US EPA, 2003). Trading is based on the fact that sources in a watershed can face very different costs to control the emission of the same pollutant. Trading programs allow facilities facing higher pollution control costs to meet their regulatory obligations by purchasing environmentally equivalent (or superior) pollution reductions from another source at a lower cost. Firms that have the financial capability and infrastructure to perform pollutant reduction below the required limit get credits for it. These credits can be sold to another firm to gain monetary benefits or banked for future use if that is a possibility.

Pollutant dischargers are mainly classified as point and non-point sources. Point sources are defined as those having direct and measurable emissions (e.g., industries, municipal waste treatment plants), while non-point sources are those with diffused emissions that are difficult to measure (e.g., agricultural or stormwater runoffs). Among the various possibilities of carrying out trading, the one between point sources is thought to be simpler and achievable. This is owing to their measurable discharges in terms of quality and quantity, and also due to the measurable assessment of pollution reduction techniques.

Point sources are regulated under the National Pollution Discharge Elimination System (NPDES) established under the Clean Water Act (CWA).

Under NPDES, all facilities that discharge pollutants from any point source into waters of the USA are required to obtain a permit, allowing them to discharge only a certain amount of the pollutant. The permit provides two levels of control: technology-based limits (based on the ability of the dischargers in the same industrial category to treat the wastewater) and water quality-based limits (if technology-based limits are not sufficient to provide protection of the water body). The permits can be an individual (specific to a company) permit or a general (applicable to a group of companies) permit. The existence of general water quality-based permits in a watershed is equivalent to the concept of Total Maximum Daily Load (TMDL) in the watershed. TMDL is established by the state for water bodies or watersheds where technology-based requirements alone are not sufficient to attain water quality goals. TMDL establishes the loading capacity of a defined watershed area, identifies reductions or other remedial activities needed to achieve water quality standards, identifies sources, and recommends waste load allocation for point (and non-point) sources.

EPA has proposed two possible trading mechanisms for point sources.

- Trades can occur in the context of individual point source permits. In this case, different point sources have individual (technology- or water quality-based) pollutant permits, and there is no common water quality based limit. Treatment cost differentials encourage trading between various point sources.
- Trades can also occur through the development of TMDL or another equivalent analytical framework. In this case, water quality-based limits, common to all the point sources in the watershed, are established. This provides a starting point to compare the costs of the baseline responsibilities necessary to achieve water quality goals with alternative allocations. Parties to the trade then negotiate within the loading capacity determined under the TMDL so as to satisfy the common discharge objective.

Various parameters that affect the economics of trading are trading ratio (how many units of pollutant reduction a source must purchase to receive credit for one unit of load reduction); transaction costs (expenses that trading participants occur only as a result of trading); the number of participants; availability of cost data; and uncertainties related to continued industry participation and data availability.

Once a TMDL has been established, each point source is assigned a particular load allocation and may need to reduce its discharge to satisfy the allocation. It has two options to accomplish this:

- The point source can implement an end of pipe treatment method, which entail certain capital and operating cost depending on the existing technology, the amount being treated and the level of reduction is achieved. This cost typically differs for different point sources.

- The point source can trade a particular amount of pollutant to another point source in the watershed that is able to reduce its discharge more than that specified by the regulation.

2.9.3 Trading Optimization Model Formulation

The formulation considers that TMDL regulation has already been developed by the state in consultation with USEPA. This translates into a specific load allocation for each point source.

Consider a set of point sources (PS_i), $i = 1, ..., N$, disposing of the pollutant containing wastewater to a common water body or watershed. The various point source specific parameters are as follows. D_i = Discharge quantity of polluted water from PS_i [volume/year].
c_i = Current pollutant discharge concentration for PS_i [mass/volume].
a_i = Current pollutant discharge quantity for PS_i [mass/year].
L_i = Load allocation to PS_i based on TMDL regulation [mass/year].
red_i = Desired pollutant quantity reduction in discharge of PS_i [mass/year].
P_i = Treatment cost incurred by PS_i to reduce discharge when trading is not permitted.
Here, the value of red_i is given by

$$red_i = D_i.c_i - L_i = a_i - L_i \tag{2.56}$$

Every PS has the option of trading or implementing a particular waste reduction technology. Let $j = 1, ..., M$ be the set of reduction technologies available to the point sources for implementation. The technology specific parameters are as follows.
TC_j = Total treatment plant cost [\$/volume].
q_j = Pollution reduction possible from the process [mass/volume].
The total plant cost TC_j is the sum of the annualized capital cost and annual operating and maintenance cost. The annualized capital cost, in turn, depends on the total equipment and setup cost. Trading is possible between all point sources. For simplicity, a single trading policy exists between all possible pairs of point sources, and a single trading ratio and the transaction fee are applicable to all the trades. Let r be the trading ratio, and F be the transaction cost (in \$/mass) to be paid by the point source trading its pollutants. The objective of the model is to achieve the desired TMDL goal at the *minimum overall cost*.

Let b_{ij} be the binary variables representing the point source-technology correlation. The variable is 1 when PS i installs technology j, and 0 otherwise. Let t_{ik} (mass/year) be the amount of pollutant traded by PS i with PS k; i.e., PS i pays PS k to take care of its own pollution. All the parameters are on an annual basis. The model is then formulated as follows.

Objective:

$$\text{Minimize} \qquad \sum_{i=1}^{N}\sum_{j=1}^{M} TC_j.D_i.\, b_{ij} \qquad\qquad (2.57)$$

Constraints:

$$t_{ii} = 0 \qquad \forall i = 1, ..., N \qquad\qquad (2.58)$$

$$red_i \leq \sum_{j=1}^{M} q_j.D_i.\, b_{ij} + \sum_{k=1}^{N} t_{ik} - r\sum_{k=1}^{N} t_{ki} \qquad \forall i = 1, ..., N$$

$$(2.59)$$

$$P_i \geq \sum_{j=1}^{M} b_{ij}.TC_j.D_i + F\left(\sum_{k=1}^{N} t_{ik} - \sum_{k=1}^{N} t_{ki}\right) \qquad \forall i = 1, ..., N$$

$$(2.60)$$

The objective function gives the sum of the technology implementation cost for all point sources. Although each PS will also spend or gain from practicing trading, the expense for one PS in a watershed is earning for one or more PS in the same watershed. As a result, for the complete watershed, trading does not contribute to the cost objective. The first set of constraints eliminates trading within the same PS. The second set of constraints ensures that all regulations are satisfied. The pollutant discharge reduction for all PS at the end of technology implementation and/or trading must be at least equal to the targeted reduction. Trading ratio r is usually set higher than 1, which requires the buyer to purchase more units of reduction than they have to achieve without trading, to include a margin of safety in the trading mechanism. This is done to account for uncertainties in the level of control needed to attain water quality standards, and to provide a buffer in case traded reductions are less effective than expected. Consequently, the PS accepting additional discharge reduction responsibility has to reduce the pollutant by an amount equal to the actual quantity traded times the trading ratio. Accordingly, in (2.59), the reduction achieved by industry i reflects the amount traded by industry k (t_{ki}) modified by the trading ratio r. The last constraint ensures that the expenses incurred by each PS with trading are not more than those without trading. This is because the trading framework by USEPA mentions that no polluter can be forced to trade. As a result, a polluter will participate in trading only if there is a financial incentive for it. In the absence of this constraint, the model solution might force some industries to spend more when trading is practiced, if such a solution results in lower overall cost. This clearly is an unlikely scenario that is avoided by this constraint.

The model given by (2.57)–(2.60) is a mixed integer linear programming problem (MILP). The decision variables in the model are binary variables b_{ij} and continuous variables t_{ik}. It should be noted that the reduction capability of the treatment process j is considered to be fixed at q_j. However, when the binary variables are fixed meaning the control technologies are already implemented; the problem becomes an LP

The model formulated above is quite general, applicable to any watershed and any pollutants. The next section discusses the application of the model on a case study of mercury waste management in the Savannah River basin in the state of Georgia, the USA.

2.9.4 Savannah River Watershed Details

This case study was performed in 2006.There were 206 advisories in Georgia in 2004. Georgia issued 178 fish consumption advisories—relating to 40 different rivers and 34 lakes and ponds. Near the Olin plant, in the Savannah River Basin, there were 24 advisories, affecting five rivers and seven lakes and ponds. TMDL was established for five contiguous segments of the Savannah River. These segments are included in the list of impaired water bodies since the tissue mercury concentration in certain fish species exceeds the Georgia Department of Natural Resources (GDNR) Fish Consumption Guidelines. Although the mid-line of the Savannah River serves as the east-west boundary between the states of Georgia and South Carolina, the TMDL does not provide waste load allocations to South Carolina NPDES facilities. This TMDL reflects an assumption that the concentrations of mercury in the South Carolina portion of Savannah River will meet the applicable Georgia water quality standards at the South Carolina-Georgia border. In order to develop the TMDL, the applicable water quality standard that gives the maximum safe concentration of mercury in water must be determined. EPA determined the applicable water quality standard for total mercury in the ambient water of the Savannah River Basin to be 2.8 ng/l (parts per trillion). At this concentration, or below, fish tissue residue concentrations of mercury will not exceed 0.4 mg/kg, which is protective of the general population from the consumption of freshwater fish. This interpretation of Georgia's water quality standard is based on site-specific data gathered for the Savannah River in 2000, specifically for the purpose of this TMDL.

In all, there are 29 significant point sources discharging mercury in Savannah River watershed. These point sources represent a wide spectrum and include 13 major municipal polluters, 12 major industrial polluters, two minor municipal polluters, and two minor industrial polluters. It should be noted that there are more point sources in this region. However, mercury discharges from those are negligible owing to their relatively low discharge volumes. One option to implement the TMDL is to apply a common WQS of 2.8 ng/lit to all point source discharges across the watershed. Therefore, under this option, the waste load allocation for each NPDES point source identified in this TMDL would be the product of 2.8 ng/l and its permitted or design flow rate. The sum of these individual waste load allocations is 0.001 kg/year, which is

significantly less than the 0.33 kg/year cumulative waste load allocation provided to all NPDES facilities. The given current discharge concentrations for
the 29 point sources are assumed. The overall reduction needed to achieve
the TMDL criteria is about 44%. The targeted overall reduction for the PS
is, therefore, taken to be 40%, and the individual discharge concentrations
are adjusted accordingly. This was taken according to USEPA guidelines for
TMDL.

Table 2.21 gives various parameter values related to the point sources. The
table also gives the values of red_i (targeted reduction) and P_i (treatment cost
without trading) for each PS at TMDL 32 Kg/year, for which the model is
solved later.

Table 2.21. Point source data for the Savannah River basin

Industry	Total Volumetric Discharge (MGD-Million Gallons per Day)	Current Discharge concentrations (ng/lit)	Targeted reduction (g/year)	Treatment cost without trading ($/year)
I_1	46.1	4.65	0.1149	1.68×10^7
I_2	1.5	3.7	0.0017	328,500
I_3	4.6	4.3	0.0092	1,679,000
I_4	1.5	3.4	0.0011	328,500
I_5	2.0	3.88	0.0028	730,000
I_6	2.24	3.7	0.0026	490,560
I_7	1.2	3.9	0.0017	438,000
I_8	27.0	4.83	0.0740	1.48×10^7
I_9	4.5	4.0	0.0072	1,642,500
I_{10}	1.0	3.1	0.00035	219,000
I_{11}	1.0	3.06	0.00029	2f 000
I_{12}	1.0	3.22	0.00052	219,000
I_{13}	2.0	3.31	0.0013	438,000
I_{14}	3.765	4.8	0.0101	2,061,337
I_{15}	18.0	4.33	0.0369	6,570,000
I_{16}	7.2	5.1	0.0224	3,942,000
I_{17}	58.6	4.87	0.1639	3.21×10^7
I_{18}	23.0	4.52	0.0532	8,395,000
I_{19}	1.152	5.05	0.0035	630,720
I_{20}	0.362	4.14	0.00064	132,130
I_{21}	108.0	4.58	0.2588	3.94×10^7
I_{22}	4.68	5.2	0.0152	2,562,300
I_{23}	28.09	4.41	0.0607	1.02×10^7
I_{24}	1.921	3.9	0.0028	701,165
I_{25}	0.544	4.5	0.0012	198,560
I_{26}	0.5	3.95	0.0008	182,500
I_{27}	0.003	3.72	3.62×10^{-6}	657
I_{28}	1.246	4.1	0.0021	454,790
I_{29}	0.054	3.4	4.14×10^{-5}	11,826

Technology Details

Three treatment technologies are considered for this model, and they are available to all point sources for implementation. These include coagulation and filtration, activated carbon adsorption, and ion exchange process. The capital requirement and reduction capability of any process are expected to be (nonlinearly) related to the capacity of the treatment plant and the form and concentration of the waste to be treated, among many other factors. For this analysis, though, such complex relationships are ignored for simplicity, and the treatment cost is only linearly related to the volume of the waste. Trading in the presence of nonlinear cost models is discussed in Chapter 3 onwards. Total plant cost data for the treatment methods are reported in USDOI as a function of the waste volume. The total plant cost includes capital as well as annual operating cost per unit volume of waste treated, and is calculated using the following equations.

$$\text{Annualized capital cost} = [\text{Total capital equipment cost} +$$
$$\text{Project related special cost}]$$
$$\times \text{ (capital recovery factor for 30 years}$$
$$\text{at 3.89\% real annual interest based}$$
$$\text{on lagged impact on interest} = 0.057) \quad (2.61)$$
$$\text{Total annual cost}(TC) = \text{Annualized capital cost} + \text{Total annual}$$
$$\text{operations and maintenance cost} \quad (2.62)$$

Since waste volumes encountered in this case study are mostly greater than 1 MGD, asymptotic values reported in USDOI are used. The treatment efficiencies depend on waste composition and concentration. In general, though, a more efficient treatment is likely to be more expensive. This criterion, along with data given in the literature, is used to decide the treatment efficiencies. Table 2.22 gives the technology data.

Table 2.22. Data for the various treatment technologies

Process	Mercury reduction capability (ng/lit)	Capital requirement ($/1000 gallons)
Coagulation and Filtration (A)	2.0	1.00
Activated carbon adsorption (B)	3.0	1.5
Ion exchange (C)	1.0	0.6

Trading Details

Trading parameters needed for problem-solving are trading ratio and trading transaction cost. For a point source-point source trading, as considered in this model, a study conducted for the USEPA for the trading of toxic pollutants in Delaware River recommends a trading ratio between 1.1 and 1.25. For the solution of this model, the ratio r is 1.1. It should be noted here that since mercury trading in water has not been practiced yet, the transaction fee is not easy to decide. However, an EPA document gives a hypothetical example of water quality trading (USEPA, 1996) that considers the transaction fee in the range of the per kg treatment cost of the pollutant. This is also observed for SO_2 trading, which is already practiced. The values of per unit marginal abatement costs and trading transaction costs for SO_2 trading are reported in literature. For this work, using the average volumetric discharge and the average desired discharge reduction for all point sources, the average treatment cost is calculated. To calculate this cost, data for the three treatment processes reported in Table 2.22 is used. This gives the treatment cost in "$/Kg of mercury." Based on this value, the transaction cost is fixed at $ 1.5 Million per Kg. However, in the later formulations, the transaction costs are ignored as their contribution to the solution is negligible.

In this chapter, we present the mercury management problem as an LP by assuming that the technology selection is already made. In the next subsection, we consider a smaller version of the problem as an LP.

2.9.5 LP Problem Details

Industry Details

Seven polluters are considered in the problem discharging the polluted water to the water body. The considered industries represent a wide spectrum and include one major municipal polluter, four major industrial polluters, and two minor industrial polluters. For this work, the average reduction in discharge concentration is assumed to be about 30%.

Table 2.23 gives the values of the various parameters related to the industries. The total mercury reduction target for the seven industries is 47.2371×10^{-5}.

Technology Details

Since the treatment volume is significantly less than the 29 sources problem, the technology linear model range is different. Table 2.24 gives the values for these processes.

Results and Discussions

Table 2.25 shows the implemented technologies (fixed binary variables, b_{ij}). This converts the problem to an LP . The results for trading volumes for this LP are shown in Table 2.26.

Table 2.23. Data for the industries

Industry	Total volumetric discharge (MGD)	Current discharge concentrations (ng/lit)	Mercury reduction requirement (Kg/year) $(\times 10^{-5})$	Available annual capital (Million $)
I_1	46.1	4.1	8.239	27.5
I_2	58.6	4.0	9.664	50
I_3	108.0	4.3	22.9853	65
I_4	3.765	3.6	0.4160	5
I_5	23.0	4.4	5.7187	15
I_6	1.246	3.7	0.1538	7.5
I_7	0.54	3.4	0.0597	6.5

Table 2.24. Data for the various treatment technologies

Process	Mercury reduction capability (ng/lit)	Capital requirement ($/1000 gallons)
Coagulation and Filtration (A)	1.15	1.0
Activated carbon adsorption (B)	1.8	1.55
Ion exchange (C)	0.8	0.65

Table 2.25. Technology selection

Industry	Technology selected
I_1	C
I_2	A
I_3	B
I_4	C
I_5	B
I_6	C
I_7	C

Table 2.26. The trading matrix in 10^{-5} Kg/year

Industry	I_1	I_2	I_3	I_4	I_5	I_6	I_7
I_1	0	0	3.1	0	0	0	0
I_2	0	0	0.4	0	0	0	0
I_3	0	0	0	0	0	0	0
I_4	0	0	0	0	0	0	0
I_5	0	0	0	0	0	0	0
I_6	0	0	0.0162	0	0	0	0
I_7	0	0	0	0	0	0	0

2.10 Summary

Linear programming problems involve linear objective function and linear constraints. The LP optimum lies at a vertex of the feasible region, which is the basis of the simplex method. LP can have 0 (infeasible), 1, or infinite (multiple) solutions. LPs do not have multiple local minima. As a general rule, LP computational effort depends on the number of constraints rather than the number of variables. Many of the LP methods are derived from the simplex method, and special classes of problems can be solved efficiently with special LP methods. The interior point method is based on the transformation of variables and using a search direction similar to nonlinear programming methods discussed in the next chapter. This method is polynomially bounded, but only large-scale problems where no prior information is available show computational savings.

Exercises

2.1 Write the following problems in standard form and solve using the simplex method. Verify your solutions graphically (where possible).

1.

$$\max 6x_1 + 4x_2$$
$$3x_1 + 2x_2 \leq 8$$
$$-4x_1 + 9x_2 \leq 20$$
$$x_1, x_2 \geq 0$$

2.

$$\max 3x_1 + 2x_2$$
$$-2x_1 + x_2 \leq 1$$
$$x_1 + 3x_2 \geq 2$$
$$x_1, x_2 \geq 0$$

3.

$$\min 2x_1 - 4x_2$$
$$3x_1 + x_2 \leq 1$$
$$-2x_1 + x_2 \geq 3$$
$$x_1, x_2 \geq 0$$

4.

$$\max x_1 + 5x_2$$
$$x_1 + 3x_2 \leq 5$$
$$2x_1 + x_2 = 4$$
$$x_1 - 2x_2 \geq 1$$
$$x_1, x_2 \geq 0$$

5.

$$\min 3x_1 + 4x_2 - x_3$$
$$x_1 + 3x_2 - x_3 \geq 1$$
$$2x_1 + x_2 + 0.5x_3 \geq 4$$
$$x_1, x_2 \geq 0; x_3 \text{ is unconstrained}$$

6.

$$\min 8x_1 - 3x_2 + 10x_3$$
$$5x_1 - 2x_2 - 4x_3 \geq 3$$
$$3x_1 + 6x_2 + 8x_3 \geq 4$$
$$2x_1 - 4x_2 + 8x_3 \geq -4$$
$$-x_2 + 5x_3 \geq 1$$
$$x_1, x_2, x_3 \geq 0$$

Also solve this problem using a dual formulation.

2.2 A refinery has two crude oil materials with which to create gasoline and lube oil:

1. Crude A costs \$28/bbl and 18,000 bbl are available.
2. Crude B costs \$38/bbl and 32,000 bbl are available.

The yield and sale price per barrel of the products and the associated markets are shown in Table 2.27.

Table 2.27. Yield and sale prices of products

Product	Yield/bbl Crude A	Yield/bbl Crude B	Sale price per bbl	Market (bbl)
gasoline	0.6	0.85	\$60	20,000
lube oil	0.4	0.15	\$130	12,000

How much crude A and B should be used to maximize the profit of the company? Formulate and solve the problem using the simplex algorithm. Verify your solution graphically. How would the optimal solution be affected if

1. The market for lube oil increased to 14,000.bbl.
2. The cost of crude A decreased to $20/bbl.

2.3   A manufacturer sells products A and B. The profit from A is $12/kg and
from B $7/kg. The available raw materials for the products are 100 kg
of C and 80 kg of D. To produce one kilogram of A the manufacturer
needs 0.5 kg of C and 0.5 kg of D. To produce one kilogram of B the
manufacturer needs 0.4 kg of C and 0.6 kg of D. The market for product
A is 60 kg and for B 120 kg. How much raw material should be used to
maximize the manufacturer's profit? Formulate and solve the problem
using the simplex algorithm. Verify your solution graphically. How would
the optimal solution be affected if
1. The availability of C were increased to 120 kg.
2. The availability of D were increased to 100 kg.
3. The market for A were decreased to 40 kg.
4. The profit of A were $10/kg.

2.4   On the bank of a river there are three neighboring cities that are discharging two kinds of pollutants A and B into the river. Now the state
government has set up a treatment plant that treats pollutants from City
1 for $15/ton which reduces pollutants A and B by the amount of 0.10
and 0.45 tons per ton of waste, respectively. It costs $10/ton to process
a ton of City 2 waste and consequentially reducing pollutants A and B
by 0.20 and 0.25 tons per ton of waste, respectively. Similarly City 3
waste is treated for $20 reducing A by 0.40 and B by 0.30 tons per ton
of waste. The state wishes to reduce the amount of pollutant A by at
least 30 and B by 40 tons. Formulate the LP that will minimize the cost
of reducing pollutants by desired amount.

2.5   Products I and II that are manufactured by a firm are sold at the rate of
$2 and $3, respectively. Both products have to be processed on machine
A and B. Product I requires 1 min on A and 2 min on B whereas Product
II requires 1 min on each machine. Machine A is not available for more
than 6 h 40 min/day, whereas machine B is not available for more than
10 h. Formulate the problem for profit maximization. Solve this problem
using the simplex method.

2.6   There are many drug manufacturers producing various combinations for
a similar ailment. Now a doctor wishes to prescribe a combination dosage
such that the cost is minimum so that it could be given to poor patients.
Drug A costs 50 cents, Drug B costs 20 cents, Drug C 30 cents, and
Drug D 80 cents per tablet, respectively. Daily requirements are 5 mg of
Medicine 1, 6 mg Medicine 2, 10 mg Medicine 3, and 8 mg Medicine 4.
The prescribed composition of each drug is given in Table 2.28. Write
the prescription that satisfies the medicinal requirements at minimum
cost.

2.7   A manufacturing firm has discontinued production of a certain profitable
product line. This created considerable excess production capacity. Management is considering devoting their excess capacity to one or more of

Table 2.28. Prescribed composition of each drug

Drug	Medicine 1	Medicine 2	Medicine 3	Medicine 4
A	4	3	2	2
B	2	2	2	4
C	1.5	0	4	1
D	5	0	4	5

three products 1, 2, and 3. The available capacity on the machines and the number of machine-hours required for each unit of the respective product are given in Table 2.29.

Table 2.29. Available machine capacities

Machine	Available time(hrs/week)	Productivity(hrs/unit)		
		Product 1	Product 2	Product 3
Milling	250	8	2	3
Lathe	150	4	3	0
Grinder	50	2	0	1

The unit profit would be $20, $6, and $8, respectively, for products 1, 2, and 3. Find how much of each product the firm should produce in order to maximize profit.

2.8 Four professors are each capable of teaching any of four different courses. Class preparation time in hours for different topics varies from professor to professor and is given in Table 2.30. Each professor is assigned only one course. Find the assignment policy schedule so as to minimize the total course preparation time for all the courses.

Table 2.30. Course preparation times in hours

Professor	LP	Queueing theory	Dynamic programming	Regression analysis
1	2	10	9	7
2	15	4	14	8
3	13	14	16	11
4	3	15	13	8

2.9 The investment opportunities are available (Table 2.31) with their cash flow and net present value (million dollars) for a firm. It at the start has 30 million dollars and estimates that at the end of 1 year it will have 15 million dollars. The firm can purchase any fraction of any investment, the cash flow, and net present value accordingly. The firm's objective is to maximize the NPV. Assumption is that any funds left over time at time zero cannot be used at time one.

Table 2.31. Investment opportunities

	Investment 1	Investment 2	Investment 3
Time 0 cash flow	$11	$297$5	
Time 1 cash flow	$3	$34	$5
NPV	$13	$39	$16

2.10 The engineering department for Alash Inc. has their computers distributed to their employees according to Table 2.32.

Table 2.32. Computer distribution

Computer (RAM)	Designer	Analysts 1	Analysts 2	Engineers
266 MHz 64 MB	10	7	7	14
200 MHz 64 MB	8	2	2	6
166 MHz 32 MB	18	2	2	34
133 MHz 32 MB	7	0	0	17
350 MHz 128 MB	0	0	0	0

The designer and analysts (grade 1) are responsible for generating engineering designs, whereas the analysts (grade 2) and engineers are responsible for generating repair item reports. Currently, all of the designers and analysts (grade 1) utilize Autocad software on their computers for generating the designs. The analysts (grade 2) and engineers utilize software M (name changed for confidentiality reasons) on their computers. Autocad software requires more Pentium and more RAM than software M. With a computer with 266 MHz Pentium and 64 MB of RAM, it takes a designer or analyst (grade 1) an average of 40 man-hours to produce one drawing. A difference in one MHz of Pentium changes the speed of producing a drawing on Autocad 0.02 % and an increase of 32 MB of RAM allows the computer 0.15% faster. With a computer with 166 MHz and 32 MB of RAM it takes an engineer an average of 20 man-hours to produce one repair item report. Find the distribution that will minimize the cost to finish the required amount of work.

Bibliography

Arbel A. (1993), *Exploring Interior-Point Linear Programming Algorithms and Software*, MIT Press, Cambridge, MA.

Carter M. W. and C. C. Price (2001), *Operations Research: A Practical Introduction*, CRC Press, New York.

Dantzig G. B.(1963), *Linear Programming and Extensions*, Princeton University Press, NJ.

Dantzig G. B. and Thapa M. N. (1996), *Linear Programming*, Springer-Verlag, New York.

Diwekar U. and Y. Shastri (2010), Green process design, green energy, and sustainability: a systems analysis perspective, *Computers and chemical Engineering* **34**, 1348.

Emmett A. (1985), Karmarkar's algorithm: A threat to simplex, *IEEE Spectrum*, December, 54.

Fiacco A. V. and G. P. McCormick (1968), *Nonlinear Programming: Sequential Unconstrained Minimization*, John Wiley and Sons, New York (reprinted by SIAM, 1990).

Narayan V.,U. Diwekar and M. Hoza (1996), Synthesizing optimal waste blends, *Industrial and Engineering Chemistry Research* **35**, 3519.

Nocedal J. and S. Wright (1999), *Numerical Optimization*, Springer Series in Operations Research, Springer-Verlag, New York.

Taha H. A. (1997), *Operations Research: An Introduction*, Sixth Edition, Prentice-Hall, Upper Saddle River, NJ.

Shastri Y. and U. Diwekar. Optimal control of lake pH for mercury bioaccumulation control, (2008), *Ecological Modelling* **216**,1.

Shastri Y. and U. Diwekar. (2009) Freshwater ecosystem conservation and management: A control theory approach. Chapter in*Freshwater Ecosystems: Biodiversity, Management and Conservation*, Nova Science Publishers.

Shastri Y., U. Diwekar and S. Mehrotra.,(2011),An innovative trading approach for mercury waste management,*International Journal of Innovation Science*,**3**, 9.

Winston W. L. (1991), *Operations Research: Applications and Algorithms*, Second Edition, PWS-KENT, Boston.

Wright S. J. (1997), *Primal-Dual Interior-Point Methods*, SIAM, Philadelphia.

Wright S. J. (1999), Algorithms and software for linear and nonlinear programming, *Foundations of Computer Aided Process'99*, Paper I07, CACHE Corporation, AIChE, New York.

3

Nonlinear Programming

In nonlinear programming[1] (NLP) problems, either the objective function, the constraints, or both the objective and the constraints are nonlinear, as shown below in Example 3.1.

Example 3.1: Consider a simple isoperimetric problem described in Chapter 1. Given the perimeter (16 cm) of a rectangle, construct the rectangle with maximum area. To be consistent with the LP formulations of the inequalities seen earlier, assume that the perimeter of 16 cm is an upper bound to the real perimeter.

Solution: Let x_1 and x_2 be the two sides of this rectangle. Then the problem can be formulated as a nonlinear programming problem with the nonlinear objective function and the linear inequality constraints given below.

$$\text{Maximize} \quad Z = x_1 \times x_2 \qquad (3.1)$$
$$x_1, \; x_2$$

subject to

$$2x_1 + 2x_2 \leq 16 \quad \text{Perimeter Constraint} \qquad (3.2)$$
$$x_1 \geq 0; x_2 \geq 0$$

Electronic Supplementary Material: The online version of this chapter (https://doi.org/10.1007/978-3-030-55404-0_3) contains supplementary material, which is available to authorized users.

---

[1]One section (constraint qualification) of this chapter is written by Dr. Yogendra Shastri, Department of Bioengineering, University of Illinois at Chicago.

U. M. Diwekar, *Introduction to Applied Optimization*, Springer Optimization and Its Applications 22, https://doi.org/10.1007/978-3-030-55404-0_3

Let us start plotting the constraints and the iso-objective (equal-area) contours in Figure 3.1. As stated earlier in the figure, the three inequalities are represented by the region on the other side of the hatched lines. The objective function lines are represented as dashed contours. The optimal solution is at $x_1 = 4\,\text{cm}$; $x_2 = 4\,\text{cm}$. Unlike LP, the NLP solution is not lying at the vertex of the feasible region, which is the basis of the simplex method.

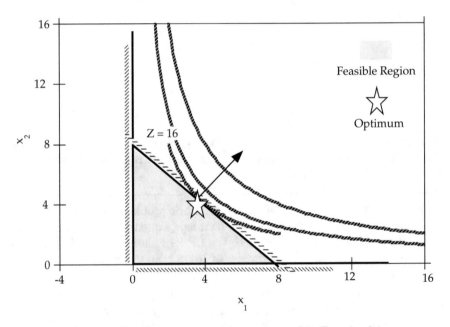

Fig. 3.1. Nonlinear programming contour plot, Exercise 3.1

The above example demonstrates that NLP problems are different from LP problems because

- An NLP solution need not be a corner point.
- An NLP solution need not be on the boundary (although in this example it is on the boundary) of the feasible region.

It is obvious that one cannot use the simplex method described in Chapter 2 for solving an NLP. For an NLP solution, it is necessary to look at the relationship of the objective function to each decision variable.

Consider the previous example. Let us convert the problem into a one-dimensional problem by assuming constraint (3.2) (isoperimetric constraint) as an equality. One can eliminate x_2 by substituting the value of x_2 in terms of x_1 using constraint (3.2).

$$\text{Maximize} \quad Z = 8x_1 - x_1^2 \tag{3.3}$$
$$x_1$$

subject to

$$x_1 \geq 0 \tag{3.4}$$

Figure 3.2 shows the graph of the objective function versus the single decision variable x_1.

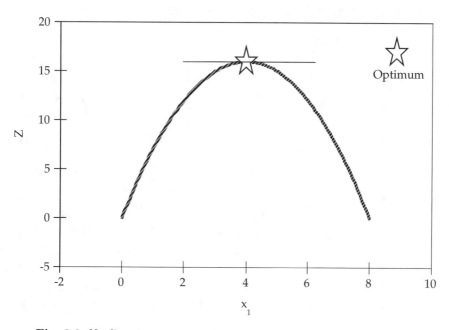

Fig. 3.2. Nonlinear programming graphical representation, Exercise 3.1

In Figure 3.2, the objective function has the highest value (maximum) at $x_1 = 4$. At this point in the figure, the x-axis is tangent to the objective function curve, and the slope dZ/dx_1 is zero. This is the first condition that is used in deciding the extremum point of a function in an NLP setting.

Is this a minimum or a maximum?

Let us see what happens if we convert this maximization problem into a minimization problem with $-Z$ as the objective function.

$$\text{Minimize} \quad -Z = -8x_1 + x_1^2 \tag{3.5}$$
$$x_1$$

subject to

$$x_1 \geq 0 \tag{3.6}$$

Figure 3.3 shows that $-Z$ has the lowest value at the same point, $x_1 = 4$. At this point in both figures, the x-axis is tangent to the objective function curve, and slope dZ/dx_1 is zero. It is obvious that for both the maximum and minimum points, the necessary condition is the same. What differentiates a minimum from a maximum is whether the slope is increasing or decreasing around the extremum point. In Figure 3.2, the slope is decreasing as you move away from $x_1 = 4$, showing that the solution is a maximum. On the other hand, in Figure 3.3 the slope is increasing, resulting in a minimum. Whether the slope is increasing or decreasing (sign of the second derivative) provides a sufficient condition for the optimal solution to an NLP.

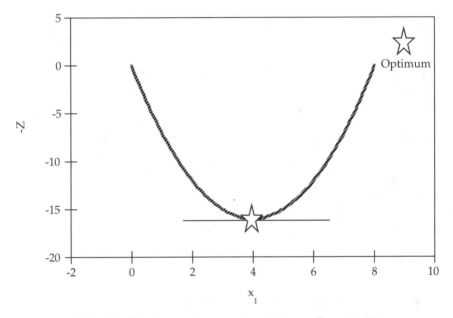

Fig. 3.3. Nonlinear programming minimum, Exercise 3.1

Figures 3.2 and 3.3 have a single optimum. However, instead of the ideal minimum shown in Figure 3.3, consider dealing with an objective function like that shown in Figure 3.4. It is obvious that for this objective function, there are two minima, one being better than the other.

This is another case in which an NLP differs from an LP, as

- In LP, a local optimum (the point is better than any "adjacent" point) is a global (best of all the feasible points) optimum. With NLP, a solution can be a local minimum.
- For some problems, one can obtain a global optimum. For example,
  - Figure 3.2 shows a global maximum of a concave function.
  - Figure 3.3 presents a global minimum of a convex function.

What is the relation between the convexity or concavity of a function and its optimum point?

The following section describes convex and concave functions and their relation to the NLP solution.

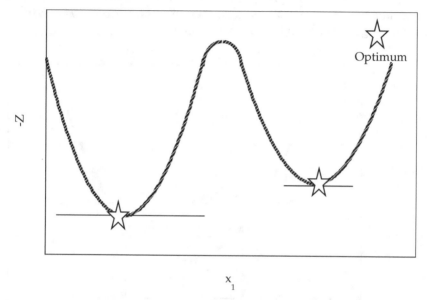

Fig. 3.4. Nonlinear programming multiple minima, Exercise 3.1

3.1 Convex and Concave Functions

A set of points S is a convex set if the line segment joining any two points in the space S is wholly contained in S. In Figure 3.5a and b are convex sets, but c is not a convex set.

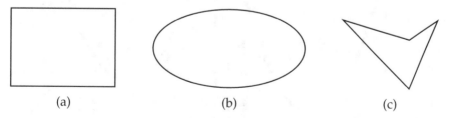

(a)   (b)   (c)

Fig. 3.5. Examples of convex and nonconvex sets

Mathematically, S is a convex set if, for any two vectors x_1 and x_2 in S, the vector $x = \lambda x_1 + (1 - \lambda)x_2$ is also in S for any number λ between 0 and 1.

Therefore, a function $f(x)$ is said to be strictly convex if, for any two distinct points x_1 and x_2, the following equation applies.

$$f(\lambda x_1 + (1-\lambda)x_2) < \lambda f(x_1) + (1-\lambda)f(x_2) \qquad (3.7)$$

Figure 3.6a describes Equation (3.7), which defines a convex function. This convex function (Figure 3.6a) has a single minimum, whereas the nonconvex function (Figure 3.6b) can have multiple minima.

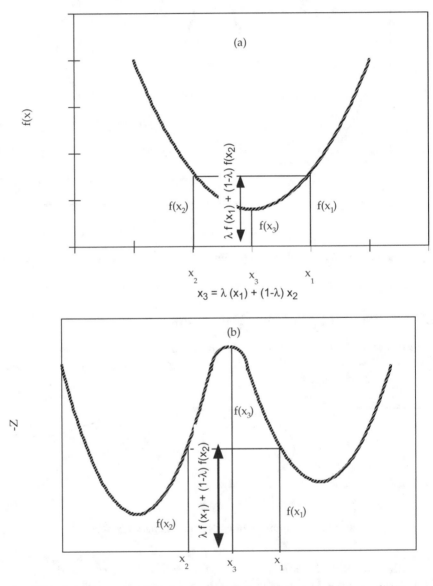

Fig. 3.6. Convex and nonconvex functions and the necessary condition

Conversely, a function $f(x)$ is strictly concave if $-f(x)$ is strictly convex. As stated earlier, Figure 3.2 is a concave function and has a single maximum.

Therefore, to obtain a global optimum in NLP, the following conditions apply.

- Maximization: The objective function should be concave and the solution space should be a convex set (as was the case in Figure 3.2).
- Minimization: The objective function should be convex and the solution space should be a convex set (Figure 3.3).

Note that every global optimum is a local optimum, but the converse is not true. The set of all feasible solutions to a linear programming problem is a convex set. Therefore, a linear programming optimum is a global optimum.

It is clear that the NLP solution depends on the objective function and the solution space defined by the constraints. The following sections describe the unconstrained and constrained NLP, and the necessary and sufficient conditions for obtaining the optimum for these problems.

3.2 Unconstrained NLP

In the following NLP optimization problem, when constraints (Equations (3.9) and (3.10)) are not present or eliminated, the results are an unconstrained nonlinear programming problem.

$$\text{Optimize} \quad Z = z(x) \qquad (3.8)$$
$$x$$

subject to

$$h(x) = 0 \qquad (3.9)$$
$$g(x) \leq 0 \qquad (3.10)$$

The first-order optimality condition for an unconstrained NLP is given by the following equation and requires the first derivative (Jacobian/gradient) of the objective function with respect to each decision variable to be zero.

The first-order necessary condition for an unconstrained optimum at x^* is

$$\nabla z(x^*) = 0 \qquad (3.11)$$

This condition is applicable for a maximum, and a minimum as well as a saddle point. The second-order necessary condition applies to the Hessian and differentiates between a maximum and a minimum. The second-order necessary condition states that the Hessian H for a strong local minimum has to be positive semidefinite and for a strong local maximum has to be negative semidefinite. The extremum point is a saddle point if the Hessian is indefinite.

Necessary Conditions:

For x^* to be a strong local minimum, *the Jacobian J must be zero and the Hessian H must be positive semidefinite.*

It should be noted that the first-order necessary condition merely identifies the extremum point without any indication about the nature of the point. The second-order necessary condition, on the other hand, distinguishes among a maximum, a minimum, and a saddle point, but is not sufficient to ascertain the presence of a strong local minimum or maximum. For example, Figure 3.6a shows a strong minimum, whereas point A in Figure 3.7 is a weak local minimum. This is because at least one point in the neighborhood of A has the same objective function value as A and the Hessian H at A is zero or positive (positive semidefinite). In the same figure, one can see that the first derivative vanishes at all extremum points, including the saddle point B in the figure (Hessian indefinite).

The presence of a strong local maximum or minimum is determined by the second-order sufficiency condition that applies to the Hessian. The sufficiency condition requires the Hessian H for a strong local minimum to be positive definite and for a strong local maximum to be negative definite. The extremum point is again a saddle point if the Hessian is indefinite. These conditions are explained in mathematical terms below.

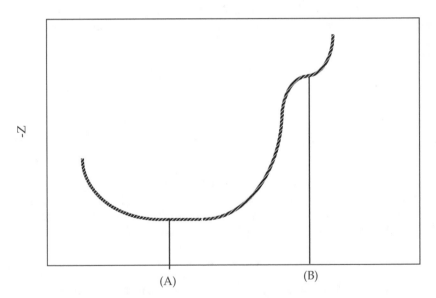

Fig. 3.7. Concept of local minimum and saddle point

Sufficiency Conditions:

x^* is a strong local minimum, if *the Jacobian J is zero and the Hessian H is positive definite*

$$J = \nabla z(x^*) = 0 \tag{3.12}$$
$$H = \nabla^2 z(x^*) \tag{3.13}$$

where

- $\nabla z(x^*) =$ the column vector of first-order partial derivatives of $z(x)$ evaluated at x^*.
- $\nabla^2 z(x^*) =$ the symmetric matrix of second-order partial derivatives of $z(x)$ evaluated at x^*, often called the Hessian matrix. The element in the ith row and jth column is $\partial^2 z/\partial x_i \partial x_j$.

The matrix H is said to be positive definite if and only if

$$Q(x) = \nabla x^T H \nabla x > 0 \big|_{x = x^*}$$

This is equivalent to saying that the eigenvalues for H are positive. Following linear algebra, the condition of H can be defined in terms of $Q(x)$, as given below.

- H is *positive definite* if for all x, $Q(x) > 0$.
- H is *positive semidefinite* if for all x, $Q(x) \geq 0$.
- H is *negative definite* if for all x, $Q(x) < 0$.
- H is *negative semidefinite* if for all x, $Q(x) \leq 0$.
- H is *indefinite* if for some x, $Q(x) > 0$
  and for other x, $Q(x) < 0$.

To determine whether H is positive definite, we evaluate $Q(x)$ or eigenvalues of H. In addition, we can also check the definiteness of H using the determinants of the principal minors of H. The ith principal minor of any matrix A is the matrix A_i constructed by the first i rows and columns.

- H is *positive definite* if for all principal minors H_i, $\det(H_i) > 0$.
- H is *positive semidefinite* if for all principal minors H_i, $\det(H_i) \geq 0$.
- H is *negative definite* if for all principal minors H_i, $\det(H_i)$ are nonzero and alternate in signs with starting $\det(H_1) < 0$.
- H is *negative semidefinite* if for all principal minors H_i, $\det(H_i)$ has alternating signs with starting $\det(H_1) \leq 0$.

For the necessary conditions for an unconstrained optimum at $x*$, the sufficiency conditions can help to check if the stationary point $x*$ is a minimum, a maximum, or a saddle point.

- Minimum if H is positive definite.
- Maximum if H is negative definite.
- Saddle point if H is indefinite.

For some special functions, the Hessian vanishes at the extremum point. For such functions, second-order conditions are not sufficient to determine the nature of the point. Instead, higher-order conditions need to be examined. This is true for single as well as multi-variable functions.

Consider the example of a single-variable function $z(x) \in x^k$ which has the following relations at the extremum point x^*.

$$\nabla z(x^*) = 0 \tag{3.14}$$
$$\nabla^{k-1} z(x^*) = 0 \tag{3.15}$$
$$\nabla^k z(x^*) \neq 0 \tag{3.16}$$

Then,

1. x^* is a point of local minimum if k is even and

$$\nabla^k z(x^*) > 0. \tag{3.17}$$

2. x^* is a point of local maximum if k is even and

$$\nabla^k z(x^*) < 0. \tag{3.18}$$

3. x^* is neither a minimum nor a maximum if k is odd.

For multi-variable functions, analysis of higher-order derivatives determines the nature of the extremum point. For a function z, if matrix M of the kth derivatives of z with respect to the dependent variables is nonzero at extremum point x^* ($k > 2$), then the extremum point is a saddle point if k is odd. If k is even, then

1. x^* is a point of local minimum if M is positive definite.
2. x^* is a point of local minimum if M is negative definite.
3. x^* is a saddle point if M is indefinite.

Example 3.2: For the given Hessian matrix, determine if the matrix is positive definite or not using (a) determinants of principal minors and (b) eigenvalues.

$$H = \begin{bmatrix} 1 & 1 & 0 \\ 1 & 2 & -1 \\ 0 & -1 & 1 \end{bmatrix}$$

(a) Determinants of principal minors (H_i): For an $n \times n$ matrix A the determinants can be calculated by the following equation.

$$|A| = \sum_{i=1}^{n} a_{i1}(-1)^{i+1} |S_{i1}|$$

where a_{ij} is an element of matrix A, and S_{i1} is a submatrix by deleting row i and column 1 of A. Therefore,

$$\det(H_1) = 1$$
$$\det(H_2) = 1 \times 2 - 1 \times 1 = 1$$
$$\det(H_3) = 1 \cdot (2 - 1) - 1 \cdot (1 - 0) + 0 \cdot (-1 + 0) = 0$$

Because all determinants of principal minors of H are either positive or zero, the Hessian matrix is positive semidefinite.

(b) Eigenvalues of H: For an $n \times n$ matrix A, the eigenvalues are given by:

$$\det(A - \rho I) = 0$$

Thus,

$$H - \rho I = \begin{bmatrix} 1 - \rho & 1 & 0 \\ 1 & 2 - \rho & -1 \\ 0 & -1 & 1 - \rho \end{bmatrix}$$
$$\det(H - \rho I) = (1 - \rho)[(2 - \rho)(1 - \rho) - 1] - 1[(1 - \rho)] = 0$$

After solving the above equation, we can get $\rho = 0, 1, 3$, which are positive or zero. Therefore, the Hessian matrix is positive semidefinite.

Example 3.3: Consider the problem in Example 3.1. The problem has two unknowns, x_1 and x_2, and one equality constraint resulting in a single degree of freedom. We eliminate x_2 by substituting it in terms of x_1 using constraint (3.2). The problem results in the unconstrained NLP shown below.

$$\text{Maximize} \quad Z = 8x_1 - x_1^2 \tag{3.19}$$
$$x_1$$

Solution: Necessary condition:

$$\frac{\partial Z}{\partial x_1} = 8 - 2x_1 = 0 \tag{3.20}$$
$$x_1 = 4.0 \tag{3.21}$$
$$x_2 = 4.0 \tag{3.22}$$
$$Z = 16 \tag{3.23}$$

To know whether this is a minimum, maximum, or a saddle point, let us look at the sufficiency condition.

Sufficiency condition check:

$$H = \frac{\partial^2 z}{\partial x_1^2} = -2 \tag{3.24}$$
$$H < 0 \tag{3.25}$$

H is negative definite, so the solution is a local and global maximum. Therefore, the maximum area rectangle is a square.

Example 3.4: Analyze the following four different functions for optimality. $f_1 = x_1{}^3 + x_2{}^3$, $f_2 = x_1{}^4 + x_2{}^4$, $f_3 = -x_1{}^4 - x_2{}^4$, and $f_4 = x_1{}^4 - x_2{}^4$.

Solution: The first-order necessary condition suggests the origin as the extremum point for all four functions. However, the Hessian vanishes at the origin for all functions. Hence, higher-order derivatives must be analyzed to determine the nature of the extremum point. The criteria mentioned for multivariable functions are used.

For f_1, $\nabla^3 z(x^*) > 0$, and $k = 3$. Because k is odd, the extremum point (origin) is a saddle point.

For f_2, f_3, and f_4, $\nabla^4 z(x^*) > 0$, and $k = 4$. For these functions the matrix $M = \nabla^4 z(x^*)$ is analyzed to determine the nature of the extremum point.

- For f_2, M is positive definite and hence the origin is the minimum.
- For f_3, M is negative definite and hence the origin is the maximum.
- For f_4, M is indefinite and hence the origin is a saddle point.

3.3 Necessary and Sufficient Conditions and Constrained NLP

The condition for NLP optimality can be explained easily by considering the example of a ball rolling in a valley, as shown in Figure 3.8.

Consider the smooth valley shown in Figure 3.8. A ball rolling in this valley will go to the lowest point in the valley due to the gradient (gravitational pull). If our objective function is represented by the surface of the valley, then the gravitational force is acting in the gradient direction shown by the arrow. At the lowest point, the stationary point for the ball, there is no force acting on this ball and hence, we have a zero gradient, $\nabla z(x^*) = 0$. We know that if we move the ball away from x^* in any direction, it will roll back. This means here the surface has a positive curvature (convex function, $\nabla^2 z(x^*) > 0$). We did not put any restriction on the ball traveling in this valley. Suppose only certain parts of the valley are free for moving the ball and are marked by the two fences shown in Figure 3.9. We know that the fences will constrain the movement of the ball by not allowing it to cross their boundaries. This can be represented by the two inequality constraints $g_1(x) \leq 0$ and $g_2(x) \leq 0$. Again, the ball rolling in the valley within the fences will roll to the lowest allowable point x^*, but at the boundary of the fence $g_1(x^*) \leq 0$, making the constraint

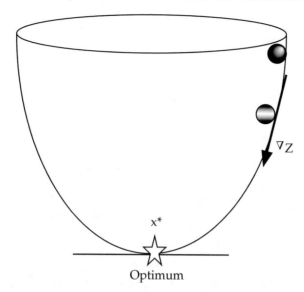

Fig. 3.8. Unconstrained NLP minimization: ball rolling in a valley

active $(g_1(x^*) \leq 0)$. At this position we no longer have $\nabla z(x^*) = 0$. Instead, we see that the ball remains stationary because of a balance of "forces": the force of "gravity" $(-\nabla z(x^*))$ and the "normal force" exerted on the ball by the fence $(-\nabla g_1(x^*))$. Also, in Figure 3.9, note that the constraint $g_2(x) \leq 0$ is inactive and does not participate in this "force balance." Again, when looking at the movement of the ball around this stationary point, if we move the ball from x^* in any direction along the fence, it will roll back (similar to that of the objective function) showing positive curvature.

Now if we want to curb the movement of the ball, we can introduce a rail in the valley which will guide the movement of the ball, as shown in Figure 3.10. Because the ball has to be on the rail all the time, this introduces an equality constraint $h(x) = 0$ into the problem. The ball rolling on the rail and within the fence will stop at the lowest point x^*. This point will also be characterized by a balance of forces: the force of gravity $(-\nabla z(x^*))$, the normal force exerted on the ball by the fence $(-\nabla g_1(x^*))$, and the normal force exerted on the ball by the rail $(-\nabla h(x^*))$. However, we see that this equality constraint is not allowing the ball to move around the direction of the fence g_1, but has a positive curvature in the direction of the rail (Hessian positive semidefinite, indicating the second derivative is zero or positive). This condition is sufficient for optimality.

To consider this force balance, let us define a new objective function by combining all the constraints as the Lagrangian function. Now the decision variables also include μ and λ.

$$L(x,\mu,\lambda) = Z(x) + g(x)^T\mu + h(x)^T\lambda \qquad (3.26)$$

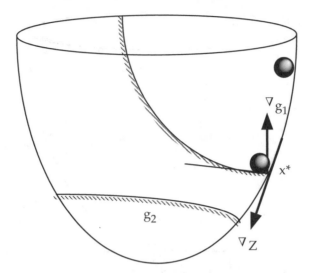

Fig. 3.9. Constrained NLP minimization with inequalities

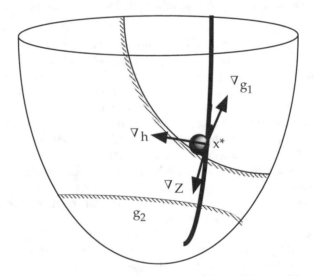

Fig. 3.10. Constrained NLP minimization with equalities and inequalities

We know that the forces in Figure 3.10 define the necessary conditions for the stationary point (optimality). These optimality conditions are referred to as the Kuhn–Tucker conditions or Karush–Kuhn–Tucker (KKT) conditions and were developed independently by Karush (1939), and Kuhn and Tucker (1951). Here, the vectors μ and λ act as weights for balancing the forces. The variables μ and λ are referred to as *dual variables, Kuhn–Tucker multipliers,* or *Lagrange multipliers.* They also represent shadow prices of the constraints.

The first-order Kuhn–Tucker conditions necessary for optimality can be written as follows.

1. Linear dependence of gradients (balance of forces in Figure 3.10):

$$\nabla L \left(x^*, \mu^*, \lambda^*\right) = \nabla Z(x^*) + \nabla g(x^*)^T \mu^* + \nabla h(x^*)^T \lambda^* = 0 \quad (3.27)$$

   where * refers to the optimum solution.
2. Feasibility of the NLP solution (within the fences and on the rail):

$$g(x^*) \leq 0 \qquad (3.28)$$

$$h(x^*) = 0. \qquad (3.29)$$

3. Complementarity condition; either $\mu^* = 0$ or $g(x^*) = 0$ (either at the fence boundary or not):

$$\mu^{*T} g(x^*) = 0. \qquad (3.30)$$

4. Nonnegativity of inequality constraint multipliers (normal force from the fence can only act in one direction):

$$\mu^* \geq 0. \qquad (3.31)$$

It should be remembered that the direction of inequality is very important here. For example, if in Figure 3.9 the fence is on the other side of the valley, as shown in Figure 3.11, then the rolling ball will not be able to reach the feasible solution from the point where it started rolling (point A). It may shuttle between the two constraints (points A and B), or it may reach the optimum (point C) if an infeasible path optimization method is used. Note that point C is a completely different stationary point from earlier because of the changed direction of the inequality constraint. The nonnegativity requirement (for a minimization problem) above ensures that the constraint direction is not violated and the solution is in the feasible region.

The signs of the inequality constraints are very important for finding the optimum solution, therefore we need to define a convention for representing an NLP, as shown below.

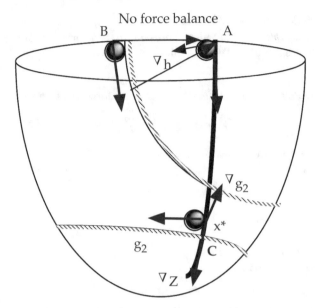

Fig. 3.11. Constrained NLP with inequality constraint

$$\text{Minimize} \quad Z \;=\; z(x) \qquad\qquad (3.32)$$
$$x$$

subject to

$$h(x) \;=\; 0 \qquad\qquad (3.33)$$
$$g(x) \;\leq\; 0 \qquad\qquad (3.34)$$

Any NLP can be converted to the above form so that the KKT conditions in the form defined above are applicable. The first KKT condition represents the linear dependence of gradients and is also referred to as the Kuhn–Tucker error. The second condition requires that NLP satisfies all the constraints. The third and fourth conditions relate to the fact that if the constraint is active ($g_i = 0$), then the corresponding multiplier (μ_i) is positive (right direction of the constraint), or if the constraint is inactive, then the corresponding multiplier is zero.

As seen above, only active inequalities and equalities need to be considered in the constrained NLP solution (Equation (3.27)). When an inequality becomes active, it is equivalent to having an additional equality. Therefore, let us first consider NLP problems with equality constraints.

NLP with Equalities

In this case, the necessary and sufficient conditions for a constrained local minimum are given by the stationary point of a Lagrangian function formed by augmenting the constraints to the objective function shown below.

$$\text{Minimize} \quad L = l(x, \lambda_j) = z(x) + \sum_j \lambda_j h_j(x) \tag{3.35}$$

$$x, \lambda_j$$

where the above problem is an unconstrained NLP (necessary and sufficient conditions are given by Equations (3.12) and (3.13)) with x and λ_j as decision variables.

Necessary conditions:

$$\frac{\partial L}{\partial x} = \nabla z(x) + \sum_j \lambda_j \nabla h_j(x) = 0 \tag{3.36}$$

$$\frac{\partial L}{\partial \lambda_j} = h_j(x) = 0 \tag{3.37}$$

Note that Equation (3.37) is the same as Equation (3.29) in the KKT conditions and is the equality constraint in the original NLP formulation. The above equations (Equations (3.36) and (3.37)) constitute a set of simultaneous equations with the number of equations equal to the number of unknowns. For relatively simple problems, the values of x and λ_j can be obtained through analytical solutions. The following example illustrates this in the context of the isoperimetric problem described earlier. For more complicated problems though, numerical methods must be used.

Example 3.5: Consider the problem in Example 3.1. Convert the perimeter constraint as an equality, as done in Example 3.3, and remove the other inequality constraints from the problem (as we know that the inequality constraints are not active). Solve the problem using the KKT conditions.

Solution: Removing the inequalities from Example 3.1 results in an NLP with the equality constraints given below.

$$\text{Maximize} \quad Z = x_1 \times x_2 \tag{3.38}$$

$$x_1, x_2$$

subject to

$$x_1 + x_2 = 8 \tag{3.39}$$

Converting the NLP into a minimization problem and formulating the augmented Lagrangian function result in the following unconstrained NLP.

$$\text{Minimize} \quad L = -x_1 \times x_2 + \lambda(x_1 + x_2 - 8) \tag{3.40}$$

$$x_1, x_2, \lambda$$

KKT Necessary condition:

$$\frac{\partial L}{\partial x_1} = -x_2 + \lambda = 0 \tag{3.41}$$

$$\frac{\partial L}{\partial x_2} = -x_1 + \lambda = 0 \tag{3.42}$$

$$\frac{\partial L}{\partial \lambda} = x_1 + x_2 - 8 = 0 \tag{3.43}$$

Solving the above three equations for the three unknowns x_1, x_2, and λ finds the following optimal solution.

$$x_1 = 4.0 \tag{3.44}$$

$$x_2 = 4.0 \tag{3.45}$$

$$\lambda = 4 \tag{3.46}$$

$$Z = 16 \tag{3.47}$$

Note that the Lagrange multiplier is positive here. However, because it is the multiplier corresponding to the equality constraint, the sign of λ is not important except for use in the sensitivity analysis.

Sufficiency condition check:

$$\frac{\partial^2 L}{\partial x_1{}^2} = 0 \tag{3.48}$$

$$\frac{\partial^2 L}{\partial x_1 x_2} = -1 \tag{3.49}$$

$$\frac{\partial^2 L}{\partial x_2 \lambda} = 1 \tag{3.50}$$

$$\frac{\partial^2 L}{\partial x_2 x_1} = -1 \tag{3.51}$$

$$\frac{\partial^2 L}{\partial x_2{}^2} = 0 \tag{3.52}$$

$$\frac{\partial^2 L}{\partial x_2 \lambda} = 1 \tag{3.53}$$

$$\frac{\partial^2 L}{\partial \lambda x_1} = 1 \tag{3.54}$$

$$\frac{\partial^2 L}{\partial \lambda x_2} = 1 \tag{3.55}$$

$$\frac{\partial^2 L}{\partial \lambda^2} = 0 \tag{3.56}$$

$$H = \begin{bmatrix} 0 & -1 & 1 \\ -1 & 0 & 1 \\ 1 & 1 & 0 \end{bmatrix} \tag{3.57}$$

$\det(H_1) = 0$

$\det(H_2) = 0 \times 0 - -1 \times -1 = -1$

$\det(H_3) = 0(0-1) + 1(o+1) + 1(-1+0) = -1$

H is indefinite or negative semidefinite. The solution is a saddle point or a local maximum. We are not sure at this point. Let us look at the eigenvalues of H.

$$H - \rho I = \begin{bmatrix} -\rho & -1 & 1 \\ -1 & -\rho & 1 \\ 1 & 1 & -\rho \end{bmatrix}$$

$$\det(H - \rho I) = -(\rho)^3 + 3\rho - 2 = 0 \qquad (3.58)$$

From Equation (3.58) the eigenvalues are $\rho = -2, 1, 1$. H is indefinite, so the solution is a saddle point. This is a surprising result given that earlier when we transformed the problem in one dimension (Example 3.3) we could find a global maximum. This can be explained by looking carefully at the objective function Z, in the two-dimensional space x_1 and x_2. This two-dimensional function is a bilinear function (nonconvex) and has multiple solutions. This is also obvious from the plots shown earlier in Figure 3.1. There is a contour of solutions for $Z = 16$ but only one lies on the constraint. This figure shows that there is a unique optimum that can be obtained for this problem due to the equality constraint. Figure 3.12 shows the 3-d plot for this function with axes x_1, x_2, and Z. The saddle can be seen in the figure. Therefore, when we eliminated this constraint and transformed the two-dimensional problem into a one-dimensional problem, we could show that the problem has a unique (global) optimal solution.

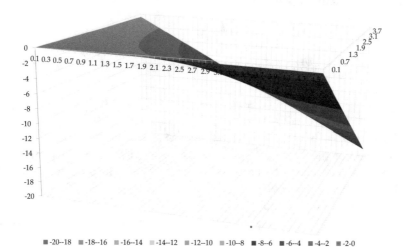

Fig. 3.12. 3-d plot for this function with axes x_1, x_2, and Z for Example 3.6

It should be remembered that the existence of a Lagrange multiplier value in an NLP constrained optimization problem is not guaranteed. There are problems where the Lagrange multiplier cannot take finite value. In general, for an NLP with Lagrange multiplier, we can write one of the necessary condition as:

$$\frac{\partial L}{\partial x} = \nabla z(x) + \sum_j \lambda_j \nabla h_j(x) = 0$$

To calculate λ from this condition, we need to take the inverse of the matrix $\frac{\partial h_j(x)}{\partial x}$. This requires the matrix to be of full rank. If the condition is not satisfied, then we cannot calculate λ. This is illustrated in the following example.

Example 3.6: Consider the problem of minimization of squared distance with one equality constraint.

$$\text{Minimize} \quad Z = x_1^2 + (x_2 - 1)^2 \tag{3.59}$$
$$x_1, \, x_2$$

subject to

$$- x_1^2 + (x_2 - 1)^3 = 0 \tag{3.60}$$

Solution: The Lagrangian function for this problem is given by

$$\text{Minimize} \quad L = - x_1^2 + (x_2 - 1)^2 + \lambda(-x_1^2 + (x_2 - 1)^3) \tag{3.61}$$
$$x_1, x_2, \lambda$$

KKT Necessary condition:

$$\frac{\partial L}{\partial x_1} = 2x_1 \, 2x_1\lambda = 0 \tag{3.62}$$

$$\frac{\partial L}{\partial x_2} = 2(x_2 - 1) + 3\lambda(x_2 - 1)^2 = 0 \tag{3.63}$$

$$\frac{\partial L}{\partial \lambda} = (x_2 - 1)^3 - x_1^2 = 0 \tag{3.64}$$

We know $x = (0, 1)$ is the optimal solution to this problem. If we calculate the elements of the matrix $\frac{\partial h_j(x)}{\partial x}$ as given below, the rank of the matrix at (0,1) is zero. Therefore, no finite value for λ. This is also obvious from the KKT conditions given above.

$$\frac{\partial h}{\partial x_1} \Big|_{(0,1)} = - 2x_1 = 0 \tag{3.65}$$

$$\frac{\partial h}{\partial x_2} \Big|_{(0,1)} = 3(x_2 - 1)^2 = 0 \tag{3.66}$$

NLP with Inequalities

For NLP with inequalities, again the problem is formulated in terms of the augmented Lagrangian function.

$$\text{Minimize} \quad L = l(x, \lambda_j, \, \mu_k)$$

$$= z(x) + \sum_j \lambda_j h_j(x) + \sum_k \mu_k g_k(x) \qquad (3.67)$$

$$x, \lambda_j, \ \mu_k$$

Necessary conditions:

$$\frac{\partial L}{\partial x} = \nabla z(x^*) + \sum_j \lambda_j^* \nabla h_j(x^*) + \sum_k \mu_k \nabla g_k(x^*) = 0 \qquad (3.68)$$

$$\frac{\partial L}{\partial \lambda_j} = h_j(x) = 0 \qquad (3.69)$$

$$\frac{\partial L}{\partial \mu_i} = g_i(x) = 0 \ \mu_i \geq 0 \qquad (3.70)$$

As seen in Figure 3.10, the inequality constraints that are not active do not contribute in the force balance, implying that the multipliers for those constraints are zero. However, to solve the above equations, one needs to find the constraints that are going to be active. For small-scale problems, the following iterative steps are generally used.

Active constraint strategy

1. Assume no active inequalities and equate all the Lagrange multipliers associated with these inequalities constraints to zero.
2. Solve the KKT conditions for the augmented Lagrangian function for all the equalities. Find the solution $x = x_{inter}^*$.
3. If all the inequalities $g_k(x_{inter}^*) \leq 0$ are satisfied and for all the active inequalities (zero for the first iteration), $\mu_k \geq 0$, then the optimal solution is reached, $x^* = x_{inter}^*$.
4. If one or more μ is negative, remove that active inequality with the largest constraint violation. Add this constraint to the active constraint list and go to Step 1.

The following example shows how to use the active constraint strategy.

Example 3.7: Consider the problem in Example 3.5 above with all the inequality constraints indicating the sides of the rectangle to be nonnegative. Impose an additional constraint that one of the sides should be less than or equal to 3 cm. Use the active constraint strategy to obtain the optimal solution.

Solution: The problem statement for Example 3.6 results in the following NLP.

$$\text{Maximize} \quad Z = x_1 \times x_2 \qquad (3.71)$$

$$x_1, \ x_2$$

subject to

$$h(x) = x_1 + x_2 - 8 = 0 \tag{3.72}$$
$$g_1(x) = x_1 - 3 \le 0 \tag{3.73}$$
$$g_2(x) = -x_1 \le 0 \tag{3.74}$$
$$g_3(x) = -x_2 \le 0 \tag{3.75}$$

Converting the NLP into a minimization problem and formulating the augmented Lagrangian function results in the following unconstrained NLP.

$$\text{Minimize} \quad L = -x_1 \times x_2 + \lambda(x_1 + x_2 - 8)$$
$$+ \mu_1(x_1 - 3) + \mu_2(-x_1) + \mu_3(-x_2) \tag{3.76}$$
$$x_1, x_2, \lambda, \mu_1, \mu_2, \mu_3$$

Active constraint strategy:

1. Assume no active constraints. KKT Necessary condition:

$$\frac{\partial L}{\partial x_1} = -x_2 + \lambda + \mu_1 - \mu_2 = 0 \tag{3.77}$$

$$\frac{\partial L}{\partial x_2} = -x_1 + \lambda - \mu_3 = 0 \tag{3.78}$$

$$\frac{\partial L}{\partial \lambda} = x_1 + x_2 - 8 = 0 \tag{3.79}$$

$$\mu_1 = 0; \mu_2 = 0; \mu_3 = 0 \tag{3.80}$$

Solving the above equations for the three unknowns x_1, x_2, and λ results in the following optimal solution.

$$x_1 = 4.0 \tag{3.81}$$
$$x_2 = 4.0 \tag{3.82}$$
$$\lambda = 4 \tag{3.83}$$
$$h(x) = 0 \tag{3.84}$$
$$g_1(x) = 4 - 3 > 0 \quad \text{Constraint violated} \tag{3.85}$$
$$g_2(x) = -4 \le 0 \tag{3.86}$$
$$g_3(x) = -4 \le 0 \tag{3.87}$$

2. The first inequality constraint is violated. This constraint is included in the augmented Lagrange function as an active constraint, and now we are solving the KKT conditions to obtain the optimal values of x_1, x_2, λ, and μ_1.
   KKT Necessary condition:

$$\frac{\partial L}{\partial x_1} = -x_2 + \lambda + \mu_1 - \mu_2 = 0 \tag{3.88}$$

$$\frac{\partial L}{\partial x_2} = -x_1 + \lambda - \mu_3 = 0 \tag{3.89}$$

$$\frac{\partial L}{\partial \lambda} = x_1 + x_2 - 8 = 0 \qquad (3.90)$$

$$\frac{\partial L}{\partial \mu_1} = x_1 - 3 = 0 \qquad (3.91)$$

Solution:

$$\mu_2 = 0; \ \mu_3 = 0 \qquad (3.92)$$

$$x_1 = 3 \qquad (3.93)$$

$$x_2 = 5 \qquad (3.94)$$

$$\lambda = 3 \qquad (3.95)$$

$$\mu_1 = 2 \qquad (3.96)$$

$$h(x) = 0 \qquad (3.97)$$

$$g_1(x) = 0 \leq 0 \qquad (3.98)$$

$$g_2(x) = -3 \leq 0 \qquad (3.99)$$

$$g_3(x) = -5 \leq 0 \qquad (3.100)$$

3. Because all the constraints are satisfied and all the Lagrange multipliers associated with the inequality constraints μ are nonnegative, the solution is reached.

In the second step, instead of causing the first inequality constraint to be active, if we make the constraint $g_3(x)$ to be active, it will result in the following solution.

KKT necessary conditions:

$$\frac{\partial L}{\partial x_1} = -x_2 + \lambda + \mu_1 - \mu_2 = 0 \qquad (3.101)$$

$$\frac{\partial L}{\partial x_2} = -x_1 + \lambda - \mu_3 = 0 \qquad (3.102)$$

$$\frac{\partial L}{\partial \lambda} = x_1 + x_2 - 8 = 0 \qquad (3.103)$$

$$\frac{\partial L}{\partial \mu_3} = -x_2 = 0 \qquad (3.104)$$

Solution:

$$\mu_1 = 0; \mu_2 = 0 \qquad (3.105)$$

$$x_1 = 8 \qquad (3.106)$$

$$x_2 = 0 \qquad (3.107)$$

$$\lambda = 0 \qquad (3.108)$$

$$\mu_3 = -8 \quad \text{negative multiplier} \qquad (3.109)$$

$$h(x) = 0 \qquad (3.110)$$

$$g_1(x) = 8 - 3 > 0 \quad \text{Constraint violated} \qquad (3.111)$$

$$g_2(x) = -3 \leq 0 \qquad (3.112)$$
$$g_3(x) = 0 \leq 0 \qquad (3.113)$$

Because the constraint g_1 is violated and the Lagrange multiplier is negative for constraint g_3, constraint g_1 is made active in the next iteration, and g_3 is made inactive, resulting in the optimal solution that is the same as the one obtained earlier. However, this strategy took one additional iteration to reach the optimum.

3.4 Constraint Qualification

Equation sets (3.36), (3.37), and (3.68)–(3.70) represent the necessary conditions for a constrained NLP with equality and inequality constraints, respectively. For these necessary conditions to be applicable, the problem must satisfy certain conditions known as constraint qualifications.

The requirement of constraint qualification is due to the first-order approximations of the objective function and constraint functions used in the necessary conditions, as well as while deciding the search direction and step size in an iterative algorithm. Because a first-order Taylor series expansion of the objective function and constraint functions is used, it is important that the linear approximations capture the essential geometric features of the feasible set near the current search point x. Constraint qualifications are assumptions about the nature of the active constraints at x that ensure the similarity of the actual constraints and their linear approximations in the neighborhood of x. Here, the active constraint set includes all equality constraints and active inequality constraints (i.e., $g(x) = 0$). The constraint qualification that is most often used states that the set of gradients of active constraints evaluated at x be linearly independent. First- and second-order necessary optimality conditions for a constrained NLP require that this condition be satisfied. However, the second-order sufficiency condition does not require constraint qualification. The condition that all active constraints be linear is another possible constraint qualification. This is neither a stronger nor a weaker condition as compared to the condition of linear independence. It must also be noted that constraint qualifications are sufficient but not necessary conditions for the linear approximations to be adequate.

3.5 Sensitivity Analysis

The sensitivity analysis information for an NLP is similar to that of an LP except that for the NLP solution, the information reflects local values around the optimum. The Lagrange multipliers in the augmented Lagrangian representation are analogous to dual prices in LP. The augmented Lagrangian representation can be used to show that the primal representation of a standard LP is equivalent to the dual representation used in the dual simplex method, as illustrated in the following example.

Example 3.8: Show that the primal and dual representation of a standard LP given in Table 3.1 are equivalent.

Table 3.1. The primal and dual representation for an LP

Primal	Dual
Maximize $Z = \sum_{i=1}^{n} C_i x_i$	Minimize $Z_d = \sum_{j=1}^{m} b_j \mu_j$
$x_i, \; i = 1, 2, \ldots, n$	$\mu_j, \; j = 1, 2, \ldots, m$
$\sum_{i=1}^{n} a_{ij} x_i \leq b_j$	$\sum_{j=1}^{m} a_{ij} \mu_j \geq C_i$
$j = 1, 2, \ldots, m$	$i = 1, 2, \ldots, n$
$x_i \geq 0$	$\mu_j \geq 0;$

Solution: Let us consider the primal representation and write it in the standard NLP form given below.

$$\text{Minimize} \quad -Z = -\sum_{i=1}^{n} C_i x_i \qquad (3.114)$$
$$x$$

subject to

$$g_j(x) = \sum_{i=1}^{n} a_{ij} x_i - b_j \leq 0 \qquad (3.115)$$

$$x_i \geq 0 \quad i = 1, 2, \ldots, n \qquad (3.116)$$

The augmented Lagrangian representation of this above problem results in the following equation.

$$\text{Minimize} \quad L = -\sum_{i=1}^{n} C_i x_i + \sum_{j=1}^{m} \mu_j \left(\sum_{i=1}^{n} a_{ij} x_i - b_j \right)$$

$$x_i, \mu_j, \upsilon_i \quad - \sum_{i=1}^{n} \upsilon_i x_i$$

$$(3.117)$$

where μ_j represents the Lagrange multiplier for the inequality constraint $g_j(x)$ and υ_i is the Lagrange multiplier corresponding to the nonnegativity constraint $x_i \geq 0$.

The KKT conditions for the above minimization problem are given below.

$$- C_i + \sum_{j=1}^{m} a_{ij}\mu_j + v_i = 0 \quad i = 1, 2, \ldots, n \tag{3.118}$$

$$\mu_j \left(\sum_{i=1}^{n} a_{ij}x_i - b_j \right) = 0 \quad j = 1, 2, \ldots, m \tag{3.119}$$

$$-v_i x_i = 0 \quad i = 1, 2, \ldots, n \tag{3.120}$$

$$x_i \geq 0 \quad i = 1, 2, \ldots, n \tag{3.121}$$

$$v_i \geq 0 \quad i = 1, 2, \ldots, n \tag{3.122}$$

$$\mu_j \geq 0 \quad j = 1, 2, \ldots, m \tag{3.123}$$

Getting the value of v_i from Equation (3.118) and substituting in Equation (3.122) results in:

$$- C_i + \sum_{j=1}^{m} a_{ij}\mu_j \geq 0 \quad i = 1, 2, \ldots, n \tag{3.124}$$

$$\sum_{j=1}^{m} a_{ij}\mu_j \geq C_i \quad i = 1, 2, \ldots, n \tag{3.125}$$

Adding m rows of Equation (3.119), interchanging the sums, and rearranging leads to:

$$\sum_{i=1}^{n} x_i \left(\sum_{j=1}^{m} a_{ij}\mu_j \right) = \sum_{j=1}^{m} b_j\mu_j \tag{3.126}$$

Multiplying Equation (3.124) by x_i, substituting the value of $v_i x_i$ from Equation (3.120), and using it in Equation (3.126) results in:

$$\sum_{i=1}^{n} C_i x_i = \sum_{j=1}^{m} \mu_j b_j \tag{3.127}$$

The right-hand side of the above equation is the dual objective function and Equation (3.125) represents the dual constraints given in Table 3.1.

The Lagrange multipliers for an NLP show change in the objective function value per unit change in the right-hand side of the constraint. The reduced gradients are analogous to reduced costs and show change in the objective function value per unit change in the decision variable. However, this information is only accurate for infinitesimal changes.

3.6 Numerical Methods

As seen in the last section, NLP involves an iterative scheme to solve the problem. In Figure 3.8, the initial position of the ball reflects the initial values of the decision variables, and the ball should move in the direction of the optimum.

It is obvious from Figure 3.8 that the ball should change its position towards the gradient direction; this is the basis of steepest ascent and conjugate gradient methods. However, these methods are slow to converge. If one looks at our earlier procedure in the last section, what we are trying to solve is the set of nonlinear equations resulting from the KKT conditions. In nonlinear equation-solving procedures, the Newton–Raphson method shows the fastest convergence if one is away from the solution. Therefore, Newton methods which use the Newton–Raphson method as their basis are faster than the gradient direction methods.

Newton–Raphson Method

Consider the following nonlinear equation.

$$f(x) = 0 \tag{3.128}$$

The procedure involves stepping from the initial values of $x = x_0$, to the next step x_1, and so on, using the derivative value. At any kth step, the next step can be obtained by using the derivative as follows.

$$f'(x_k) = \frac{\partial f}{\partial x} = \frac{f(x_{k+1}) - f(x_k)}{x_{k+1} - x_k} \tag{3.129}$$

Because we want $f(x) = 0$, in the next step substitute $f(x_{k+1}) = 0$ in the above equation, resulting in the Newton–Raphson step.

$$x_{k+1} = x_k - \alpha \frac{f(x_k)}{f'(x_k)} \tag{3.130}$$

Here α shows the step size and for the conventional Newton–Raphson method $\alpha = 1$. Figure 3.13 shows the Newton–Raphson procedure for solving the above nonlinear equation. One note of caution about the Newton–Raphson method is that, depending on the starting point and the step length, Newton–Raphson can diverge, as illustrated for point B.

In the optimization problem, the KKT condition states that the first derivative of the Lagrangian function is zero ($f(x) = \nabla L = 0$). Following the Newton–Raphson method described above, it is necessary to have information about the second derivative (or Hessian) to take the next step. Also, these Newton methods demand the Hessian to be positive definite (for minimization). The calculation of this Hessian is computationally expensive. The frustration of Dr. Davidon, a physicist working at Argonne National Laboratory with the Newton methods for large-scale problems, was due to continuous computer crashing before the Hessian calculation was complete. This led to the first idea behind the quasi-Newton methods that are currently so popular.

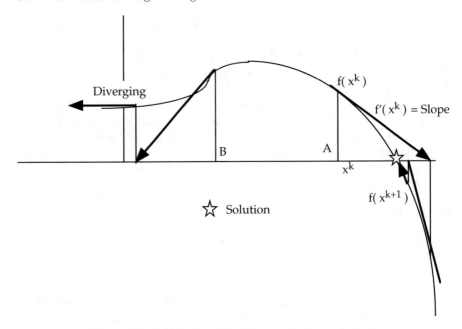

Fig. 3.13. Schematic of the Newton–Raphson method

The quasi-Newton methods use approximate values of the Hessian obtained from the gradient information and avoid the expensive second derivative calculation. The Hessian can be calculated numerically using the two values of gradients at two very close points. This is what Davidon used to obtain the derivative from the last two optimization iterations. However, because the steps are not as close as possible, this information is very approximate, and must be updated at each iteration. Furthermore, the question of the starting value of the Hessian needs to be addressed. In general, the starting value can be taken as the identity matrix. Although Dr. Davidon proposed the first step towards the quasi-Newton algorithms, his paper (written in the mid 1950s) was not accepted for more than 30 years until it appeared in the first issue of the *SIAM Journal on Optimization*(1991).

The most popular way of obtaining the value of the Hessian is to use the BFGS update, named for its discoverers Broyden, Fletcher, Goldfarb, and Shanno. The following procedure illustrates basic steps involved in the BFGS updating.

BFGS Updating

Given that we want to solve the KKT condition as $\nabla L = 0$ to obtain the decision variables x.

1. Specify the initial values of the decision variables.
2. Find the partial derivatives and calculate the KKT conditions.

3. Denote that the approximate Hessian at any step k is B_k. Initially assume B_1 to be the identity matrix (positive definite).
4. Use the approximate Hessian to calculate the Newton step.
5. Update the Hessian as follows.
   - Let s_k be the Newton step given by $s_k = x_{k+1} - x_k$ and $b_k = \nabla L_{k+1} - \nabla L_k$.
   - Then the BFGS update for the Hessian is given by:

$$B_{k+1} = B_k - \frac{B_k s_k (B_k s_k)^T}{s_k^T B_k s_k} + \frac{b_k b_k^T}{b_k^T s_k}. \tag{3.131}$$

6. If the KKT conditions are not satisfied, then go to Step 4; else stop.

The BFGS updating guarantees a "direction matrix" that is positive definite and symmetric, which can be numerically "better" than a poorly behaved Hessian.

Example 3.9: Demonstrate how the BFGS updating can be used to solve the following simple minimization problem.

$$\underset{x}{\text{Minimize}} \quad Z = x^3/3 + x^2 - 3x$$

Solution: The KKT condition results in the following nonlinear algebraic equation.

$$\nabla L = x^2 + 2x - 3 = 0$$

We can solve this equation analytically and obtain the values of x as $x = -3$ where H will be negative definite and $x = 1$ where H is positive, and hence the solution for the minimization problem. The quasi-Newton steps leading to the same solution are shown in Table 3.2 using the BFGS updating.

Table 3.2. Newton steps in BFGS updating for Example 3.8

Iteration	x	∇L	B_k	H_k
1	0.000	−3.000	1.000	2.000
2	3.000	12.000	5.000	8.000
3	0.600	−1.440	5.600	3.200
4	0.857	−0.551	3.457	3.714
5	1.016	0.0066	3.874	4.033
6	1.000	0.0002	4.016	3.999
7	1.000	0.0000	4.000	4.000

Quasi-Newton Methods

Currently, the two major methods for NLP commonly used in various commercial packages are: (1) the generalized reduced gradient (GRG) method and (2) the sequential quadratic programming (SQP). These two are quasi-Newton methods.

GAMS uses MINOS, a particular implementation of the GRG method. The basic idea behind the reduced gradient methods is to solve a sequence of subproblems with linearized constraints, where the subproblems are solved by variable elimination. This is similar to two-level optimization (please refer to the next chapter on mixed integer nonlinear programming). The outer problem takes the Newton step in the reduced space, and the inner subproblem is the linearly constrained optimization problem. SQP, on the other hand, takes the Newton step using the KKT conditions. It can be easily shown that the Newton step (if the exact Hessian is known) results in a quadratic programming problem (objective function quadratic, constraints linear) containing the Newton direction vector as decision variables. For the quasi-Newton approximation, the BFGS update is generally used to obtain the Hessian. GRG methods are best suited for problems involving a significant number of linear constraints. On the other hand, SQP methods are useful for highly nonlinear problems. Extension of the successful interior-point methods for LPs to NLP is the subject of ongoing research.

3.7 Global Optimization and Interval Newton Method

As stated earlier, nonlinear programming methods can have multiple optima as shown in Figure 3.4. The methods and algorithms that are used to reach a global optimum are interval mathematical programming methods, dynamic programming, probabilistic methods such as simulated annealing and evolutionary algorithms. Dynamic programming is described in Chapter 7, probabilistic methods are presented in Chapter 4. Here we present the interval Newton method as an example of mathematical programming algorithms.

Interval Newton Method

The advantage of the interval Newton method as compared to other methods described in Chapters 4 and 7 is that it is faster to converge and interval methods avoid roundoff errors. In summary, the interval Newton method has the following advantages.

1. Relatively faster convergence to the global optimum, without restarting the search.

2. If the search space is large, then the interval Newton method is faster than the traditional Newton method.

3. It can avoid divergence without changing the individual step size (multiplication factor, a in Equation (3.129) as is normally used in the basic Newton method.

The interval Newton method was derived first by Moore in 1966 in the following manner. Originally Moore assumed that $0 \in f'(X_k)$.

If X_k is an inclusion of zero, then an improved interval X_{k+1} may be computed by

$$x_k = m(X_k) \tag{3.132}$$

$$X_{k+1} = (x_k - \frac{f(x_k)}{f'(X_k)}) \cap X_k \tag{3.133}$$

where $m(X_k)$ is a point within the interval X, usually the midpoint.

3.8 What to Do When NLP Algorithm is Not Converging

There is a number of reasons algorithms do not converge after specified iterations. These are given below.

Optimality tolerance: The most common problem is the optimality tolerance. That is, the KKT condition (or Kuhn–Tucker error) tolerance is too small.

Feasibility tolerance: For some methods, one has to specify feasibility tolerance. One can play with this tolerance value to obtain convergence.

Perturbation Size: Many algorithms use perturbation derivatives. The perturbation size should be small enough to calculate the derivatives but large enough to circumvent numerical errors.

Scaling: The performance of most of the algorithms can be enhanced if the scale of the variables, constraints, and objective function is within a few orders of magnitude (absolute values) of each other.

Starting point: The solution to most of the NLP problems is greatly dependent on the initial values of the decision variables. It is good practice to solve the problem using multiple initializations.

3.9 Hazardous Waste Blending: An NLP

The nuclear waste blending problem presented in the LP chapter eliminated constraints related to durability, viscosity, and electrical conductivity. These constraints need to be included in the formulation to solve the blending problem. These constraints are nonlinear, making the blending problem an NLP (linear objective function with nonlinear constraints). This NLP formulation is presented below.

The NLP problem can be derived from the LP presented in Chapter 2.

$$MinG \equiv Min \sum_{i=1}^{n} f^{(i)} \tag{3.134}$$
$$w_{ij}, W^{(i)}, f^{(i)}, g^{(i)}, G, p^{(i)}$$

where i (1–15) corresponds to the component ID and j (1–3) corresponds to tank ID.

Subject to the following constraints:

1. Definition of decision variables:

$$W^{(i)} = \sum_{j=1}^{3} w_{ij} \tag{3.135}$$

$$g^{(i)} = W^{(i)} + f^{(i)} \tag{3.136}$$

$$G = \sum_{i=1}^{n} g^{(i)} \tag{3.137}$$

$$p^{(i)} = g^{(i)}/G \tag{3.138}$$

2. Component bounds:
   (a) $0.42 \le p^{(SiO_2)} \le 0.57$
   (b) $0.05 \le p^{(B_2O_3)} \le 0.20$
   (c) $0.05 \le p^{(Na_2O)} \le 0.20$
   (d) $0.01 \le p^{(Li_2O)} \le 0.07$
   (e) $0.0 \le p^{(CaO)} \le 0.10$
   (f) $0.0 \le p^{(MgO)} \le 0.08$
   (g) $0.02 \le p^{(Fe_2O_3)} \le 0.15$
   (h) $0.0 \le p^{(Al_2O_3)} \le 0.15$
   (i) $0.0 \le p^{(ZrO_2)} \le 0.13$
   (j) $0.01 \le p^{(other)} \le 0.10$
3. Five glass crystallinity constraints:
   (a) $p^{(SiO_2)} > p^{(Al_2O_3)} * C_1$
   (b) $p^{(MgO)} + p^{(CaO)} < C_2$
   (c) $p^{(Fe_2O_3)} + p^{(Al_2O_3)} + p^{(ZrO_2)} + p^{(Other)} < C_3$
   (d) $p^{(Al_2O_3)} + p^{(ZrO_2)} < C_4$
   (e) $p^{(MgO)} + p^{(CaO)} + p^{(ZrO_2)} < C_5$
4. Solubility Constraints:
   (a) $p^{(Cr_2O_3)} < 0.005$
   (b) $p^{(F)} < 0.017$
   (c) $p^{(P_2O_5)} < 0.01$
   (d) $p^{(SO_3)} < 0.005$
   (e) $p^{(Rh_2O_3)} + P^{(PdO)} + P^{(Ru_2O_3)} < 0.025$
5. Viscosity constraints:
   (a) $\sum_{i=1}^{n} \mu_a^i * p^{(i)} + \sum_{j=1}^{n} \sum_{i=1}^{n} \mu_b^{ij} * p^{(i)} * p^{(j)} > \log(\mu_{min})$
   (b) $\sum_{i=1}^{n} \mu_a^i * p^{(i)} + \sum_{j=1}^{n} \sum_{i=1}^{n} \mu_b^{ij} * p^{(i)} * p^{(j)} < \log(\mu_{max})$
6. Conductivity Constraints:
   (a) $\sum_{i=1}^{n} k_a^i * p^{(i)} + \sum_{j=1}^{n} \sum_{i=1}^{n} k_b^{ij} * p^{(i)} * p^{(j)} > \log(k_{min})$
   (b) $\sum_{i=1}^{n} k_a^i * p^{(i)} + \sum_{j=1}^{n} \sum_{i=1}^{n} k_b^{ij} * p^{(i)} * p^{(j)} < \log(k_{max})$

7. Dissolution rate for boron by PCT test (DissPCTbor):
$$\sum_{i=1}^{n} Dp_a^i * p^i + \sum_{j=1}^{n} \sum_{i=1}^{n} Dp_b^{ij} * p^{(i)} * p^{(j)} < \log\left(D_{max}^{PCT}\right)$$
8. Dissolution rate for boron by MCC test (DissMCCbor):
$$\sum_{i=1}^{n} Dm_a^i * p^i + \sum_{j=1}^{n} \sum_{i=1}^{n} Dm_b^{ij} * p^{(i)} * p^{(j)} < \log\left(D_{max}^{MCC}\right)$$
9. Nonnegativity Constraint:
   (a) $f^{(i)} \leq 0$

where μ, k and D_p, D_m are the property constants. The GAMS file and data for this problem are presented online on Springer website with the book link. When solved using the iterative solution procedure, the solution to the NLP is given in Table 3.3. Thus Hanford should add 590 kgs of frit to the blend of these three tanks. Note that the objective function (amount of frit) is the same as that of the LP solution although the decision variable values are different. Looking at the limiting constraints, it is obvious that in the NLP formulation, the nonlinear constraints are not active, essentially resulting in the same solution as LP (Please see the GAMS output files on the attached CD).

Table 3.3. Composition for the optimal solution

Component	Mass in the waste, $W^{(i)}$	Mass in frit $f^{(i)}$
SiO_2	11.2030	355.3436
B_2O_3	2.4111	51.9439
Na_2O	34.1980	127.4987
Li_2O	0.0000	14.7568
CaO	5.5436	30.1039
MgO	2.8776	10.6249
Fe_2O_3	89.0097	
Al_2O_3	45.5518	
ZrO_2	11.4111	
Other	41.7223	
Total	243.9281	590.2718

Let us see what happens if we tighten some of the nonlinear constraints. This strategy does not have any effect on the LP solution as these constraints are not present in the LP formulation. To achieve this, we have changed some of the parameters in the conductivity constraints and tightened the viscosity, conductivity, and dissolution of boron by PCT. This formulation is presented in Appendix A and the GAMS files are available on the attached CD. With this formulation, the LP solution remains the same, but the NLP solution changes to the solution given in Table 3.4.

The frit mass requirement for this alternative formulation, where some of the nonlinear constraints are active, is increased from 590 kg to 737 kg. The

Table 3.4. The optimal solution for alternative formulation

Component	Mass in the waste, $W^{(i)}$	Mass in frit $f^{(i)}$
SiO_2	11.2030	424.6195
B_2O_3	2.4111	116.2736
Na_2O	34.1980	75.7986
Li_2O	0.0000	16.9259
CaO	5.5436	69.9114
MgO	2.8776	
Fe_2O_3	89.0097	
Al_2O_3	45.5518	25.7268
ZrO_2	11.4111	7.3591
Other	41.7223	
Total	243.9281	736.6449

reason that the frit mass is increased from the LP solution is that the problem is more constrained than the LP problem.

3.10 Sustainable Mercury Management: An NLP

The basic optimization model assuming that all information is deterministically known is presented in the previous chapter as a MILP. However, the technology cost, which was assumed to be linear, is an approximation. Therefore, the generalized form of the model without a functional form is presented below.

Objective:

$$\text{Minimize} \quad \sum_{i=1}^{N} \sum_{j=1}^{M} f_j(\phi_j, D_i). \, b_{ij} \tag{3.139}$$

Constraints:

$$t_{ii} = 0 \qquad \forall i = 1, ..., N \tag{3.140}$$

$$red_i \leq \sum_{j=1}^{M} q_j. D_i. \, b_{ij} + \sum_{k=1}^{N} t_{ik} - r \sum_{k=1}^{N} t_{ki} \qquad \forall i = 1, ..., N \tag{3.141}$$

$$P_i \geq \sum_{j=1}^{M} b_{ij}. f_j(\phi, D_i) + F \left(\sum_{k=1}^{N} t_{ik} - \sum_{k=1}^{N} t_{ki} \right) \qquad \forall i = 1, ..., N \tag{3.142}$$

The meanings of various symbols and explanation of the constraints can be found in Section 2.9.3. The difference in the formulation presented above is in the representation of technology cost functions. The cost function for technology j is represented here as $f_j(\phi_j, D_i)$. Here, ϕ_j is the set of design parameters of technology j. Thus, the cost of technology implementation is considered to be a function (linear or nonlinear) of the design parameters ϕ_j (specific to technology j) and treated volume D_i (specific to industry i).

For a linear model considered in the last chapter, the cost function is given as:

$$f_j(\phi_j, D_i) = TC_j.D_i \qquad (3.143)$$

Here, TC_j represents cost per unit volume for technology j [\$/volume]. With linear cost functions, a mixed integer linear programming (MILP) problem is formulated.

As mentioned in the introduction though, the assumption of linear cost models might not be reasonable in many situations. Hence the formulation must be modified to include nonlinear models. For a nonlinear model, the cost function f_j is a nonlinear function of design parameters ϕ_j and volume D_i. The details of this nonlinear model for the three technologies are presented in Appendix B. This leads to the formulation of a mixed integer nonlinear programming (MINLP) problem. Apart from the modifications in the cost function, the rest of the optimization model is the same for linear and nonlinear model formulation. When the binary variables b_{ij} are assumed fixed, this formulation results in an NLP.

3.11 Summary

Nonlinear programming problems involve either the objective function or the constraints, or both the objective function and the constraints are nonlinear. Unlike LP, an NLP optimum need not lie at a vertex of the feasible region. NLP can have multiple local optima. The NLP local optimum is a global minimum if the feasible region and the objective function are convex, and is a global maximum if the feasible region is convex and the objective function is concave. Karush–Kuhn–Tucker conditions provide necessary conditions for an NLP solution and are used in numerical methods to solve the problem iteratively. Currently, the two most popular methods, reduced gradient methods and successive quadratic programming methods, are based on the idea of quasi-Newton direction proposed by Davidon in 1950. SQP is suitable for highly nonlinear problems, and GRG is best suited when there are a large number of linear constraints present. Interior-point methods for NLP are the current area of algorithmic research in nonlinear programming.

Exercises

3.1   Determine if the point provided is an optimal point. Also, where possible, determine if the point is: (1) local or global and (2) a minimum or maximum.

1.
$$3x_1^2 + 2x_1 + 2x_2 + 7, \mathbf{x} = (2, 1)$$

2.
$$0.1x_1^2 + x_2^2 + x_1x_2 + x_1 - 10, \mathbf{x} = (4, 1)$$

3.
$$(x_1 - 2)^2 + x_2^2, \mathbf{x} = (1, 1)$$

4.
$$x_1x_2 - x_1^2 - (x_2 - 3)^2, \mathbf{x} = (2, 4)$$

5.
$$x_1^3 + 2x_1^2 + x_2^2 + 3x_2 - 5, \mathbf{x} = (0, -1.5)$$

6.
$$3x_1 - x_1^3 - (x_2 - 2)^3 + 12x_2 + 3, \mathbf{x} = (1, 4)$$

3.2   Given the following functions, find if the function is convex, concave, or a saddle point.
$$2 - x_1^2 + 2x_1 + 2x_2 - x_2^2$$
$$x_1^2 - x_2^2$$
$$-x_1^2 - x_2^2$$
$$x_1^2 + x_2^2$$

3.3   Given the following optimization problem,

$$\text{Minimize} \quad f(x) = x_1{}^2 + x_2{}^2$$

subject to
$$2x_1 + x_2 - 2 \leq 0$$
$$x_2 - x_1 \geq 0$$
$$x_2 \leq 2$$

Plot the contours for $f(x)$ for $f(x) = 0, 0.5, 1, 2, 3, 4$, and the feasible region. From inspection, what is the optimal solution?

3.4 Solve the following quadratic programming (quadratic objective function and linear constraints) problem using an active constraint strategy. Plot the gradients of the objective and active constraints at the optimum, and verify geometrically the Kuhn–Tucker conditions. Determine also whether the optimal solution is unique.

$$\text{Minimize} \quad f(x) = 0.5(x_1{}^2 + x_2{}^2) - 3x_1 - x_2$$

subject to

$$x_1 + 0.5x_2 - 2 \leq 0$$

$$x_2 - x_1 \leq 0$$

$$-x_2 \leq 0$$

3.5 For the following problems: (1) solve the problem using the active constraint method, (2) perform two iterations of the Newton–Raphson method, and (3) perform two iterations of the BFGS quasi-Newton method. Use the starting point provided.

1.

$$\min 2x_1^2 + x_2^2 - 6x_1 - 2x_1x_2, x_0 = (1,1)$$

2.

$$\min 20x_1^2 + 10x_2^2 - 5x_1 - 2x_2, x_0 = (3,1)$$

3.

$$\min x_1^2 + x_2^2 - 4x_1 - 2x_2, x_0 = (1,1),$$

$$x_1 + x_2 \geq 4$$

$$x_1, x_2 \geq 0$$

Validate graphically.

4.

$$\min x_1^2 + x_2^2 - 4x_1 - 4x_2, x_0 = (1,1),$$

$$x_1 + 2x_2 - 4 \leq 0$$

$$3 - x_1 \leq 0$$

$$x_1, x_2 \geq 0$$

Validate graphically.

3.6 Design a beer mug to hold as much beer as possible. The height and radius of the mug should be no more than 20 cm. The mug must have a radius of at least 5 cm. The surface area of the sides of the mug must not exceed 900 cm² (we are ignoring the surface area related to the bottom of the mug). The ratio of the height to radius should be between 2.4 and 3.4. Formulate and solve this optimal design problem.

3.7 Design a circular tank, closed at both ends, with a volume of 200 m³.
The cost is proportional to the surface area of material, which is priced
at \$400/m². The tank is contained within a shed with a sloping roof,
thus the height of the tank h is limited by $h \leq 12 - d/2$, where d is
the tank diameter. Formulate the minimum cost problem and solve the
design problem.

3.8 Consider three cylindrical objects of equal height but with different radii,
$(r_1 = 1\,\mathrm{cm}, r_2 = 2\,\mathrm{cm}, r_3 = 5\,\mathrm{cm})$ as shown in Figure 3.14. What is
the box with the smallest perimeter that will contain these three cylin-
ders? Formulate and analyze this nonlinear programming problem. Using
GAMS find a solution. Turn the box by a right angle and check the solution.

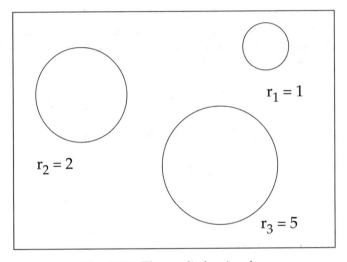

Fig. 3.14. Three cylinders in a box

3.9 The following reaction is taking place in a batch reactor, $A \rightarrow R \rightarrow S$,
where R is the desired product and A is the reactant, with an initial
concentration $CA_0 = 10$ moles/volume. The rate equations for this
reaction are provided below. Draw the graph of concentrations of A, R,
and S (CA, CR, and CS) with respect to time t, when time changes
in increments of 0.04 h. Solve the problem to obtain (1) maximum concentration of R, (2) maximum yield of R, (3) maximum profit, where
profit = 100 × concentration of product R - 10 × concentration of raw
material, (4) maximum profit, where profit = value of product less cost
of removing impurities A and S. Product value is the same as given for
(3) but the cost of removing A is 1 unit, whereas the cost of removing
S is 25 units (5) maximum profit, where profit = product value—raw
material cost (same as (3))—labor cost (25 units per unit time).

$$CA_t = CA_0 \exp(-k_1 t)$$

$$CR_t = CA_0 k_1 \left[\frac{\exp(-k_1 t)}{k_2 - k_1} - \frac{\exp(-k_2 t)}{k_2 - k_1} \right]$$

$$CS_t = CA_0 \left[1 + \frac{\exp(-k_1 t)}{k_2 - k_1} - \frac{\exp(-k_2 t)}{k_2 - k_1} \right]$$

where $k_1 = 10$ per hour, $k_2 = 0.1$ per hour
Now include the temperature (T) effects on the reaction in terms of the reaction constants k_1 and k_2 given below and resolve the above five optimization problems.

$$k1_{(T)} = 19531.2 \exp(-2258.0/T)$$
$$k2_{(T)} = 382700. \exp(-4517.0/T)$$

3.10 Use the interval Newton method and find the solution of the three humpback camel function given below.

$$f(x_1, x_2) = 2x_1^2 - 1.05x_1^4 + 1/6x_1^6 - x_1 x_2 + x_2^2 = 0$$

Bibliography

Arora J. S. (1989), *Introduction to Optimum Design*, McGraw-Hill, New York.

Bazaraa M., H. Sherali, and C. Shetty (1993), *Nonlinear Programming, Theory and Applications*, Second Edition, John Wiley & Sons, New York.

Beightler C. S., D. T. Phillips, and D. J. Wilde (1967), *Foundations of Optimization*, Prentice-Hall, Englewood Cliffs, NJ.

Biegler L., I. Grossmann, and A. Westerberg (1997), *Systematic Methods of Chemical Process Design*, Prentice-Hall, Englewood Cliffs, NJ.

Carter M. W. and C. C. Price (2001), *Operations Research: A Practical Introduction*, CRC Press, New York.

Davidon W. C. (1991), Variable matric method for minimization, *SIAM Journal on Optimization* **1**, 1.

Edgar T. F., D. M. Himmelblau, and L. S. Lasdon (2001), *Optimization of Chemical Processes*, Second Edition, McGraw-Hill, New York.

Fletcher R. (1987), *Practical Methods of Optimization*, Wiley, New York.

Gabasov R. F., F.M. Kirillova (1988), *Methods of optimization*, Optimization Software, Publications Division, New York.

Gill P. E., W. Murray, M. A. Saunders, and M. H. Wright (1981), *Practical Optimization*, Academic Press, New York.

Hansen E. and G. W. Walster(2004), *Global Optimization Using Interval Analysis, Second Edition*, Marcel Decker, New York.

Karush N. (1939), MS Thesis, Department of Mathematics, University of Chicago.

Kuhn H. W. and Tucker A. W. (1951), Nonlinear Programming, In J. Neyman (Ed.), *Proceedings of the Second Berkeley Symposium on Mathematical Statistics and Probability*, University of California Press, Berkeley.

Narayan, V., U. M. Diwekar and M. Hoza (1996), Synthesizing optimal waste blends, *Industrial and Engineering Chemistry Research* **35**, 3519.

Nocedal J. and S. J. Wright, (2006), *Numerical Optimization*, Springer Series in Operations Research, New York.

Reklaitis G. V., A. Ravindran, and K. M. Ragsdell (1983), *Engineering Optimization : Methods and Applications*, Wiley, New York.

Taha H. A. (1997), *Operations Research: An Introduction*, Sixth Edition, Prentice-Hall, Upper Saddle River, NJ.

Winston W. L. (1991), *Operations Research: Applications and Algorithms*, Second Edition, PWS-KENT, Boston.

4

Discrete Optimization

Discrete optimization problems involve discrete decision variables as shown below in Example 4.1.

Example 4.1: Consider the isoperimetric problem solved in Chapter 3 to be an NLP. This problem is stated in terms of a rectangle. Suppose we have a choice among a rectangle, a hexagon, and an ellipse, as shown in Figure 4.1.

Fig. 4.1. Isoperimetric problem discrete decisions, Exercise 4.1

Draw the feasible space when the perimeter is fixed at 16 cm and the objective is to maximize the area.

Solution: The decision space in this case is represented by the points corresponding to different shapes and sizes as shown in Figure 4.2.

Electronic Supplementary Material The online version of this chapter (https://doi.org/10.1007/978-3-030-55404-0_4) contains supplementary material, which is available to authorized users.

U. M. Diwekar, *Introduction to Applied Optimization*, Springer
Optimization and Its Applications 22,
https://doi.org/10.1007/978-3-030-55404-0_4

Discrete optimization problems can be classified as integer programming (IP) problems, mixed integer linear programming (MILP), and mixed integer nonlinear programming (MINLP) problems. Now let us look at the decision variables associated with this isoperimetric problem. We need to decide which shape and what dimensions to choose. As seen earlier, the dimensions of a particular figure represent continuous decisions in a real domain, whereas selecting a shape is a discrete decision. This is an MINLP as it contains both continuous (e.g., length) and discrete decision variables (e.g., shape), and the

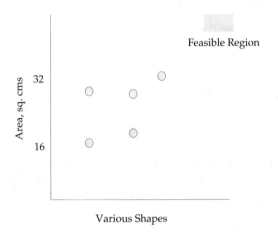

Fig. 4.2. Feasible space for discrete isoperimetric problem

objective function (area) is nonlinear. For representing discrete decisions associated with each shape, one can assign an integer for each shape or a binary variable having values of 0 and 1 (1 corresponding to yes and 0 to no). The binary variable representation is used in traditional mathematical programming algorithms for solving problems involving discrete decision variables. However, probabilistic methods such as simulated annealing and genetic algorithms which are based on analogies to a physical process such as the annealing of metals or to a natural process such as genetic evolution, may prefer to use different integers assigned to different decisions.

Representation of the discrete decision space plays an important role in selecting a particular algorithm to solve the discrete optimization problem. The following section presents the two different representations commonly used in discrete optimization.

4.1 Tree and Network Representation

Discrete decisions can be represented using a tree representation or a network representation. Network representation avoids duplication and each node corresponds to a unique decision. This representation is useful when one is using methods like discrete dynamic programming (Chapter 7 describes dynamic programming for continuous path optimization). Another advantage of the network representation is that an IP problem that can be represented appropriately using the network framework can be solved as an LP (see Chapter 2). Examples of network models include transportation of supply to satisfy a demand, flow of wealth, assigning jobs to machines, and project management. The tree representation shows clear paths to final decisions; however, it involves duplication. The tree representation is suitable when the discrete decisions are represented separately, as in the branch-and-bound method. This method is more popular for IP than the discrete dynamic programming method in the mathematical programming literature due to its easy implementation and generalizability. The following example from Hendry and Hughes (1972) illustrates the two representations.

Example 4.2: Given a mixture of four chemicals A, B, C, D for which different technologies are used to separate the mixture of pure components. The cost of each technology is given in Table 4.1. Formulate the problem as an optimization problem with tree and network representations.

Table 4.1. Cost of separators in 1000 $/year

Separator	Cost
A/BCD	50
AB/CD	170
ABC/D	110
A/BC	40
AB/C	69
B/CD	228
BC/D	40
A/B	144
B/C	50
C/D	329

Solution: Figure 4.3 shows the decision tree for this problem. In this representation, we have multiple representations of some of the separation options. For example, the binary separators A/B, B/C, C/D appear twice in the terminal nodes. We can avoid this duplication by using the network representation shown in Figure 4.4. In this representation, we have combined the branches that lead to the same binary separators. The network representation has 10 nodes, and the tree representation has 13 nodes. The optimization problem is to find the path that will separate the mixture into pure components for a minimum cost. From the two representations, it is very clear that the decisions involved here are all discrete decisions. This is a pure integer programming problem. The mathematical programming method commonly used to solve this problem is the branch-and-bound method. This method is described in the next section.

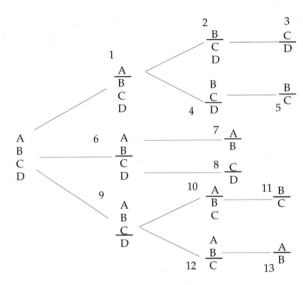

Fig. 4.3. Tree representation, Example 4.2

4.2 Branch-and-Bound for IP

Having developed the representation, the question is how to search for the optimum. One can go through the complete enumeration, but that would involve evaluating each node of the tree. The intelligent way is to reduce the search space by implicit enumeration and evaluate as few nodes as possible. Consider the above example of separation sequencing. The objective is to minimize the cost of separation. If one looks at the nodes for each branch, there are an initial node, intermediate nodes, and a terminal node. Each node is the sum of the costs of all earlier nodes in that branch. Because this cost increases monotonically as we progress through the initial, intermediate, and final nodes, we can define the upper bound and lower bounds for each branch.

- The cost accumulated at any intermediate node is a lower bound to the cost of any successor nodes, as the successor node is bound to incur additional cost.

- For a terminal node, the total cost provides an upper bound to the original problem because a terminal node represents a solution that may or may not be optimal.

The above two heuristics allow us to prune the tree for cost minimization. If the cost at the current node is greater than or equal to the upper bound defined earlier (either from one of the prior branches or known to us from experience), then we do not need to go further in that branch. These are the

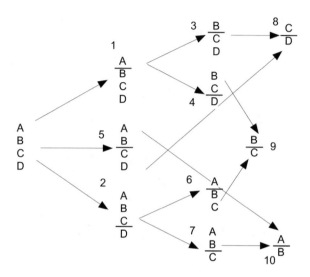

Fig. 4.4. Network representation, Example 4.2

two common ways to prune the tree based on the order in which the nodes are enumerated:

- Depth-first: Here, we successively perform one branching on the most recently created node. When no nodes can be expanded, we backtrack to a node whose successor nodes have not been examined.
- Breadth-first: Here, we select the node with the lowest cost and expand all its successor nodes.

The following example illustrates these two strategies for the problem specified in Example 4.2.

Example 4.3: Find the lowest cost separation sequence for the problem specified in Example 4.2 using the depth-first and breadth-first branch-and-bound strategies.

Solution: Consider the tree representation shown in Figure 4.5 for this problem.

First, let us examine the depth-first strategy, as shown in Figure 4.6 and enumerated below.

- Branch from Root Node to Node 1: Sequence Cost = 50.
- Branch from Node 1 to Node 2: Sequence Cost = 50 + 228 = 278.
- Branch from Node 2 to Node 3: Sequence Cost = 278 + 329 = 607.
  - Because Node 3 is terminal, current upper bound = 607.
  - Current best sequence is (1, 2, 3).

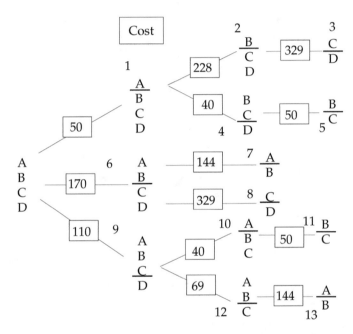

Fig. 4.5. Tree representation and cost diagram, Example 4.3

- Backtrack to Node 2.
- Backtrack to Node 1.
- Branch from Node 1 to Node 4: Sequence Cost = 50 + 40 = 90 < 607.
- Branch from Node 4 to Node 5: Sequence Cost = 90 + 50 = 140 < 607.
  - Because Node 5 is terminal and 140 < 607, current upper bound = 140.
  - Current best sequence is (1, 4, 5).
- Backtrack to Node 4.
- Backtrack to Node 1.
- Backtrack to Root Node.
- Branch from Root Node to Node 6: Sequence Cost = 170.
  - Because 170 > 140, prune Node 6.
  - Current best sequence is still (1, 4, 5).

- Backtrack to Root Node.
- Branch from Root Node to Node 9: Sequence Cost = 110.
  - Branch from Node 9 to Node 10: Sequence Cost = 110 + 40 = 150.
  - Branch from Node 9 to Node 12: Sequence Cost = 110 + 69 = 179.
  - Because 150 > 140, prune Node 10.
  - Because 179 > 140, prune Node 12.
  - Current best sequence is still (1, 4, 5).

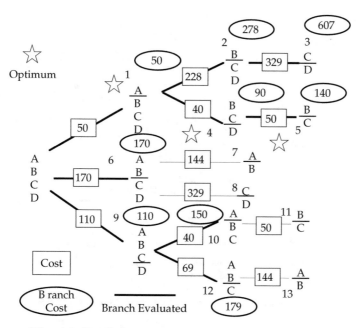

Fig. 4.6. Depth-first strategy enumeration, Example 4.2

- Backtrack to Root Node.
- Because all the branches from the Root Node have been examined, stop.
- Optimal Sequence (1, 4, 5), Minimum Cost = 140.

Note that with the depth-first strategy, we examined 9 nodes out of 13 that we have in the tree. If the separator costs had been a function of continuous decision variables, then we would have had to solve either an LP or an NLP at each node, depending on the problem type. This is the principle behind the depth-first branch-and-bound strategy.

The breadth-first strategy enumeration is shown in Figure 4.7. The steps are elaborated below.

- Branch from Root Node to:
  - Node 1: Sequence Cost = 50.
  - Node 6: Sequence Cost = 170.
  - Node 9: Sequence Cost = 110.

- Select Node 1 because it has the lowest cost.
- Branch Node 1 to:
    - Node 2: Sequence Cost = 50 + 228 = 278.
    - Node 4: Sequence Cost = 50 + 40 = 90.
- Select Node 4 because it has the lowest cost among 6, 9, 2, 4.

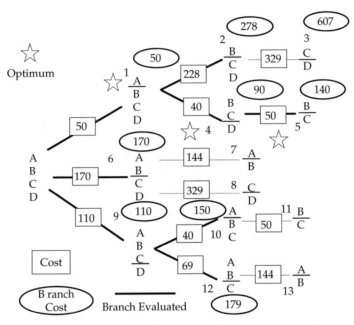

Fig. 4.7. Breadth-first strategy enumeration, Example 4.2

- Branch Node 4 to:
    - Node 5: Sequence Cost = 90 + 50 = 140.
- Because Node 5 is terminal, current best upper bound = 140 with the current best sequence (1, 4, 5).
- Select Node 9 because it has the lowest cost among 6, 9, 2, 5.
- Branch Node 9 to:
    - Node 10: Sequence Cost = 110 + 40 = 150.
    - Node 12: Sequence Cost = 110 + 69 = 179.
- From all the available nodes 6, 2, 5, 10, and 12, Node 5 has the lowest cost, so stop.
- Optimal Sequence (1, 4, 5), Minimum Cost = 140.

Note that with the breadth-first strategy, we only had to examine 8 nodes out of 13 nodes in the tree, one node less than the depth-first strategy.

In general, the breadth-first strategy requires the examination of fewer nodes and no backtracking. However, depth-first requires less storage of nodes because the maximum number of nodes to be stored at any point is the number of levels in the tree. For this reason, the depth-first strategy is commonly used. Also, this strategy has a tendency to find the optimal solution earlier than the breadth-first strategy. For example, in Example 4.3, we had reached the optimal solution in the first few steps using the depth-first strategy (seventh step, with five nodes examined).

4.3 Numerical Methods for IP, MILP, and MINLP

In Example 4.3, we could carry out the branch-and-bound for IP using graphical representation. However, for large-scale problems, it is impossible to solve the problem using the graphical way of enumerating the branch-and-bound steps. We need an algebraic representation of the graphical problem above for a numerical procedure. The following example presents the algebraic representation of the problem in Example 4.3.

Example 4.4: Provide the algebraic representation of the problem specified in Example 4.2 for the numerical branch-and-bound procedure.

Solution: Consider the tree representation for this problem to be as shown in Figure 4.8. As stated earlier, the decision variables associated with each node can be represented by binary variables y_i, for mathematical programming techniques. The figure also shows the binary variable associated with each node, representing the logic that if the node were present in the sequence, the binary variable associated with that node would be equal to one, else it would be zero. C_i denotes the cost of each node, and y_i represent the binary variable associated with each node. They are numbered (as subscripts) according to the nodes shown in the figure (e.g., y_9 corresponds to Node 9). Let us translate the tree structure into logical constraints. The objective function is the minimization of total costs, the cost of each node present in the final sequence. Because we do not know which node will be selected, we can write the objective function in terms of the cost of each node multiplied by the binary variable. Given that node not appearing in the sequence, the corresponding binary variable y will go to zero.

$$\text{Min} \quad z = \sum_{i=1}^{13} C_i y_i$$

$$y_i$$

subject to:

At the Root Node we can only select one of the three nodes.

$$y_1 + y_6 + y_9 = 1$$

Node 2 or Node 4 will exist if Node 1 is considered.

$$y_2 + y_4 = y_1$$

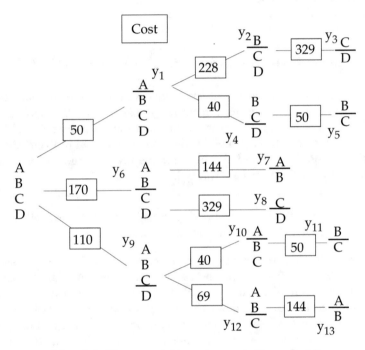

Fig. 4.8. Binary variable assignment, Example 4.2

Node 3 will exist if Node 2 is considered.

$$y_3 = y_2$$

Node 5 will exist if Node 4 is considered.

$$y_5 = y_4$$

Node 7 will exist if Node 6 is present and Node 8 will exist only if Node 7 is considered.

$$y_7 = y_6$$
$$y_8 = y_7$$

Node 10 or Node 12 will exist if Node 9 is considered.

$$y_{10} + y_{12} = y_9$$

Node 11 will exist if Node 10 is present and Node 13 will exist if Node 12 is considered.

$$y_{11} = y_{10}$$

$$y_{13} = y_{12}$$

It is obvious from the above example that once the discrete variables are assigned, it is possible to write logical constraints. Typical examples are:

1. Multiple Choice Constraints:
   - Select only one item:
   $$\sum_i y_i = 1$$
   - Select at most one item:
   $$\sum_i y_i \leq 1$$
   - Select at least one item:
   $$\sum_i y_i \geq 1$$

2. Implication Constraints:
   - If item k is selected, item j must be selected, but not necessarily vice versa:
   $$y_k - y_j \leq 0$$
   - If binary variable y is zero, an associated continuous variable x must be zero:
   $$x - Uy \leq 0$$
   $$x \geq 0$$

   where U is an upper limit to x.

3. Either-or constraints:
   - Either constraint $g_1(x) \leq 0$ or constraint $g_2(x) \leq 0$ must hold:
   $$g_1(x) - Uy \leq 0$$
   $$g_2(x) - U(1 - y) \leq 0$$

   where U is a large value.

As can be seen above, the IP problem can be represented by the following generalized form.

$$\text{Optimize} \quad Z \;=\; z(y) \;=\; \sum_i c_i y_i \;=\; C^T y \qquad (4.1)$$
$$y_i$$

where $y_i \,\epsilon\, 0, 1$.
subject to

$$h(y) \;=\; A^T y + B \;=\; 0 \qquad (4.2)$$
$$g(y) \;=\; D^T y + E \;\leq\; 0 \qquad (4.3)$$

As can be seen from the above example and the generalized representation, an IP problem tends to be linear. One way to solve these problems is to relax the constraint on the binary variables by making them continuous variables and then solving the LP. Figure 4.9 shows the feasible region of a two-dimensional problem where the IP is converted to an LP. What happened to the feasible region that consisted of discrete points? It became a continuous region, and the size of the feasible region increased. We have seen in the hazardous waste problem in earlier chapters that the solution to a less-constrained problem is as good as or better than the constrained solution. If the relaxed solution to an LP is a pure integer set, then the solution of the IP is reached. If the LP solution is not the IP solution, then the LP solution provides a lower bound to the (less constrained) IP solution for a minimization problem. The advantage of getting a lower bound to a branch a priori is that if the current upper bound is lower than the lower bound of the respective branch, then one does not have to enumerate that branch at all. Normally, the relaxed LP solution is used as a starting point for the branch-and-bound method.

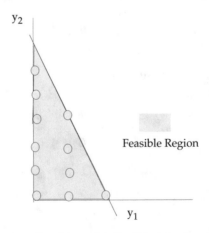

Fig. 4.9. Feasible region for IP and relaxed LP

As in the branch-and-bound algorithm, the cutting plane algorithm also starts with a relaxed LP solution. However, rather than using branching and bounding, it finds the solution by successively adding specially constructed cuts (constraints) to the problem. The added cuts do not eliminate any of the feasible integer points but must pass through at least one feasible or one infeasible integer point. Figure 4.10 shows the basic concepts behind the cutting plane method. In the figure (a) shows the relaxed LP solution, (b) the solution after one cut, and (c) the final cut and the integer solution.

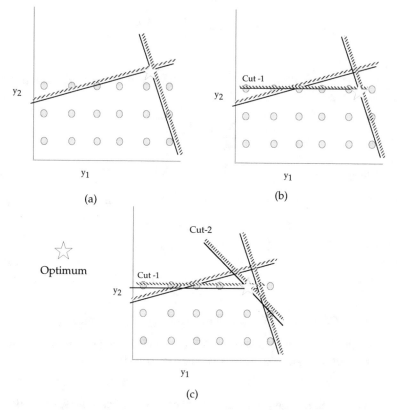

Fig. 4.10. The cutting plane method conceptual iterations

The idea behind the cutting plane method is that if the LP relaxation problem solution is an integer, then we are done. If not, then a valid inequality is found that separates the fractional solution or cuts it off. With the inclusion of this cutting plane, the former solution is forbidden; that is, it is tabu and will not be encountered within subsequent steps of the search. The cutting plane method was first proposed by Gomory in 1958. Therefore, the cut proposed by this method is called the Gomory cut. This is briefly described below.

Gomory cut: We know that the solution to the integer program in the cutting plane method starts with a relaxed LP solution. If the solution is not an integer solution, then we start adding cuts. For example, consider a general constraint at this solution given by $a_i y_i = a_0$, with y_i as integers, then we can derive the Gomory cut for this constraint as follows.

- Divide both sides of the constraint by a_j, as shown below.

$$y_j + \sum_{i \neq j} \frac{a_i}{a_j} y_i = \frac{a_o}{a_j} \tag{4.4}$$

$$y_j + \sum_{i \neq j} a_{i,j} y_i = a_{0,j} \tag{4.5}$$

where $a_{i,j} = \frac{a_i}{a_j}$.

- If $\lfloor \ \rfloor$ denotes rounding, then we can write the following expression from Equation 4.5.

$$y_j + \sum_{i \neq j} \lfloor a_{i,j} \rfloor y_i \leq \lfloor a_{0,j} \rfloor \tag{4.6}$$

- Subtracting Equation 4.5 from Equation 4.6 results in

$$\sum_{i \neq j} (\lfloor a_{i,j} \rfloor - a_{i,j}) y_i \leq \lfloor a_{0,j} \rfloor - a_{0,j} \tag{4.7}$$

- Let $f_{ij} = a_{i,j} - \lfloor a_{i,j} \rfloor$, Equation 4.7 becomes

$$-\sum_{i \neq j} f_{ij} y_i \leq -f_{0j} \tag{4.8}$$

This is the Gomory cut for the constraint $a_i y_i = a_0$, with y_i.

In the following example, we revisit Example 2.1 from Chapter 2 solution and how we can find a cut to forbid the earlier solution.

Example 4.5: Start the problem at the solution of Example 2.1 and derive a Gomory cut to cut out the previous solution.

Solution: In Example 2.2, we reached the solution $x = (0.5)$ with the following final Simplex tableau. This is also shown in Figure 4.11.

From Table 4.2, we can write the three constraints as follows.

$$2x_1 + s_1 - s_2 = 3 \tag{4.9}$$
$$x_2 + s_2 = 5 \tag{4.10}$$
$$x_1 + s_2 + s_3 = 9 \tag{4.11}$$

Table 4.2. The final simplex tableau, Example 2.2

Row	$-Z$	x_1	x_2	s_1	s_2	s_3	RHS	Basic	Ratio
0	1	4	0	0	1	0	5	$-Z = 5$	–
1	0	2	0	1	−1	0	3	$s_1 = 3$	–
2	0	0	1	0	1	0	5	$x_2 = 5$	–
3	0	1	0	0	1	1	9	$s_3 = 9$	–

Consider the constraint given by Equation 4.9 for deriving the Gomory cut. By dividing both sides of constraint by 2, and then rounding results in the following inequality.

$$x_1 + 0s_1 - 0s_2 \leq 1 \tag{4.12}$$

Subtracting Equation 4.9 after dividing by 2 from Equation 4.12, we get the following constraint (Gomory cut).

$$s_1 - s_2 \leq 1 \tag{4.13}$$

Substituting Equation 4.13 in Equation 4.9 results in the following equation. This Gomory cut shown in Figure 4.11 is cutting the previous solution.

$$x_1 \geq 1 \tag{4.14}$$

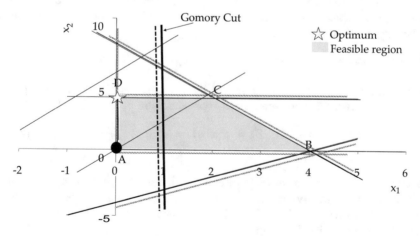

Fig. 4.11. Gomory Cut

The two disadvantages of the cutting plane method are for some problems a large number of cuts are needed to find the solution, and if the rounding is not done properly, the cuts can reduce the feasible region.

This same concept of forbidding the previous solutions is used in the tabu search. The basic concept of tabu search, as described by Glover (1986), is a meta-heuristics superimposed on another heuristic. The overall approach is to avoid entrainment in cycles by forbidding or penalizing moves that take the solution, in the next iteration, to points in the solution space previously visited (hence "tabu"). The tabu search begins by marching to a local minimum. To avoid retracing the steps used, the method records the moves in one or more tabu lists. At initialization, the goal is to make a coarse examination of the solution space, known as "diversification," but as candidate locations are identified the search is more focused to produce local optimal solutions in a process of "intensification."

MILP Problems

The mixed integer linear programming problems are of the form given below.

$$\text{Optimize} \quad Z = z(x, y) = a^T y + C^T x \qquad (4.15)$$
$$x, y_i$$

where $y_i \in 0, 1$ and x is a set of continuous variables. Note that the IP part in the objective function is again linear.
subject to

$$g(x, y) = -By + A^T x \leq 0 \qquad (4.16)$$

Branch-and-bound is a commonly used technique for solving MILP problems, where at each node, instead of looking at the fixed costs as we have seen in Example 4.3, an LP is solved.

MINLP Problems

What happens when you have a mixed integer nonlinear programming problem? The following is the generalized representation of an MINLP problem.

$$\text{Optimize} \quad Z = z(x, y) = a^T y + f(x) \qquad (4.17)$$
$$x, y_i$$

where $y_i \in 0, 1$ and x is a set of continuous variables. The first term represents a linear function involving the binary variables y and the second term is a nonlinear function in x. This formulation avoids nonconvexities and bilinear terms in the objective function.
Similarly, for the constraints the following formulation is used.
subject to

$$h(x) = 0 \qquad (4.18)$$

$$g(x, y) = -B^T y + g(x) \leq 0 \qquad (4.19)$$

Branch-and-bound for MINLP is a direct extension of the linear case (MILP). This method starts by relaxing the integrality requirements of the

0–1 variables, which leads to a continuous NLP optimization problem. It then continues by performing the tree enumeration where a subset of 0–1 variables is successively fixed at each node and an NLP problem is solved at each node.

The major disadvantage of the branch-and-bound method is that it may require the solution of a relatively large number of huge NLP problems, making this method computationally expensive. The relaxed NLP can lead to singularities and convergence problems. On the other hand, if the MINLP has a tight NLP relaxation, the number of nodes enumerated may be modest. In the limiting case where the NLP relaxation exhibits 0–1 solutions for the binary variable (convex hull representation), only one single NLP problem needs to be solved. A convex hull is a smallest convex set containing all the points.

The alternatives to branch-and-bound for MINLP are the generalized Bender's decomposition (GBD) and outer-approximation (OA) algorithms. These algorithms consist of solving an NLP subproblem (with all 0–1 variables) and an MILP master problem at each major iteration, as shown in Figure 4.12. The NLP subproblem has the role of optimizing the continuous variables, and the MILP master problem provides the 0–1 variables at each iteration. The master problem represents the linearized representation of the NLP and hence provides the lower bound to the MINLP. The following paragraph explains the linearization procedure and why the master problem provides a lower bound.

Consider the nonlinear objective function shown in Figure 4.13. As can be seen, this is a convex function and the problem is to locate the minimum of this function, as given below.

$$\text{Minimize}_{x_1} \quad Z \; = \; z(x_1) \; = \; -8x_1 + x_1^2 \qquad (4.20)$$

$$x_1 \; \geq \; 0 \qquad (4.21)$$

The linearization of this problem at the point in Figure 4.13 resulted in a tangent at that point (point k). It is obvious that the line provides a boundary to the function and hence is represented by an inequality where the objective function has to lie on the other side of the hashed line in Figure 4.13. So the linearized optimization problem can be represented as shown below.

Weak LP representation:

$$\text{Minimize}_{x_1} \quad Z \; = \; z(x_1) \; = \; \alpha \qquad (4.22)$$

Using the Taylor series expansion:

$$\alpha \; \geq \; -8x_1^{\,k} \; + \; (x_1^{\,k})^2 \; + \; (-8 + 2x_1^{\,k})\,(x_1 - x_1^{\,k}) \qquad (4.23)$$

$$x_1 \; \geq \; 0 \qquad (4.24)$$

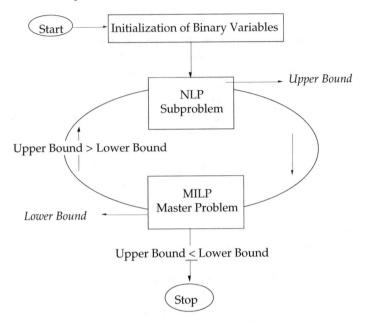

Fig. 4.12. Main steps in GBD and OA algorithms

As can be seen in the figure, the optimum solution lies lower than the original NLP. This linearization is a weak representation of the original function. To represent the NLP, we need to add linearization at several points, as shown in Figure 4.14, leading to the same optimum solution. This LP representation will have several binding constraints such as the one represented above (Equation (4.23)), one for each line.

In GBD-OA algorithms (Figure 4.12), the MILP problem is a linearized representation of the MINLP calculated at the previous NLP solution points (with fixed binary variables). The linearization is based on the above principle. As can be seen above, the MILP solution would provide a lower bound to the MINLP. At each iteration, the binary variables are calculated by the MILP master problem. For these fixed binary variables, the NLP is solved and linearizations are obtained. If the NLP solution (upper bound) crosses or is equal to the lower bound predicted by the MILP, then stop, else the iteration continues and a new linearized representation is added to the MILP. In GBD, the Lagrangian, or the dual representation of the problem is used for linearization, whereas in OA the linearizations are carried out, keeping the original (primal) representation of the problem.

Let us first look at the GBD linearization for the following generalized representation of the MINLP.

MINLP:

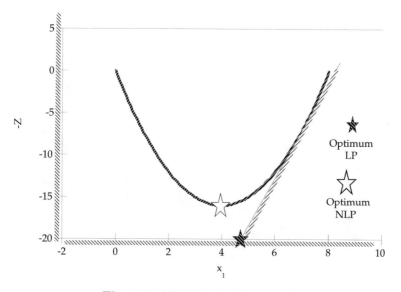

Fig. 4.13. NLP linearization, step 1

$$\underset{x,\,y_i}{\text{Minimize}} \quad Z \;=\; z(x,y) \;=\; a^T y \;+\; f(x) \tag{4.25}$$

where $y_i \,\epsilon\, 0,1$ and x is a set of continuous variables.
  subject to

$$h(x) \;=\; 0 \tag{4.26}$$

$$g(x,y) \;=\; g(x) \;-\; B^T y \;\le\; 0 \tag{4.27}$$

Lagrangian or dual representation of the above MINLP:

$$\begin{aligned}
\underset{x,\,y,\,\lambda_j,\,\mu_i}{\text{Minimize}} \quad L \;&=\; l(x,y,\lambda_j,\,\mu_i) \\
&=\; a^T y \;+\; f(x) \;+\; \sum_j \lambda_j h_j(x) \\
&\quad +\; \sum_i \mu_i(g_i(x) \;-\; B^T y)
\end{aligned} \tag{4.28}$$

where λ_j and μ_i are Lagrangian constraint multipliers.

  For the kth iteration of the master problem, which results in the binary variables solution $y = y^k$, the GBD linearization can be obtained as follows.
  MILP at the kth GBD iteration:

$$\underset{\alpha,\,y}{\text{Minimize}} \quad \alpha \tag{4.29}$$

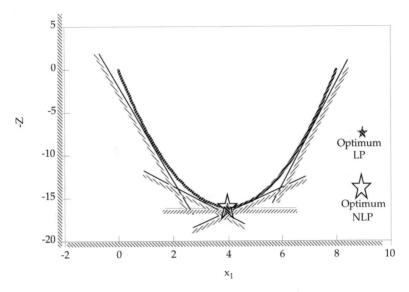

Fig. 4.14. NLP linearization

subject to

$$\alpha \geq a^T y + f(x^k) + \sum_i \mu_i^k (g_i(x^k) - B^T y) \tag{4.30}$$

On the other hand, OA uses the original representation for linearization. OA in its original form could not handle equality constraints. A variant of OA called outer-approximation/Equality Relaxation (OA/ER) was proposed later to handle equalities. If we eliminate the equality constraint in the MINLP formulation, then OA linearization results in the following.

MILP for OA:

$$\text{Minimize} \quad Z = \alpha \tag{4.31}$$
$$\alpha, x, y_i$$

subject to

$$\alpha \geq a^T y + f(x^k) + \nabla f(x^k)(x - x^k) \tag{4.32}$$

$$g(x, y) = -B^T y + g(x^k) + \nabla g(x^k)^T (x - x^k) \leq 0 \tag{4.33}$$

Both GBD and OA master problems accumulate new constraints as the iterations proceed. However, GBD accumulates one additional constraint, whereas OA accumulates a set of linear constraints per iteration. The master problem of OA is richer in information than the GBD, so it requires fewer iterations than the GBD. It should be noted that the GBD master problem only predicts discrete variables, and is an IP. OA, on the other hand, is an

MILP problem and may require more computational efforts to solve the master problem as compared to the GBD. The following two MINLP examples (one simple and one complex) demonstrate the GBD and OA algorithms.

Example 4.6: Consider the problem of minimization of a curve similar to the shown in Figure 4.13 but in two dimensions x_1 and x_2. We will also introduce an integer variable y, which becomes 1 when $x_2 > 0$; otherwise, it is zero. x_2 also has an upper bound $U = 100$. We can formulate this problem as follows.

$$\underset{x_1}{\text{Minimize}} \quad Z = z(x_1) = -8x_2 + x_1^2 \tag{4.34}$$

$$x_2 \leq x_1 \tag{4.35}$$
$$x_2 - Uy \leq 0 \tag{4.36}$$
$$x_1 \geq 0 \quad x_2 \geq 0 \tag{4.37}$$

Solution: Let us start the value of integer variable $y = 0$.

- First NLP problem:
Using Lagrangian formulation, the NLP results in

Minimize L
$$L = -8x_2 + x_1^2 + \mu_1(z_2 - 100y) + \mu_2(x_2 - x_1) + \mu_3(-x_1) + \mu_4(-x_1) \tag{4.38}$$

This NLP results in the following solution.

$$x_1 = 0$$
$$x_2 = 0$$
$$\mu_1 = 8$$
$$\mu_2 = 0$$
$$\mu_3 = 0$$
$$\mu_4 = 0$$
$$Z_{up} = 0$$

- First MILP problem using GBD:

$$\text{Minimize} \ \alpha \tag{4.39}$$
$$\alpha \geq 8(0 - 100y) \tag{4.40}$$

This results in $y = 1$, $Z_{lower} = -800$.

- First MILP problem using OA:

$$\text{Minimize } \alpha \tag{4.41}$$
$$\alpha \geq 8(x_2) \tag{4.42}$$
$$x_2 \leq x_1 \tag{4.43}$$
$$x_2 - Uy \leq 0 \tag{4.44}$$

$$x_1 \geq 0 \quad x_2 \geq 0 \tag{4.45}$$

This results in $y = 1, x_2 = 100, x_1 = 100, Z_{lower} = -800$.
- Second NLP at $y = 1$. Again, using the same Lagrangian formulation given in Equation 4.38, the NLP results in the following solution.

$$x_1 = 4$$
$$x_2 = 4$$
$$\mu_1 = 0$$
$$\mu_2 = 8$$
$$\mu_3 = 0$$
$$\mu_4 = 0$$
$$Z_{up} = -16$$

- Second MILP using GBD:

$$\text{Minimize } \alpha \tag{4.46}$$
$$\alpha \geq -16 \tag{4.47}$$
$$\alpha \geq 8(0 - 100y) \tag{4.48}$$

This results in $y = 1, \ Z_{lower} = -16$.
- Second MILP problem using OA:

$$\text{Minimize } \alpha \tag{4.49}$$
$$\alpha \geq -16 - 8(x_2 - 4) + 8(x_1 - 4) \tag{4.50}$$
$$\alpha \geq 8(x_2) \tag{4.51}$$
$$x_2 \leq x_1 \tag{4.52}$$
$$x_2 - Uy \leq 0 \tag{4.53}$$

$$x_1 \geq 0 \quad x_2 \geq 0 \tag{4.54}$$

This results in $y = 1, x_2 = 4, x_1 = 4, \ Z_{lower} = -16$. Since $z_{up} = z_{lower}$, the solution is reached.

Example 4.7: Consider the three objects shown in Figure 4.15. Each object shows the maximum area that is allowed to be covered by that kind of figure. It is given that the length of the square is equal to the radius of each circular object and is limited by an upper bound of 4 cm. Formulate the problem as an MINLP to find the object that will provide the maximum area. Use OA and GBD algorithms to solve this problem.

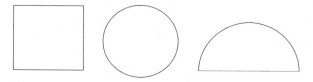

Fig. 4.15. Maximum area problem, Example 4.5

Solution: Let us first define the decision variables.

A_i area corresponding to object i

x maximum allowable length or radius for each object

y_i binary variable corresponding to object i; if y_i is 1, object i is selected, else y_i is zero.

MINLP formulation

In order to avoid nonconvexity, the following formulation is used.

$$\text{Maximize} \quad Z = z(A, x, y) = A_1 + A_2 + A_3 \qquad (4.55)$$
$$A_i, x, y_i$$

or

$$\text{Minimize} \quad Z = z(A, x, y) = -A_1 - A_2 - A_3 \qquad (4.56)$$
$$x, y_i$$

subject to

$$A_1 \leq x^2 \qquad (4.57)$$
$$A_2 \leq \pi x^2 \qquad (4.58)$$
$$A_3 \leq \pi/2 \, x^2 \qquad (4.59)$$
$$0 \leq x \leq 4 \qquad (4.60)$$

If binary variable y_i disappears, corresponding A_i vanishes.

$$A_i \leq U y_i \qquad i = 1, 2, 3. \qquad (4.61)$$

$$\sum_{i=1}^{3} y_i = 1 \qquad (4.62)$$

U is an arbitrary large number. We assume $U = 100$.

Outer-Approximation (OA)

Let us start the first iteration with $y^0 = (1, 0, 0)$.

- First NLP subproblem:

$$\underset{A, x}{\text{Minimize } Z} = z(A, x, y^0) = -A_1 - A_2 - A_3 \quad (4.63)$$

subject to

$$A_1 \leq x^2 \quad (4.64)$$
$$A_2 \leq \pi x^2 \quad (4.65)$$
$$A_3 \leq (\pi/2) x^2 \quad (4.66)$$
$$0 \leq x \leq 4 \quad (4.67)$$

$$A_1 \leq U \quad (4.68)$$
$$A_2 \leq 0 \quad (4.69)$$
$$A_3 \leq 0 \quad (4.70)$$

NLP solution: $A_2 = A_3 = 0$, $x = 4$, and $A_1 = 16$, $Z = -16$.
- First MILP master problem using linearization:

$$\underset{\alpha, A_i, x, y_i}{\text{Minimize } Z} = z(A, x, y) = \alpha \quad (4.71)$$

subject to

$$\alpha \geq -A_1 - A_2 - A_3 \quad (4.72)$$
$$A_1 \leq (4)^2 + 2(4)(x - 4) \quad (4.73)$$
$$A_2 \leq \pi(4)^2 + 2(4)\pi(x - 4) \quad (4.74)$$
$$A_3 \leq (\pi/2)(4)^2 + 2(4)(\pi/2)(x - 4) \quad (4.75)$$
$$0 \leq x \leq 4 \quad (4.76)$$

where constraints (4.73)–(4.75) represent linearizations at $x = 4$, and A = (16, 0, 0).

$$A_i \leq U y_i \quad i = 1, 2, 3. \quad (4.77)$$

$$\sum_{i=1}^{3} y_i = 1 \quad (4.78)$$

MILP solution: $y^1 = (0, 1, 0)$, $A_1 = A_3 = 0$, $x = 4$, and $Z = \alpha = -16\pi$.

- Second NLP iteration:

$$\text{Minimize} \quad Z \; = \; z(A, x, y^1) \; = \; - A_1 \; - \; A_2 \; - \; A_3 \qquad (4.79)$$
$$A_i, x$$

subject to

$$A_1 \; \leq \; x^2 \qquad (4.80)$$
$$A_2 \; \leq \; \pi x^2 \qquad (4.81)$$
$$A_3 \; \leq \; \pi/2 x^2 \qquad (4.82)$$

$$A_1 \; \leq \; 0 \qquad (4.83)$$
$$A_2 \; \leq \; U \qquad (4.84)$$
$$A_3 \; \leq \; 0 \qquad (4.85)$$

NLP solution: $A_1 \; = \; A_3 \; = 0$, $x \; = \; 4$, and $A_2 \; = \; 16\pi$, $Z \; = \; -16\pi$.
- Because $Z_{NLP} \; \leq \; Z_{MILP}$, the solution is reached in two NLP and one MILP iterations.

Remember that the branch-and-bound solution for this problem will take three NLP iterations.

Generalized Bender's Decomposition (GBD)

Initial binary variables with $y^0 \; = \; (1, 0, 0)$.

- Lagrangian or dual representation of the MINLP:

$$\text{Minimize} \qquad L \; = \; - A_1 \; - \; A_2 \; - \; A_3 \; + \; \mu_1(A_1 \; - \; x^2)$$
$$A_i, \mu_i, \mu_{1i}, \mu_0, x, y_i \quad + \; \mu_2(A_2 \; - \; \pi x^2)$$

$$+ \; \mu_3(A_3 \; - \; (\pi/2)x^2) \; + \; \sum_{i=1}^{3} \mu_{1i}(A_i \; - \; Uy_i)$$

$$+ \; \mu_0(x - 4) \; + \; \mu_{00}(-x) \qquad (4.86)$$

- First NLP solution from the KKT conditions:
Considering only active inequality constraints (corresponding to μ_1, μ_0, μ_{12}, μ_{13} as μ_{00}, μ_2, μ_3, μ_{11} are equal to zero).

$$\nabla L \; = \; 0 \qquad (4.87)$$

$$\frac{\partial L}{\partial A_1} \; = \; -1 \; + \; \mu_1 \; = 0 \qquad (4.88)$$

$$\frac{\partial L}{\partial A_2} \; = \; -1 \; + \; \mu_2 \; + \; \mu_{12} = 0 \qquad (4.89)$$

$$\frac{\partial L}{\partial A_3} \; = \; -1 \; + \; \mu_3 \; + \; \mu_{13} = 0 \qquad (4.90)$$

$$\frac{\partial L}{\partial x} = -2x\mu_1 + \mu_0 = 0 \tag{4.91}$$

$$\frac{\partial L}{\partial \mu_1} = A_1 - x^2 = 0 \tag{4.92}$$

$$\frac{\partial L}{\partial \mu_0} = x - 4 = 0 \tag{4.93}$$

$$\frac{\partial L}{\partial \mu_{12}} = A_2 = 0 \text{ because } y_2 = 0 \tag{4.94}$$

$$\frac{\partial L}{\partial \mu_{13}} = A_3 = 0 \text{ because } y_3 = 0 \tag{4.95}$$

Nonactive constraints:

$$\mu_{00} = 0 \tag{4.96}$$

$$\mu_2 = 0 \tag{4.97}$$

$$\mu_3 = 0 \tag{4.98}$$

$$\mu_1 1 = 0 \tag{4.99}$$

NLP Solution: $\mu_0 = 8$, $\mu_1 = \mu_{12} = \mu_{13} = 1$, $Z = -16$
- MILP master problem:

$$\text{Minimize}_{y_i} \quad Z = z(y) = \alpha \tag{4.100}$$

subject to

$$\alpha \geq -16 - Uy_2 - Uy_3 \tag{4.101}$$

$$\sum_{i=1}^{3} y_i = 1 \tag{4.102}$$

MILP solution: $y^1 = (0,1,0)$, $Z = -116$
- Table 4.3 shows solution steps and the MILP and NLP iteration summary for the GBD algorithm.

Table 4.3. GBD solution summary

Iteration	Z_{NLP}	Z_{MILP}	y	x
0	–	–	(1,0,0)	–
1	-16	-116	(0,1,0)	4
2	-16π	-16π	(0,1,0)	4

Because the binary variables obtained in two consecutive iterations are the same, the solution is reached in two NLP iterations. The following was the final MILP master problem.

Second MILP master iteration:

$$\text{Minimize} \quad Z = z(y) = \alpha \qquad (4.103)$$
$$y_i$$

subject to

$$\alpha \geq -16 - Uy_2 - Uy_3 \qquad (4.104)$$
$$\alpha \geq -16\pi - Uy_1 - Uy_3 \qquad (4.105)$$
$$\sum_{i=1}^{3} y_i = 1 \qquad (4.106)$$

The MINLP algorithms described above are designed for open equation systems where the information is transparent for problem solving. Furthermore, they encounter difficulties when functions do not satisfy convexity conditions, for systems having large combinatorial explosion, or when the solution space is discontinuous. Probabilistic methods such as simulated annealing and genetic algorithms provide an alternative to mathematical programming techniques such as the branch-and-bound, GBD, and OAs.

4.4 Probabilistic Methods

Simulated annealing (SA) and genetic algorithms (GA) are combinatorial methods based on ideas from the physical world. These are probabilistic combinatorial methods. Table 4.4 illustrates the key features of these algorithms and highlights marked differences and similarities between the two approaches. The following paragraphs describe the details of the two algorithms.

Simulated Annealing:

Simulated annealing is a heuristic combinatorial optimization method based on ideas from statistical mechanics (Kirkpatrick et al., 1983). The analogy is to the behavior of physical systems in the presence of a heat bath: in physical annealing, all atomic particles arrange themselves in a lattice formation that minimizes the amount of energy in the substance, provided the initial temperature is sufficiently high and the cooling is carried out slowly. At each temperature T, the system is allowed to reach thermal equilibrium, which is characterized by the probability (P_r) of being in a state with energy E given by the Boltzmann distribution:

$$P_r(Energy\ state = E) = \frac{1}{Z(t)} \exp\left(-\frac{E}{K_b T}\right) \qquad (4.107)$$

where K_b is Boltzmann's constant ($1.3806 \times 10^{23}\ J/K$) and $1/Z(t)$ is a normalization factor (Collins et al., 1988). See Figure 4.16.

Table 4.4. SA and GA comparison: theory and practice

	Simulated annealing	Genetic algorithms
In Theory		
Analogous Physical Phenomena	Statistical mechanics	Biological evolution and natural selection
Nature of Algorithm	Probabilistic	Probabilistic
Objective Function	Minimize the energy	Maximize the fitness of a generation
Mode of Operation	Works on a single solution string at any time	Works on a population of solution strings at any time
Initialization	Random or heuristic set of decision variables	Random population generated initially
Change in Decision Variables for Subsequent Iteration	Random perturbation	Crossover, mutation, and immigration
Stopping Criteria	Low temperature No improvement for consecutive iterations	Desired average fitness No improvement for consecutive generations
Key Algorithm Parameters	Temperature, Decrement factor, No. of moves at each temperature	No. of solution strings in a population, Percentage of reproduction, crossover, and mutation
In Practice		
Type of Optimization Problems That Can Be Solved	Large-scale, discrete, combinatorial, black-box, and nonconvex problems	Large-scale, discrete, combinatorial, black-box, and nonconvex problems
Global Optimization	Asymptotically converges to global optima if move sequences are Markov chains	No proof for optimal convergence
Optimization of Nonconvex Objective Function	Yes and does not require objective function gradient information	Yes and does not require objective function gradient information
Avoidance of Local Optima	Yes. By accepting moves by Metropolis criterion	Yes. By crossover and mutation techniques
Some Applications	Heat exchanger networks (Chauduri et al., 1997), Multidatabase systems (Subramanian and Subramanian, 1998), DNA structure (Guarnieri and Mezei, 1996)	Molecular design (Tayal and Diwekar, 2001), Aircraft design (Dunn, 1997), Internet (Joseph and Kinsner, 1997), Virology and AIDS (Shapiro and Wu, 1997), Truss design (Vazquez-Espi et al., 1997), Market simulation (Price, 1997)

In SA, the objective function (usually cost) becomes the energy of the system. The goal is to minimize the cost (energy). Simulating the behavior of the system then becomes a question of generating a random perturbation that displaces a "particle" (moving the system to another configuration). If the configuration that results from the move has a lower energy state, the move is accepted. However, if the move is to a higher energy state, the move is accepted according to the Metropolis criteria (accepted with probability $= \exp(-\Delta E/K_b T)$; VanLaarhoven and Aarts, 1987).

This implies that at high temperatures, a large percentage of uphill moves is accepted. However, as the temperature gets colder, a small percentage of uphill moves is accepted. After the system has evolved to thermal equilibrium at a given temperature, the temperature is lowered and the annealing process continues until the system reaches a temperature that represents "freezing." Thus, SA combines both iterative improvements in local areas and random jumping to help ensure that the system does not get stuck in a local optimum. The general SA is as follows (VanLaarhoven and Aarts, 1987).

1. Get an initial solution configuration S.
2. Get an initial temperature, $T = T_{initial}$.
3. While not yet frozen ($T > T_{froze}$) perform the following.

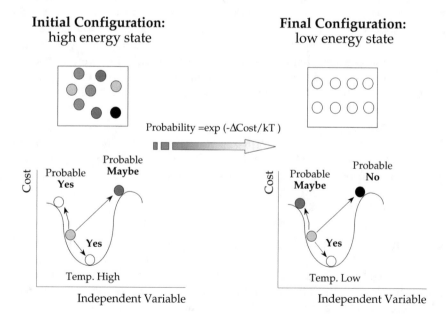

Fig. 4.16. Simulated annealing, basic concepts

(a) Perform the following loop K times until equilibrium is reached (K is the number of the moves per temperature level and is a function of moves accepted at that temperature level).
- Generate a move S' by perturbing S.
- Let $\Delta = Cost(S') - Cost(S)$.
- If $\Delta \leq 0$ (accept downhill move for minimization), then set $S = S'$ (accept the move), else, if $\Delta > 0$, it is an uphill move, accept the move with probability $\exp(-\Delta/T)$.
- Update number of accepts and rejects.
- Determine K and return to Step (a).

(b) No significant change in last C steps. Go to Step (4).

(c) Decrease T and go to Step (3).

4. Optimum solution is reached.

A major difficulty in the application of simulated annealing is defining the analogues to the entities in physical annealing. Specifically, it is necessary to specify the following: the configuration space, the cost function, the move generator (a method of randomly jumping from one configuration to another), the initial and final temperatures, the temperature decrement, and the equilibrium detection method. All of the above are dependent on problem structure. The initial and final temperatures, in combination with the temperature decrement scheme and equilibrium detection method, are generally referred to as the cooling schedule. Collins et al. (1988) have produced a very comprehensive bibliography on all aspects of SA including cooling schedule, physical analogies, solution techniques for specific problem classes, and so on. What follows is a brief summary of recommendations for developing a representation for the objective function, configuration space, cooling schedule, and move generator.

Objective Function

The objective function is a performance measure that the designer wishes to optimize. Because the analogy of the objective or cost function in annealing is energy, the problem should be defined so that the objective is to be minimized. That is, a maximization problem should be multiplied by -1 to transform it into a minimization problem.

Initial Temperature

If the initial annealing temperature is too low, the search space is limited and the search becomes trapped in a local region. If the initial temperature is too high, the algorithm spends a lot of time "boiling around" and wasting CPU time. The idea is to initially have a high percentage of moves that are accepted. Therefore, to determine the initial temperature, the following criteria should be satisfied (Kirkpatrick et al., 1983).

1. Take an initial temperature $T_{initial} > 0$.
2. Perform N sample moves according to the annealing schedule.

3. If the acceptable moves are $< 80\%$ of the total sampled, $(N_{succ}/N < 0.8)$ then set $T_{init} = 2T_{init}$ and go back to step 2.
4. If the acceptable moves are $> 80\%$ of the total sampled, $(N_{succ}/N > 0.8)$ then the initial temperature for the SA is T_{init}.

Final Temperature and Algorithm Termination
The annealing process can be terminated when one of the following conditions holds.

1. The temperature reaches the freezing temperature, $T = T_{freeze}$.
2. A specified number of moves have been made.
3. No significant changes have been made in the last C consecutive temperature decrements (C usually is fairly small, that is 5–20 temperature decrements).

Equilibrium Detection and Temperature Decrement
If the temperature decrement is too big, the algorithm quickly quenches and could get stuck in a local minimum with not enough thermal energy to climb out. On the other hand, if the temperature decrement is very small, much CPU time is required. Some rules (annealing schedule) for setting the new temperature at each level are:

1. $T_{new} = \alpha T_{old}$ where $0.8 \leq \alpha \leq 0.994$.
2. $T_{new} = T_{old}(1 + (1+\gamma)T_{old}/3\sigma)^{-1}$. This annealing schedule was developed by VanLaarhoven and Aarts and is based on the idea of maintaining quasi-equilibrium at each temperature (VanLaarhoven and Aarts, 1987). σ is the standard deviation of the cost at the annealing temperature T_{old}, and γ is the parameter that governs the speed of annealing (usually very small).
3. $T_{new} = T_{old} \exp(average(\Delta cost) \times T_{old}/\sigma^2)$
   This schedule was developed by Huang and is based upon the idea of controlling the average change in cost at each time step instead of taking a fixed change in the $\log T$ as in schedule 1. This allows one to take more moves in the region of lower variability, so that one takes many small steps at the cooler temperature when σ is low (Huang et al., 1986).

Note that the number of moves at a particular temperature N should be set in consideration of the annealing schedule. For example, many implementations chose a fairly large N (on the order of 100–1000) with large temperature decrements ($\alpha = 0.9$).

SA needs to reach quasi-equilibrium at each state or it is not truly annealing. It is difficult to detect equilibrium, but there are some crude methods, such as:

1. Set $N =$ number of states visited at each temperature.
2. Set a ratio of the number of accepted moves over the number of rejected moves.

Configuration Space
As with other discrete optimization methods, representation is one of the critical issues for successful implementation of SA and GA. In general, assigning integer values to the decision variable space instead of binary representation is better for SA and GA.

Move Generator
A move generator produces a "neighbor" solution (S' from S) from a given solution. The creation of a move generator is difficult because a move needs to be "random" yet results in a configuration that is in the vicinity of the previous configuration.

Genetic Algorithms

Genetic algorithms are search algorithms based on the mechanics of natural selection and natural genetics. Based on the idea of survival of the fittest, they combine the fittest string structures with a structured yet randomized information exchange to form a search algorithm with some of the innovative flair of human search (Goldberg, 1989). Genetic Algorithms were first developed by John Holland and his colleagues at the University of Michigan in the 1960s and 1970s, and the first full, systematic treatment was contained in Holland's book *Adaptation in Natural and Artificial Systems* published in 1975. The consistent growth in interest since then has increased markedly during the last 15 years. Applications include diverse areas, such as biological and medical science, finance, computer science, engineering and operations research, machine learning, and social science.

A GA is a search procedure modeled on the mechanics of natural selection rather than a simulated reasoning process. Domain knowledge is embedded in the abstract representation of a candidate solution, termed an organism, and organisms are grouped into sets called populations. Successive populations are called generations. A general GA creates an initial generation (a population or a discrete set of decision variables) $G(0)$, and for each generation $G(t)$, generates a new one $G(t+1)$.

The general genetic algorithm is described below.

At $t = 0$,

- Generate initial population, $G(t)$.
- Evaluate $G(t)$.
- While termination criteria are not satisfied, do

$t = t + 1$,

- Select $G(t)$.
- Recombine $G(t)$.
- Evaluate $G(t)$.

until solution is found.

In most applications, an organism consists of a single chromosome. A chromosome, also called a solution string of length n, is a vector of the form

$y_1, y_2, \ldots, y_n$ where each y_i is an allele, or a gene representing a set of decision variable values.

Initial Population

The initial population $G(0)$ can be chosen heuristically or randomly. The populations of the generation $G(t+1)$ are chosen from $G(t)$ by a randomized selection procedure, which is composed of four operators: (1) reproduction, (2) crossover, (3) mutation, and (4) immigration. Figure 4.17 shows GA strategies for developing the next generation using crossover, mutation, and immigration techniques.

Reproduction

Reproduction is a process in which individual strings are copied according to their objective function or fitness (f). Objective function f can be some measure of profit or goodness that we want to maximize. Alternatively, the objective function can represent process cost or the effluent pollutant level that we want to minimize. In the process of reproduction, only solution strings with high fitness values are reproduced in the next generation. This means that the solution strings that are fitter, and which have shown better performance, will have a higher chance of contributing to the next generation.

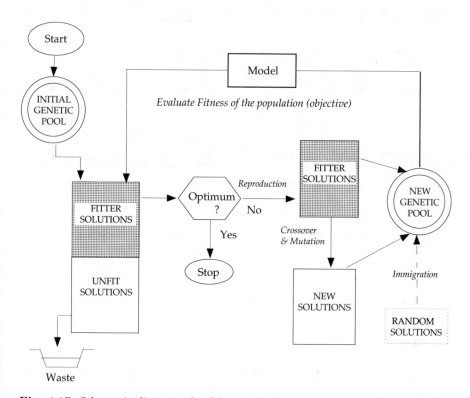

Fig. 4.17. Schematic diagram of a GA with different strategies for developing the next generation using crossover, mutation, and immigration techniques

Crossover

The crossover operator randomly exchanges parts of the genes of two parent solution strings of generation $G(t)$ to generate two child solution strings of generation $G(t + 1)$. Crossover serves two complementary search functions. First, crossover can provide new information about the hyperplanes already represented earlier in the population, and by evaluating new solution strings, GA gathers further knowledge about these hyperplanes. Second, crossover introduces representatives of new hyperplanes into the population. If this new hyperplane is a high-performance area of the search space, the evaluation of new population will lead to further exploration in this subspace.

Figure 4.18 shows three variants of crossover: one-point crossover, two-point crossover, and single-gene crossover. In a simple one-point crossover, a random cut is made and genes are switched across this point. A two-point crossover operator randomly selects two crossover points, and then exchanges genes in between. However, in a single-gene crossover, a single gene is exchanged between chromosomes of two parents at a random position.

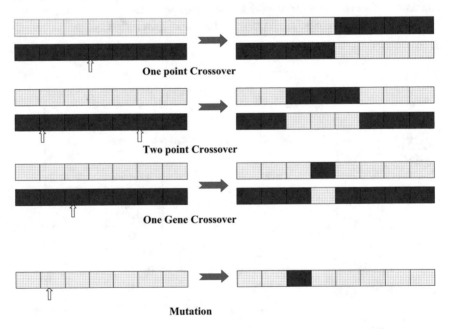

Fig. 4.18. Crossover and mutation techniques in genetic algorithms

Mutation

Mutation is a secondary search operator, and it increases the variability of the population. As shown in Figure 4.18, GA randomly selects a gene of the chromosome or solution string, and then changes the value of this gene in its permissible range. A low level of mutation serves to prevent any given bit position from remaining fixed indefinitely (forever converged) to a single value in the entire population. A high level of mutation yields essentially a random search.

Immigration

Immigration is a relatively new concept in GA and is based on immigration occurring between different societies in nature. In such scenarios, the fitness of immigrants from one society and their impact on the overall fitness of the new society to which they migrated becomes of crucial importance. It is analogous to the migration of intelligent individuals from rural areas to metropolitan in search of better prospects, and how they integrate and proliferate in the new society (being fitter) and effect the enrichment (fitness) of this new society. Thus immigration is the process of adding new, fitter individuals who will replace some existing members in the current genetic pool. Two criteria for selecting immigrants are that they should be fit and they should be quite different from the native population. Usually, immigration occurs between different populations, but can be incorporated in a single population as well (Ahuja and Orlin, 1997). Immigration offers an alternative to mutation and is usually employed when there is a danger of premature or local convergence.

Stopping Criteria

Termination criteria of the GA may be triggered by finding an acceptable approximated solution, by fixing the total number of generations to be evaluated, or by some other special criterion depending on the different approaches employed.

Key GA Parameters

The key GA parameters, which are common to all strategies explained above, are the population size in each generation (NPOP), the percentage of the population undergoing reproduction (R), crossover (C), mutation (M), and number of generations (NGEN). These can be crucial for customizing the GAs and can affect computational time significantly. These parameters govern the implementation of the algorithm to real-life optimization problems, and must be determined a priori before the procedure is applied to any given problem.

As stated earlier, SA and GA provide alternatives to the traditional mathematical programming techniques. There are a number of new probabilistic methods that appeared in the literature, like the particle swarm optimization and ant colony optimization. All these methods are based on natural phenomena. These methods were originally developed for discrete optimization where continuous variables or constraints were not present. There are various ways of dealing with this problem. For example, one can use explicit penalties for constraint violation (Painton and Diwekar, 1994), infeasible path optimization, or a coupled simulated annealing-nonlinear programming (SA-NLP) or

GA-NLP approach where the problem is divided into two levels, similar to the MINLP algorithms described above. The outer level is SA (GA), which decides the discrete variables. The inner level is NLP for continuous variables and can be used to obtain a feasible solution to the outer SA (GA). This approach is demonstrated in the following nuclear waste problem.

4.5 Hazardous Waste Blending: A Combinatorial Problem

Chapter 2 described the nuclear waste blend problem as an LP for single blend when some constraints were eliminated. Subsequently, it was converted to an NLP in Chapter 3 when all constraints were added. This chapter presents the nuclear waste problem as a discrete optimization problem.

The objective in this phase is to select the combination of blends so that the total amount of frit used is minimized. The number of possible combinations is given by the formula:

$$\frac{N!}{B!(T!)^B} \tag{4.108}$$

where N represents the total number of tanks of waste, B is the number of blends, and T is the number of tanks in each blend.

The formula indicates the complexity of the problem. To put this in perspective, if there are 6 tanks that have to be combined to form 2 blends by combining 3 tanks each, there are 10 possible combinations. If the number of individual waste tanks is 24 and 4 blends are to be formed by combining any 6 tanks, the number of possible combinations is 96,197,645,544. If the number of wastes is further quadrupled to 96 while maintaining the ratio of blends to the number of wastes in a blend at 2/3, the number of possible combinations is approximately 8.875×10^{75}. Clearly, any approach that is required to examine every possible combination to guarantee the optimum will very quickly be overwhelmed by the number of possible choices. Furthermore, note that a change in the ratio of the blends available to the number of wastes combining to form a blend affects the number of possible combinations. Figure 4.19 shows this variation when the number of wastes is 128. On the x-axis, the number of blends formed increases from left to right and the number of wastes in a blend decreases from left to right. The y-axis represents the log of possible combinations. Notice that the number of combinations first increases and then decreases and is skewed somewhat to the right.

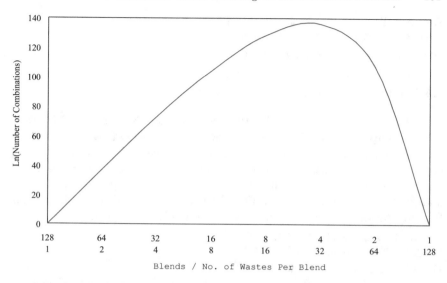

Fig. 4.19. Combinatorial complexity versus number of blends

For the purposes of this study, we have selected 21 tanks to be partitioned into three blends. The information about chemical composition and the amount of waste in each tank was obtained from Narayan et al. (1996) and is presented in Appendix A. The GAMS and other input files for this problem and the solutions can be online on Springer website with the book link. From the above formula, for a problem with 21 wastes to be partitioned into three blends, there are 66, 512, 160 possible combinations to examine. Clearly examining all possible combinations is a very onerous task and nearly impossible for larger problems. We, therefore, have to resort to either a heuristic approach or use combinatorial optimization methods such as mathematical programming techniques (GBD or OA) or simulated annealing.

In a heuristic approach to solving the discrete blending problem, we first determined the limiting constraint for a total blend of all tanks being considered (21, in this case). Then we tried to formulate blends such that all blends would have the same limiting constraint. If this can be achieved, the frit required would be the same as for the total blend. This approach, however, was very difficult to implement; rather, we formulated blends to try to ensure that all blends were near the limiting value for the limiting constraint. Using this approach, the best solution obtained was 11,736 kg of frit with the following tank configurations in each blend.

Blend 1 Tanks = [5 8 11 12 14 15 17]
Blend 2 Tanks = [4 6 7 13 18 19 20]
Blend 3 Tanks = [1 2 3 9 10 16 21]

4.5.1 The OA-based MINLP Approach

One possible approach for solving the above problem is using a MINLP with OA-based approach. GAMS uses this technique to solve MINLP problems.

The GAMS-based MINLP solution was very dependent on the starting conditions for the calculation. The conditions specified were the initial distribution of each tank among the blends (for the relaxed initial optimization) and the frit composition of each of the blends. The best MINLP solution was found to be 12,342 kg of frit with the following blend composition.

Blend 1 Tanks = [4 8 9 12 13 19 21]
Blend 2 Tanks = [1 2 7 14 15 17 18]
Blend 3 Tanks = [3 5 6 10 11 16 20]

The GAMS-based MINLP model failed to find the global optimal solution because the problem is highly nonconvex with the presence of several bilinear constraints.

For the particular problem on hand, we also developed a branch-and-bound procedure. Because this procedure was specific to the three-blend problem and also computationally intensive, it was used to check the global optimality of the simulated annealing solution procedure. Hence, it is presented as a separate section.

4.5.2 The Two-Stage Approach with SA-NLP

The optimal waste blending problem that we have addressed here is the discrete blending problem, where the amount of frit required to meet the various constraints is minimized by blending optimal configurations of tanks and blends. We have used a 2-loop solution procedure based on simulated annealing and nonlinear programming. In the inner loop, nonlinear programming is used to ensure constraint satisfaction by adding frit. In the outer loop, the best combination of blends is sought using simulated annealing so that the total amount of frit used is minimized. Figure 4.20 shows the schematic of this procedure used for solving the discrete blend problem.

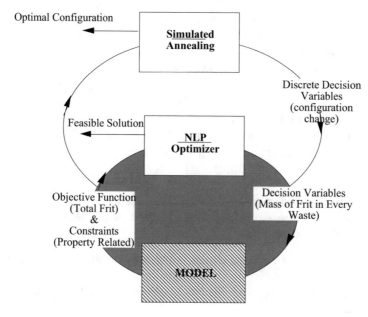

Fig. 4.20. Optimum discrete blending problem: solution procedure

We have used the inner loop NLP to solve each single- blend problem for both the two-stage approach and the branch-and-bound approach. The inner loop returns the minimum amount of frit required to satisfy all constraints given in Chapter 3. We have used two different procedures for the outer loop: (1) the simulated annealing procedure and (2) the branch-and-bound algorithm. Due to the problem characteristics, the solution from the inner loop cannot be guaranteed to be globally optimal. However, we are using the same NLP inner loop for the two-stage and the branch-and-bound approaches to find the discrete decision variables, that is, the configuration of each blend. The branch-and-bound method provides a guaranteed global optimum for the search of the discrete variables.

Simulated Annealing:

A major difficulty in the application of simulated annealing is defining the analogue to the entities in physical annealing. Specifically, it is necessary to specify the following: the objective function, the configuration space and the move generator, and the annealing schedule. All of the above are dependent on the problem structure. So for the discrete blending problem, we use the following specifications.

Objective

The objective for simulated annealing is identical to the objective given in Equation (3.134), which is to minimize the total mass of frit used over a given combination of blends.

Configuration Space and the Move Generator

Consider the problem where we have the 21 wastes shown in Appendix A (indexed by 1,2,...,21) and we wish to form three blends with these wastes. Our objective is to find the best combination of blends.

Suppose an initial state is such that: Blend 1 = [1, 2, 3, 4, 5, 6, 7];

Blend 2 = [8, 9, 10, 11, 12, 13, 14];
Blend 3 = [15, 16, 17, 18, 19, 20, 21].

A neighbor to this state can be defined as the state that can be reached by the application of a single operator. For a problem with three blends, we can devise three simple operators:

1. Swap (1,2)—where we swap elements between Blend 1 and Blend 2 (1/3 probability).
2. Swap (2,3)—where we swap elements between Blend 2 and Blend 3 (1/3 probability).
3. Swap (1,3)—where we swap elements between Blend 1 and Blend 3 (1/3 probability).

We need two more operators to decide which two elements from the two blends are to be swapped. For these studies, we have kept an equiprobable chance for one of the seven elements to be chosen from each of the two blends.

Temperatures Schedule

- Initial Temperature: If the initial temperature is too low, the search space is limited and the search becomes trapped in a local region. If the temperature is too high, the algorithm spends a lot of time jumping around, wasting CPU time. A rule of thumb for this is to select an initial temperature where a high percentage of moves is accepted. We have chosen 1000 as the initial temperature.

- Final Temperature: The final temperature is chosen so that the algorithm terminates after 10 successive temperature decrements with no change in the optimal state.

- Temperature Decrement: We have used the following simple rule with the value of α to be 0.95.

$$T_{new} = \alpha T_{old}$$

Solution:

The simulated annealing procedure provided a solution of 11,028 kg of frit (Table 4.5 provides the frit composition in each blend), which we were able to later confirm to be the global optimum using a branch-and-bound procedure. The composition of the blends was as follows.

Table 4.5. Frit composition in optimal solution

	Mass in Frit $f^{(i)}$		
Component	Blend 1	Blend 2	Blend 3
SiO_2	293.78	680.950	4550.6
B_2O_3	31.350	2.186	1212.4
Na_2O	38.683	375.06	1130.3
Li_2O	43.890	64.709	302.97
CaO	0.000	11.466	344.20
MgO	0.000	66.866	485.78
Fe_2O_3	0.000	0.000	502.11
Al_2O_3	0.000	0.000	640.96
ZrO_2	0.000	0.000	0.000
Other	0.000	0.000	250.07

Blend 1

   Tanks = [20 3 9 4 8 6 5]

Blend 2

   Tanks = [21 12 11 10 19 16 1]

Blend 3

   Tanks = [17 15 14 2 18 13 7]

4.5.3 A Branch-and-Bound Procedure

In order to find a guaranteed optimal solution among all possible combinations of wastes, each combination must be examined. Consider the example in Figure 2.10. There is a set of four wastes that has to be partitioned into two blends of two wastes each. Clearly we have three possible combinations:

$$[1, 2][3, 4], [1, 3][2, 4], [1, 4][2, 3]$$

Notice that we are indifferent to the ordering within a blend and also the ordering of blends within a possible combination. That is,

$$[1, 2][3, 4] \equiv [4, 3][2, 1]$$

This reduces the number of combinations we need to examine. For each of the three possible blend combinations, the amount of frit required for each blend must be found by the NLP. Thus, the enumerative procedure, like simulated annealing, is composed of two procedures. The outer loop is an enumerative procedure that supplies the inner loop with the wastes which might be combined to form a blend. In the inner loop, the NLP informs the outer loop about the amount of frit necessary to form glass. Although this method finds a guaranteed global optimal solution, unfortunately the number of possible combinations we need to examine grows exponentially with the number of wastes available, as given in Equation (4.108).

Objective

The objective is to minimize the total amount of frit as given by Equation (3.134).

Bounds

As mentioned before, the test problem with 21 wastes to be partitioned into three blends has 66,512,160 possible combinations to examine. The number of combinations that must be explicitly examined to verify optimality can be reduced by using a branch-and-bound method. The initial configuration is used as the starting upper bound. In the case of the test problem, the lower bound can be obtained in the following manner.

1. Fix the wastes for the first blend and calculate the amount of frit.
2. Relax the requirement that the remaining wastes must form two blends and assume that they form a single blend. In other words, we remove the binary variables y_{ij} for the two remaining blends. Now calculate the amount of frit required for this relaxation.
3. The total of the frit for the first blend and the relaxation is now a valid lower bound of the original problem.

If the lower bound is greater than the current best upper bound, then any combination where the composition of one of the blends is the same as that of the first blend cannot be optimal. All these combinations can be eliminated and can be considered as implicitly examined. This bounding method was sufficiently strong enough for us to solve the test problem to optimality. However, it still took about 3 days of computation (on a DEC-ALPHA 400 machine), as compared to an average of 45 min of CPU time using the two-stage annealing approach.

Solution Procedure

Figure 4.21 is a flowchart of the branch-and-bound procedure for a problem in which three blends are to be formed. An initial solution using any method serves as the initial upper bound. Within a loop which essentially enumerates every possible combination, the procedure first fixes the composition of the first blend. The amount of frit needed for this blend is then determined. By assuming that the remaining two blends form a single blend, the amount of frit for the composite blend is then determined. If the total amount of frit needed for this configuration of blends is greater than the current best solution or the upper bound, then we need not examine any combination of blends where one of the blends is identical to the first blend. But if this is not so, we examine all the possible combinations of the remaining two blends to determine which particular configuration requires the smallest amount of frit. The upper bound is updated if the best configuration found during enumeration is better than the upper bound. This continues until all possible combinations are either explicitly or implicitly examined. The better the lower bound estimates the eventual solution, the lower the amount of explicit enumeration that will have to be performed.

The branch-and-bound procedure can be implemented using different strategies. We implemented the procedure using a depth-first strategy because of its minimal memory requirements. The depth-first strategy is also relatively easy to implement as a recursive function. Any branch-and-bound method starts off with a start node. With reference to our problem, the start node is the assignment of waste 1 to the first blend, as shown in Figure 4.22. Because we are indifferent to the ordering of wastes within a blend and the ordering between blends, it is apparent that this assignment is a valid starting point. In the nodes succeeding the starting node, different wastes that can be combined with waste 1 to form the first blend are considered. As can be seen in Figure 4.22, there are five nodes that succeed the starting node. If we choose to expand the nodes in the order they are generated, the strategy is called *breadth-first*. If, on the other hand, we choose to expand the most recently generated nodes first, the strategy is called *depth-first*. We see that the depth-first algorithm pushes along one path until it reaches the maximum depth, then it begins to consider alternative paths of the same or less depth that differ only in the last step, then those that differ in the last two steps, and so on. As a result of this property, the number of nodes that have to be retained in memory is very small as compared to the breadth-first algorithm wherein the number of nodes residing in the computer's memory can grow exponentially with the size of the problem. In Figure 4.22, the dotted arrows indicate the direction in which the search proceeds.

Before going to the third level, a lower bound is computed and compared to the current best solution. If the lower bound is greater than the current best solution, then that part of the tree can be pruned and considered implicitly examined. The search is complete when all branches of the tree are either explicitly or implicitly examined. The better the lower bound is as an estimate of the final solution, the fewer the number of branches that have to be explicitly examined.

Optimal solution

The branch-and-bound procedure found the optimal solution to be 11,028 kg of frit, which is identical to the solution found by simulated annealing. This confirms that the two-stage SA-NLP approach provided the global optimum with respect to the configuration decisions. As before, the composition of the blends that required the minimum amount of frit was found to be

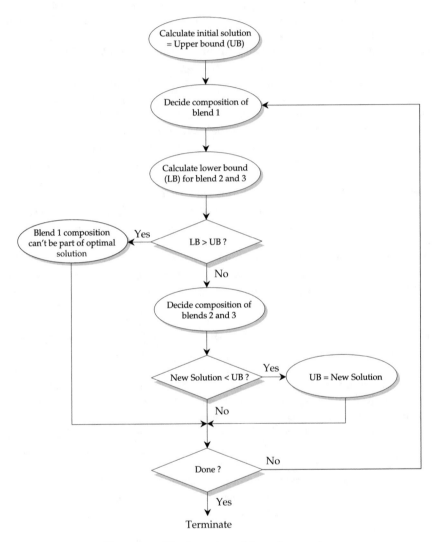

Fig. 4.21. The branch-and-bound procedure

Blend 1
   Tanks = [20 3 9 4 8 6 5]
Blend 2
   Tanks = [21 12 11 10 19 16 1]

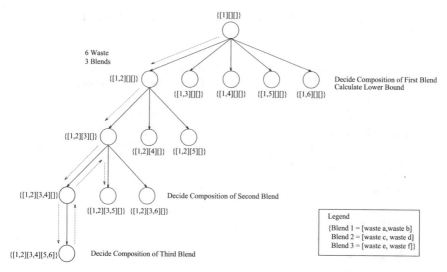

Fig. 4.22. Branch-and-bound using a depth-first strategy

Blend 3

Tanks = $[17\ 15\ 14\ 2\ 18\ 13\ 7]$

Comparison

The purpose of this case study was to develop a method that would help to determine which combination of wastes would minimize the amount of frit needed to convert the waste into glass. The benefit of reducing the amount of frit used is in reduced material costs and the reduced bulk of the glass formed, which in turn reduces the disposal costs. The search space grows exponentially with an increase in parameters defining the problem, making it almost impossible to find optimal solutions for realistically sized problems. We compared different combinatorial optimization techniques such as the GAMS-based MINLP algorithm, branch-and-bound, and the simulated annealing and heuristics approach to find the optimal solution. We have found that a two-stage approach combining simulated annealing and NLP algorithms is a computationally effective means of obtaining the global optimal or near global optimal solutions with reasonable amounts of computational effort. Both the heuristic approach and the GAMS-based MINLP result in local minima. The branch-and-bound procedure leads to a global optimum but requires a significantly longer computational time than the coupled simulated annealing-NLP approach.

4.6 Sustainable Mercury Management: A Combinatorial Problem

In Chapter 2, the generalized representation of the trading problem as a MILP is presented. Chapter 3 introduced the nonlinear cost function in the problem, making it an MINLP. Optimal solutions to the Savannah River watershed trading case study as a MILP and an MINLP are presented here.

For the sake of clarity, we are presenting the MILP and MINLP problem again here. The symbols are the same as described in Chapters 2 and 3.

The MILP Formulation:

Objective:

$$\text{Minimize} \quad \sum_{i=1}^{N} \sum_{j=1}^{M} TC_j.D_i.\, b_{ij} \tag{4.109}$$

Constraints:

$$t_{ii} = 0 \quad \forall i = 1, ..., N \tag{4.110}$$

$$red_i \leq \sum_{j=1}^{M} q_j.D_i.\, b_{ij} + \sum_{k=1}^{N} t_{ik} - r \sum_{k=1}^{N} t_{ki} \quad \forall i = 1, ..., N \tag{4.111}$$

$$P_i \geq \sum_{j=1}^{M} b_{ij}.TC_j.D_i + F\Big(\sum_{k=1}^{N} t_{ik} - \sum_{k=1}^{N} t_{ki}\Big) \quad \forall i = 1, ..., N \tag{4.112}$$

The MINLP Formulation:

Objective:

$$\text{Minimize} \quad \sum_{i=1}^{N} \sum_{j=1}^{M} f_j(\phi_j, D_i).\, b_{ij} \tag{4.113}$$

Constraints:

$$t_{ii} = 0 \quad \forall i = 1, ..., N \tag{4.114}$$

$$red_i \leq \sum_{j=1}^{M} q_j.D_i.\, b_{ij} + \sum_{k=1}^{N} t_{ik} - r \sum_{k=1}^{N} t_{ki} \quad \forall i = 1, ..., N \tag{4.115}$$

$$P_i \geq \sum_{j=1}^{M} b_{ij}.f_j(\phi, D_i) + F\Big(\sum_{k=1}^{N} t_{ik} - \sum_{k=1}^{N} t_{ki}\Big) \quad \forall i = 1, ..., N \tag{4.116}$$

The MILP and MINLP formulations above were solved for two options in the absence of trading and in the presence of trading. In the subsequent text, analysis when trading is not permitted is referred to as "technology option," while the analysis when trading is permitted is referred to as "trading option."

Tables 4.6 and 4.7 present the results of the MILP and MINLP problems for the considered TMDL range (26 Kg/year–36 Kg/year). The annual saving is computed as the difference between the total cost for technology option and total cost for trading options for a particular model under consideration. It is observed that approximate linear models underestimate the annual savings. The differences between linear deterministic and nonlinear deterministic model results are significant enough, and hence should not be ignored.

Table 4.6. Savannah River watershed trading: Solution for the MILP

| | Cost | | | Total | Technology distribution | | | | | |
| | | | | | Technology trading | | | | | |
TMDL	Technology	Trading	Savings	technology	A	B	C	A	B	C
26	188676712.5	154886655	36328632.5	0.083596104	13	13	3	15	7	0
27	187545212.5	148762155.8	40767762.5	0.082022565	13	12	4	14	6	0
28	178906903.5	142569967.3	37831337.5	0.070365225	13	10	6	15	4	1
29	178807623.5	136426688.8	43389557.5	0.062216495	14	9	6	14	3	1
30	154900123.5	129972247.8	25252671	0.042307902	16	7	6	18	4	1
31	146486873.5	124098923.3	22388990.5	0.045518446	17	6	6	18	0	2
32	145940395.5	118746216.5	27195164.5	0.060451949	14	6	9	12	0	1
33	140325783	113408420	26917728	0.076697728	16	4	9	11	1	0
34	128883617	108165925	20717692	0.099988353	14	3	12	9	0	0
35	128153617	102874264.5	25295084	0.115957863	12	3	14	11	0	2
36	126009461	97662812.75	28354295	0.142758527	13	1	13	8	0	3

TMDL for this analysis is 32 Kg/year, and the total mercury reduction target for all point sources is 0.892 g/year. From Table 4.6, it can be seen that the implementation of trading leads to about an 18% reduction in the compliance cost (technology implementation and/or trading), which amounts to around 27 Million $ annually. This supports the expected result that trading will reduce the overall expenditure. However, while satisfying TMDL, the total mercury discharge for the trading option is higher by about 19%. The technology only solution results in the implementation of technology A by 14 polluters, technology B by six polluters, and technology C by nine polluters. In comparison, when trading is permitted, 12 industries implement technology A, one industry implements technology C, while technology B is not implemented at all. Sixteen polluters trade all their reduction quantity to some other industries, while five polluters implement technology and also trade some portion of their discharge. Nine industries accept trades and thereby take care of excess discharge from other industries. The total quantity of mercury traded is 0.06 g/year, which is about 7% of the desired reduction. The

Table 4.7. Savannah River watershed trading: solution for the MINLP

	Cost			Total	Technology distribution			Technology trading		
								Technology trading		
TMDL	Technology	Trading	Savings	technology	A	B	C	A	B	C
26	167039837.7	115869556.7	51170281	0.151420021	7	19	3	5	3	1
27	164912119.7	111537713.5	53374406.18	0.1747659	7	18	4	4	3	2
28	154214576.5	106252350	47962226.52	0.165929784	8	15	6	5	2	0
29	154214576.5	101182339.9	53032236.6	0.175127431	8	15	6	4	2	1
30	139494653.5	103716924.3	35777729.19	0.122540613	10	13	6	5	2	0
31	132221260.2	98486954.01	33734306.19	0.094301779	11	12	6	7	1	0
32	130452235.1	93673310.48	36778924.57	0.115512261	9	12	8	6	1	1
33	124052801.1	89358434.86	34694366.27	0.147545346	11	10	8	5	1	0
34	113365337.7	85502031.54	27863306.14	0.121150689	10	8	11	7	0	0
35	112201768.6	80878225.72	31323542.85	0.144482586	9	7	13	6	0	1
36	108602898.9	75473166.8	33129732.1	0.150590739	10	5	12	5	0	0

results show a trend towards avoiding expensive technology options, and satisfying part of the pollutant reduction through trading. Also observed is a significant preference towards one technology (technology A) after trading is permitted.

The tables also show the number of times each technology is implemented over the complete TMDL range (summation over all TMDL values). It can be seen that there are definite implications on technology selection. With linear technology models, various small industries (industries with small volumetric discharge rates) implement technologies along with large industries (industries with large volumetric discharge rates). However, when nonlinear technology models are used, large industries implement most of the technologies, and smaller industries satisfy the regulations by trading with these large industries. The tables also show that nonlinear deterministic model involves considerably higher amount of trading (particularly at smaller TMDL values) since the smaller industries prefer to trade rather than implement technologies. The distribution of technology selection is observed to be similar for both models. For both models, coagulation and filtration is the technology most commonly implemented, followed by granular activated carbon process and ion exchange process, respectively.

4.7 Summary

Discrete optimization involves integer programming (IP), mixed integer linear programming (MILP), and mixed integer nonlinear programming (MINLP) problems. The commonly used solution method for solving IP and MILP problems is the branch-and-bound method. It uses the relaxed LP as a starting point and a lower bound for the branch-and-bound method. The branch-and-bound approach to MINLP can encounter problems of singularities and in-feasibilities, and can be computationally expensive. GBD and OA algorithms tend to be more efficient than branch-and-bound, and are commonly employed to solve MINLPs. However, these algorithms are designed for open equation systems and encounter difficulties when functions do not satisfy convexity conditions, for systems having a large combinatorial explosion, or when the solution space is discontinuous. Probabilistic methods such as simulated an-nealing and genetic algorithms provide an alternative to these algorithms. However, to obtain the best results, coupled SA-NLP or GA-NLP approaches need to be used.

Exercises

4.1 Two plants manufacture soybean oil. Plant A has six truckloads ready for shipment. Plant B has twelve truckloads ready for shipment. The products will be delivered to three warehouses: warehouse 1 needs seven truckloads; warehouse 2 needs five truckloads; and warehouse 3 needs six truckloads. Shipping a truckload of soybean oil from plant A to warehouses 1, 2, and 3 costs \$25, \$21, and \$27, respectively, and shipping a truckload of soybean oil from plant B to warehouses 1, 2, and 3 costs \$23, \$24, and \$22, respectively. The cost can be reduced by shipping more than one truckload to the same warehouse, and the discounted cost is given by

$$C_n = \frac{C}{n^{\frac{1}{3}}} \tag{4.117}$$

where C is the cost for only one truckload used for shipping to a warehouse and n is the number of truckloads from a plant to the same warehouse. A total of 18 truckloads are available at points of origin for delivery. Determine how many truckloads to ship from each plant to each warehouse to meet the needs of each warehouse at the lowest cost.

4.2 There are eight cities in Alloy Valley County. The county administration is planning to build fire stations. The planning target is to build the minimum number of fire stations needed to ensure that at least one fire station is within 18 min (driving time) of each city. The times (in minutes) required to drive between the cities in Alloy Valley County are shown in Table 4.8. Determine the minimum number of fire stations and also where they should be located.

4.3 This is a cellular network design problem. The potential customers in the planning region were grouped into small clusters based on their geograph-ical locations. Table 4.9 shows the 20 clusters in the planning region. Each cluster was characterized by the computed coordinate, the farthest dis-tance in the cluster from the centroid, and peak traffic demand. Table 4.10 shows the specifications and costs of four types of base stations available for planning. The capacity and coverage requirements for the setup of radio networks are:

- Each cluster must be served by at least one base station.

Table 4.8. Driving times (minutes) between cities in Alloy Valley County

City	1	2	3	4	5	6	7	8
1	0	18	13	16	9	29	20	25
2	18	0	25	35	22	10	15	19
3	13	25	0	15	30	25	18	20
4	16	35	15	0	15	20	25	28
5	9	22	30	15	0	17	14	29
6	29	10	25	20	17	0	25	10
7	20	15	18	25	14	25	0	15
8	25	19	20	28	29	10	15	0

- The total peak traffic of all clusters served by each base station must be within the capacity limit.
- The farthest point in a cluster must be within the coverage radius of the base station serving that cluster.

The planner wants to find the optimal base station planning method that minimizes the total cost.

Note: The coordinate of the centroid of the cluster is based on a V and H coordinate system which is commonly used in the telephone network. The distance in miles between two points is given by

$$\sqrt{[(v_1 - v_2)^2 + (h_1 - h_2)^2]/10}$$

4.4 And God said to Noah,

I have determined to make an end of all flesh, for the earth is filled with violence because of them; now I am going to destroy them along with the earth. Make yourself an ark of cypress wood; make rooms in the ark, and cover it inside and out with pitch. This is how you are to make it: the length of the ark three hundred cubits, its width fifty cubits, and its height thirty cubits. Make a roof for the ark, and finish it to a cubit above; and put the door of the ark in its side; make it with lower, second, and third decks. For my part, I am going to bring a flood of waters on the earth, to destroy from under heaven all flesh in which is the breath of life; everything that is on the earth shall die. But I will establish my covenant with you; and you shall come into the ark, you, your sons, your wife, and your sons' wives with you. And of every living thing, of all flesh, you shall bring two of every kind into the ark; they shall be male and female—(Genesis 6:13–19, New Revised Standard Version).

Suppose eight humans will be on the ark: Noah, his wife, their sons Shem, Ham, and Japheth, and each of the sons' wives. We put them into one room, which should be at least 320 square feet for their basic life, and allow

Table 4.9. 20 Planning clusters

V	H	Farthest point (mile)	Peak traffic (Mbps)
7121	8962	1.61	206.983886
7119	8952	1.33	151.1853099
7115	8951	1.49	60.76715757
7117	8961	1.56	291.965782
7129	8948	3.06	166.9175212
7117	8959	1.36	107.1509535
7119	8953	1.36	138.2899501
7119	8948	1.41	60.92519017
7122	8958	1.36	141.4339383
7118	8942	1.61	92.51767376
7116	8953	1.42	243.6052474
7127	8959	1.44	174.8569325
7116	8955	1.32	107.1905736
7121	8952	1.61	62.1673109
7128	8941	3.03	102.1435541
7121	8953	1.61	45.99242408
7116	8960	1.47	222.5078821
7128	8956	1.45	95.57552618
7126	8961	1.51	42.55302978
7120	8956	1.50	154.7507397

Table 4.10. Available four types of base stations

Base	Capacity (Mbps)	Coverage Radius	Cost ($)
Type 1	65	18	1,500,000
Type 2	130	5	2,000,000
Type 3	260	5	2,500,000
Type 4	520	5	3,000,000

them free roaming space (about 80 square feet for each deck) aboard all three decks. If we bring one pair of every living thing as ordered by God, the herbivores take a space of 625 square feet, and the carnivores take 319 square feet. Height is not a constraint because 45 feet for three decks is an average of 15 feet per deck but all of our tallest creatures are on the top deck. Assume that all these animals require a circular space, and use the insights derived from Exercise 3.8, Chapter 3. Formulate the problem as an MINLP for maximizing species and species variety so as to minimize the risk that a species may die (one cubit $= 1.5$ feet) (Katcoff J., and F. Wu, *All Creatures Great and Small: An Optimization of Space on Noah's Ark*, Course presentation, Carnegie Mellon University, 19–703, (1999)).

4.5 Consider the following small mixed integer nonlinear programming (MINLP) problem:

$$\text{Minimize} \quad z = -y + 2x_1 + x_2$$

Subject to

$$x_1 - 2.\exp\left(-x_2\right) = 0$$

$$-x_1 + x_2 + y \leq 0$$

$$0.5 \leq x_1 \leq 1.4$$

$$y\epsilon[0, 1]$$

Solve the two NLPs by fixing $y = 0$ and $y = 1$. Locate the optimum. For the above problem

- Eliminate x_2 and write down the iterative solution procedure using OA. Write down iterative solution procedure using GBD.
- Now instead of eliminating x_2, eliminate x_1. Assume $y = 0$ as a starting point. Find the solution using OA.

4.6 Given a mixture of four components A, B, C, D for which two separation technologies (given in Table 4.11) are to be considered.

Table 4.11. Cost of separators in $/year

Separator	Cost
A/BCD	50,000
AB/CD	56,000
ABC/D	21,000
A/BC	40,000
AB/C	31,000
B/CD	38,000
BC/D	18,000
A/B	35,000
B/C	44,000
C/D	21,000

1. Determine the tree and network representation for all the alternative sequences.
2. Find the optimal sequence with the depth-first strategy.
3. Find the optimal sequence with the breadth-first strategy.

4. Heuristics has provided a good upper bound for this problem. The cost of the separation should not exceed $91,000. Use the depth-first strategy to find the solution.

5. Compare the solution obtained by using the lower bound with the solution obtained without the lower bound information.

4.7 As a student in your senior year, you realize the cost of textbooks varies depending on where you buy them. With the help of Internet price comparison engines, you have been able to create a table (Table 4.12) for the books you need next term.

Table 4.12. Costs of books

Store	Idiots guide to optimization	Engineering for mathematicians	History of liechtenstein
Campus Store	$17.95	$75.75	$45.15
Nile.com	$15.50	$71.65	$47.20
buyyourbookhere.com	$16.25	$73.00	$41.50
Anotherbookstore.net	$15.99	$69.99	$42.99
dasBuch.li	$25.00	$90.00	$28.75

Although the prices are sometimes lower for the online bookstores, you realize that shipping and handling costs are not included in the price for the books. Table 4.13 provides the relevant information about these costs.

Table 4.13. Costs for shipping and handling

Store	Shipping and handling
Campus store	$0.00
Nile.com –	$4.95 + $1.00 for each additional book
buyyourbookhere.com	$4.00 for 1-2 books, add $1.00 for each set of 3 more
Anotherbookstore.net	$7.99 for 1-2 books, $10.99 for 3-7
dasBuch.li	$17.25 + $5.75 for each additional book

(a) If the local sales tax is 6.25%, what is the optimal place for purchasing all your required books from one store only? There are no taxes paid on Internet purchases.

(b) If you can buy from multiple stores, what books should you buy at which store?

(c) Three years after you graduate, a friend ends up taking the same classes. However, as all three books have new editions, she has to buy all new books. The prices remain the same, but the sales tax has increased by 0.5%. Have either optimal solutions changed in 3 years?

4.8 The following simple cost function is derived from the Brayton cycle example (Painton and Diwekar, 1994) to illustrate the use of simulated annealing and genetic algorithms.

$$\text{Min} \quad Cost = \sum_{i=1}^{N_1} \left[(N_1 - 3)^2 + (N_2(i) - 3)^2 + (N_3(i) - 3)^2 \right]$$

where N_1 is allowed to vary from one to five and both N_2^i and N_3^i can take any value from one to five.

1. How many total combinations are involved in the total enumeration?
2. Set up the problem using simulated annealing and genetic algorithms.
3. For simulated annealing, assume the initial temperature to be 50, the number of moves at each temperature to be 100, and the freezing temperature to be 0.01. Use the temperature decrement formula $T_{new} = \alpha T_{old}$ where α is 0.95.
4. Use the binary string representation for genetic algorithms and solve the problem.
5. Compare the solution obtained by the binary representation above with the solution obtained using the natural representation consisting of the vector $N = (N_1, N_2(i), i = 1, 2, \ldots, 5, N_3(i) \; i = 1, 2, \ldots, 5)$.

Bibliography

Ahuja R. K. and J. B. Orlin (1997), Developing fitter genetic algorithms, *INFORMS Journal of Computing* **9(3)**, 251.

Beale E. M. (1977), *Integer Programming: The State of the Art in Numerical Analysis*, Academic Press, London.

Biegler L., I. E. Grossmann, and A. W. Westerberg (1997), *Systematic Methods of Chemical Process Design*, Prentice-Hall, Upper Saddle River, NJ.

Chauduri P., U. M. Diwekar, and J. F. Logsdon, An automated approach to optimal heat exchanger design (1997), *Ind. Eng. Chem. Res.*, **36**, 2685.

Chiba, T., S. Okado, and I. Fujii (1996), Optimum support arrangement of piping systems using genetic algorithm, *Journal of Pressure Vessel Technology* **118**, 507.

Clerc M.(2006), *Particle Swarm Optimization*, Wiley.

Collins N. E., R. W. Eglese, and B. L. Golden (1988), Simulated annealing—An annotated biography, *American Journal of Mathematical and Management Science* **8(3)**, 209.

Diwekar U. M., I. E. Grossmann, and E. S. Rubin (1991), An MINLP process synthesizer for a sequential modular simulator, *Industrial and Engineering Chemistry Research* **31**, 313.

Dorigo M. and T. Stutzle (2004), *Ant Colony Optimization*, MIT Press.

Dunn, S.A. (1997), Modified genetic algorithm for the identification of aircraft structures, *Journal of Aircraft* **34**, 251.

Glover F. (1986), Future paths for integer programming and links to artificial intelligence, *Computers and Operations Research* **5**, 533.

Goldberg D.E. (1989), *Genetic Algorithms in Search, Optimization and Machine Learning*, Addison-Wesley, Reading MA.

Gomory R.E. (1958), Outline of an algorithm for integer solutions to linear programs, *Bull. American Math. Soc.* **64**, 275.

Guarnieri F. and M. Mezei (1996), Simulated annealing of chemical potential: A general procedure for locating bound waters. Application to the study of the differential hydration propensities of the major and minor grooves of DNA, *Journal of the American Chemical Society* **118**, 8493.

Hendry J. E. and R. R. Hughes (1972), Generating separation flowsheets, *Chemical Engineering Progress* **68**, 69.

Holland J. H. (1975), *Adaptation in Natural and Artificial Systems*, University of Michigan Press, Ann Arbor.

Holland J. H. (1992), Genetic algorithms, *Scientific American*, July, 66.

Huang M. D., F. Romeo, and A. L. Sangiovanni-Vincetelli (1986), An efficient general cooling schedule for simulated annealing, *Proceedings of IEEE Conference on Computer Design*, 381.

Joseph D., W. Kinsner (1997) Design of a parallel genetic algorithm for the Internet, IEEE WESCANEX 97 Communications, Power and Computing. Conference Proceedings, 333.

Kershenbaum A. (1997), When genetic algorithms work best, *INFORMS Journal of Computing* **9(3)**, 254.

Kirkpatrick S., C. Gelatt, and M. Vecchi (1983), Optimization by simulated annealing, *Science* **220(4598)**, 670.

Lettau M. (1997), Explaining the facts with adaptive agents: The case of mutual fund flows, *Journal of Economic Dynamics and Control* **21(7)**, 1117.

Levine D. (1997), Genetic algorithms: A practitioner's view, *INFORMS Journal of Computing* **9(3)**, 256.

Narayan V., U. Diwekar, and M. Hoza (1996), Synthesizing optimal waste blends, *Industrial and Engineering Chemistry Research* **35**, 3519.

Painton L. and U. M. Diwekar (1994), Synthesizing optimal design configurations for a Brayton cycle power plant, *Computers & chemical Engineering* **18**, 369.

Price T. C. (1997), Using co-evolutionary programming to simulate strategic behavior in markets, *Journal of Evolutionary Economics* **7(3)**, 219.

Reeves C. R. (1997), Genetic algorithms: No panacea, but a valuable tool for the operations researcher, *INFORMS Journal of Computing* **9(3)**, 263.

Ross P. (1997), What are genetic algorithms good at?, *INFORMS Journal of Computing* **9(3)**, 260.

Shapiro B. A. and J. C. Wu (1997) Predicting RNA H-type pseudoknots with the massively parallel genetic algorithm, *Comput. Appl. Biosci.*, **13(4)**, 459.

Subramanian D.K. and K. Subramanian (1998), Query optimization in multidatabase systems, *Distributed and Parallel Databases* **6(2)**, 183.

Taha H. A. (1997), *Operations Research: An Introduction*, Sixth Edition, Prentice-Hall, Upper Saddle River, NJ.

Tayal M. and U. Diwekar (2001), Novel sampling approach to optimal molecular design under uncertainty: A polymer design case study, *AIChE Journal* **47(3)**, 609.

VanLaarhoven P. J. M. and E. H. Aarts (1987), *Simulated Annealing Theory and Applications*, D. Reidel, Holland.

Vazquez-Espi, C.; Vazquez, M. Sizing, shape and topology design optimization of trusses using genetic algorithm. Journal of Structural Engineering, 1997, 123, 375–7.

Winston W. L. (1991), *Operations Research: Applications and Algorithms*, Second Edition, PWS-KENT, Boston.

5

Optimization Under Uncertainty

Change is certain, future is uncertain.

–Bertrand Russell

In previous chapters, we looked at various optimization problems. Depending on the decision variables, objectives, and constraints, the problems were classified as LP, NLP, IP, MILP, or MINLP. However, as stated above, the future cannot be perfectly forecast but instead should be considered random or uncertain. Optimization under uncertainty refers to this branch of optimization where there are uncertainties involved in the data or the model, and is popularly known as stochastic programming or stochastic optimization problems. In this terminology, stochastic refers to the randomness, and programming refers to the mathematical programming techniques such as LP, NLP, IP, MILP, and MINLP. In the discrete optimization chapter, we came across probabilistic techniques such as simulated annealing and genetic algorithms; these techniques are sometimes referred to as stochastic optimization techniques because of the probabilistic nature of the method. In general, however, stochastic programming and stochastic optimization involve optimal decision-making under uncertainty. For example, consider the LP example stated in Chapter 2 where, instead of having a fixed maximum supply of chemical X_2, the supply can be uncertain, as shown in the following stochastic programming (optimization) problem.

Electronic Supplementary Material: The online version of this chapter (https://doi.org/10.1007/978-3-030-55404-0_5) contains supplementary material, which is available to authorized users.

Example 5.1: Consider Example 2.1 from Chapter 2. In this example, the chemical manufacturer was using chemicals X_1 and X_2 to obtain minimum cost solvents. This problem had constraints due to storage capacity, safety requirements, and availability of materials. We formulated the problem as the following LP.

$$\text{Minimize} \quad Z = 4x_1 - x_2 \tag{5.1}$$
$$x_1, \ x_2$$

subject to

$$2x_1 + x_2 \leq 8 \quad \text{Storage Constraint} \tag{5.2}$$
$$x_2 \leq 5 \quad \text{Availability Constraint} \tag{5.3}$$
$$x_1 - x_2 \leq 4 \quad \text{Safety Constraint} \tag{5.4}$$
$$x_1 \geq 0; \ x_2 \geq 0$$

Let us include the uncertainties associated with the supply of X_2. A distribution of supply for a particular week is shown in Table 5.1. Find the optimum value of raw materials the manufacturer needs to buy to reduce the average cost to a minimum. How is the feasible region of operation changing with uncertainty?

Table 5.1. Weekly supply

i	Day	Supply, u_i
1	Monday	5
2	Tuesday	7
3	Wednesday	6
4	Thursday	9
5	Friday	10
6	Saturday	8
7	Sunday	4

Solution: Given that the supply of X_2 that is 5 tons per day is uncertain, the availability constraint is going to change. Our first attempt to solve this problem was to find the average supply (i.e., $u_{avg} = 7$ in this case) and change the problem formulation accordingly. This formulation is given below.

$$\text{Minimize} \quad Z = 4x_1 - x_2 \tag{5.5}$$
$$x_1, \ x_2$$

subject to

$$2x_1 + x_2 \leq 8 \quad \text{Storage Constraint} \tag{5.6}$$

$$x_2 \leq 7 \qquad \text{Availability Constraint} \qquad (5.7)$$

$$x_1 - x_2 \leq 4 \qquad \text{Safety Constraint} \qquad (5.8)$$

$$x_1 \geq 0; \; x_2 \geq 0$$

Obviously the optimal solution for this case is $x_1 = 0$ and $x_2 = 7$. Let us see whether this represents an optimal solution to the problem. Consider the distribution of supply for a typical week given earlier. The manufacturer can only buy an amount of chemical X_2 equal to the supply u if the supply is less than 7 tons, otherwise his decision to buy 7 tons of chemical X_2 remains unchanged. This results in the following cost function.

$$\begin{aligned} Cost_p(u) \quad &= \quad 4x_1 - x_2 \quad \text{if} \quad x_2 \leq u \\ &= \quad 4x_1 - u \quad\;\; \text{if} \quad x_2 \geq u \end{aligned}$$

Table 5.2 shows the cost function calculation for three sets of decision variables, one of them being the average value $x = (0, 7)$. It is obvious that $x = (0, 8)$ is a better solution than the other two, showing that the optimal solution obtained by taking an average of the input uncertain variable is not necessarily an optimum.

Table 5.2. Evaluating cost under uncertainty

i	Day	Supply, u_i	$Cost_p$		
			$x = (0,5)$	$x = (0,7)$	$x = (0,8)$
1	Monday	5	-5	-5	-5
2	Tuesday	7	-5	-7	-7
3	Wednesday	6	-5	-6	-6
4	Thursday	9	-5	-7	-8
5	Friday	10	-5	-7	-8
6	Saturday	8	-5	-7	-8
7	Sunday	4	-4	-4	-4
$Cost_{avg}$	$= \sum_i Cost_p/7.0$		-4.86	-6.14	-6.57

The other alternative is for the manufacturer to change his decisions according to the supply. If the manufacturer knows the supply curve for each week a priori (Table 5.1), then he can change the decisions x_1 and x_2 on a daily basis. This can be achieved by using the following formulation in terms of the uncertain variable u_i for each day i.

$$\text{Minimize} \quad Z_i = 4x_1 - x_2 \qquad (5.9)$$
$$x_1, \; x_2$$

subject to

$$2x_1 + x_2 \leq 8 \qquad \text{Storage Constraint} \qquad (5.10)$$

$$x_2 \leq u_i \qquad \text{Availability Constraint} \qquad (5.11)$$

$$x_1 - x_2 \leq 4 \qquad \text{Safety Constraint} \qquad (5.12)$$

$$x_1 \geq 0; \; x_2 \geq 0$$

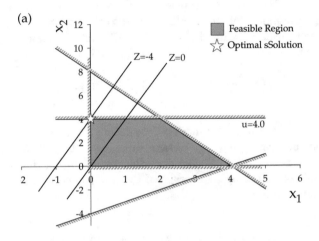

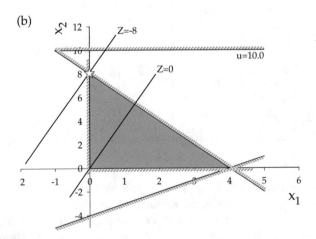

Fig. 5.1. Change in feasible region as the uncertain variable changes

The feasible region of operation is changing with the change in the uncertain variable as shown in Figure 5.1.

Table 5.3 shows the optimal decision variables for each day with the daily average minimum cost equal to $\$-7.0$.

Table 5.3. Weekly decisions

i	Day	u_i	$x = (x_1, x_2)$	Cost $, Z_i
1	Monday	5	(0,5)	-5
2	Tuesday	7	(0,7)	-7
3	Wednesday	6	(0,6)	-6
4	Thursday	9	(0,9)	-9
5	Friday	10	(0,10)	-10
6	Saturday	8	(0,8)	-8
7	Sunday	4	(0,4)	-4
Total Minimum Cost			$-49	
Average Minimum Cost			$-7.0	

However, as stated in the problem statement, the supply scenario given in Table 5.1 is a likely scenario for a particular week, but the manufacturer may not exactly know the daily situation. The information available from Table 5.1 can be translated into probabilistic information as shown in Table 5.4. In this case, the manufacturer would like to find the amount of each chemical on average, given the supply distribution in Table 5.4, to minimize the average daily cost.

Table 5.4. Weekly supply uncertainty distribution

j	Supply, u	Probability
1	4	1/7
2	5	1/7
3	6	1/7
4	7	1/7
5	8	1/7
6	9	1/7
7	10	1/7

Let us choose x_1 and x_2 to be the average amount of each chemical ordered or purchased by the manufacturer per week. This is the action the manufacturer is taking without knowing the exact daily supply data. If the supply on a specific date u_j is less than this average purchase x_2, then the manufacturer can only buy the supply amount. This is reflected in the following formulation.

$$\text{Minimize} \quad Z = Cost_{avg}(u) \tag{5.13}$$
$$x_1,\ x_2$$

$$Cost_{avg}(u) = \int_0^1 Cost_p(u)\ dp = \sum_j p_j Cost_{p(j)} \tag{5.14}$$

$$Cost_p(u) = 4x_1 - x_2 \quad \text{if } x_2 \leq u$$
$$= 4x_1 - u \quad\ \ \text{if } x_2 \geq u$$

subject to

$$2x_1 + x_2 \leq 8 \tag{5.15}$$
$$x_1 - x_2 \leq 4 \tag{5.16}$$
$$x_1 \geq 0;\ x_2 \geq 0$$

where p reflects the probability distribution of the uncertain variable u.

We can see that the problem is no longer an LP because the cost function is nonlinear and non-smooth as shown in Figure 5.2. There are special methods required to solve this problem which are described later. At this stage, we can evaluate this function using different decision variables and find the optimum cost by inspection. Table 5.5 presents the results of this exercise. The solution is -6.57.

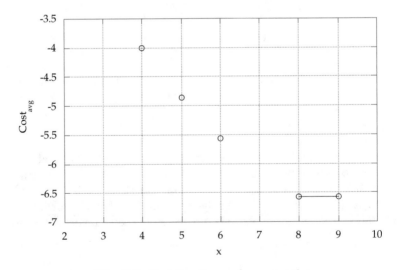

Fig. 5.2. Cost function under uncertainty

Table 5.5. Evaluating cost under uncertainty

u	Probability, p_i	$Cost_p$		
		$x = (0,5)$	$x = (0,7)$	$x = (0,8)$
4	1/7	−4	−4	−4
5	1/7	−5	−5	−5
6	1/7	−5	−6	−6
7	1/7	−5	−7	−7
8	1/7	−5	−7	−8
9	1/7	−5	−7	−8
10	1/7	−5	−7	−8
$Cost_{avg} = \sum_i p_i Cost_p$		−4.86	−6.14	−6.57

The difference between taking the average value of the uncertain variable as the solution as compared to using stochastic analysis (propagating the uncertainties through the model and finding the effect on the objective function as shown in Table 5.5) is defined as the value of stochastic solution, VSS. The VSS for this problem reflects a cost savings of $6.57 − 6.14 = 0.43$ per day.

We see that the average cost for both of the formulations (Tables 5.3 and 5.5) is similar, but in one case, the manufacturer had the perfect information and could change the decisions as the supply changed (Table 5.3). However, in the second case the manufacturer has to take action before he has the perfect information. The value of getting more accurate information about the uncertainty in this case is zero.

In general, the difference between the solution obtained when perfect information is available and the optimum solution obtained considering uncertainties is the expected value of perfect information, EVPI. The EVPI measures the maximum amount a decision-maker would be ready to pay in return for complete accurate information. For this problem the cost savings by having perfect information is $EVPI = \$7.0 − 6.57 = 0.43$. As can be expected, this value is always greater than or equal to zero. The next example shows this clearly.

5.1 Types of Problems and Generalized Representation

The need for including uncertainty in complex decision models arose early in the history of mathematical programming. The first model forms, involving action followed by observation and reaction (or recourse), appear in Beale (1955) and Dantzig (1955). In the above problem, there was action (decisions $x = (0,8)$), followed by observation ($Cost_{avg} = \$-6.57$) but the problem did not have any recourse action. A commonly used example of a recourse problem is the news vendor or the newsboy problem described below. This problem has a rich history that has been traced back to the economist (Edgeworth, 1888), who applied a variance to a bank cash-flow problem. However, it was not until the 1950s that this problem, as did many other OR/MS models seeded by the war effort, became a topic of serious and extensive study by academicians (Petruzzi and Dada, 1999).

Example 5.2: The simplest form of a stochastic program may be the news vendor (also known as the newsboy) problem. In the news vendor problem, the vendor must determine how many papers (x) to buy now at the cost of c cents for a demand which is uncertain. The selling price is s_p cents per paper. For a specific problem, whose weekly demand is shown below, the cost of each paper is $c = 20$ cents and the selling price is $s_p = 25$ cents. Solve the problem, if the news vendor knows the demand uncertainties (Table 5.6) but does not know the demand curve for the coming week (Table 5.7) a priori. Assume no salvage value $s = 0$, so that any papers bought in excess of demand are simply discarded with no return.

Table 5.6. Weekly demand uncertainties

j	Demand, d_j	Probability, p_j
1	50	5/7
2	100	1/7
3	140	1/7

Table 5.7. Weekly demand

i	Day	Demand, (u) d_i
1	Monday	50
2	Tuesday	50
3	Wednesday	50
4	Thursday	50
5	Friday	50
6	Saturday	100
7	Sunday	140

Solution: In this problem, we want to find how many papers the vendor must buy (x) to maximize the profit. We know that any excess papers bought are just thrown away. Let r be the effective sales and w be the excess that are going to be thrown away. As stated earlier, this problem falls under the category of stochastic programming with recourse where there is action (x), followed by observation $(profit)$, and reaction (or recourse) $(r$ and $w)$. Again the deterministic way to solve this problem is to find the average demand and find the optimal supply x corresponding to this demand. Because the average demand from Table 5.6 is 70 papers, $x = 70$ should be the solution. Let us see if this represents the optimal solution for the problem. Table 5.8 shows the observation (profit function) for this action.

Table 5.8. Supply and profit

i	Day	Supply, x_i	Profit, cents
1	Monday	70	-150
2	Tuesday	70	-150
3	Wednesday	70	-150
4	Thursday	70	-150
5	Friday	70	-150
6	Saturday	70	350
7	Sunday	70	350
Average weekly		$-$	-50

From Table 5.8, it is obvious that if we take the average demand as the solution, then the news vendor will have a loss of 50 cents per week. This probably is not the optimal solution. Can we do better? For that we need to propagate the uncertainty in the demand to see the effect of uncertainty on the objective function and then find the optimum value of x. This formulation is shown below.

$$\text{Maximize } Z = \text{Profit}_{\text{avg}}(u)$$
$$x$$

subject to

$$\text{Profit}_{\text{avg}}(u) = \int_0^1 [-cx + \text{Sales}(r, w, p(u))]dp$$
$$= \sum_j p_j \, \text{Sales}(r, w, d_j) - cx$$
$$\text{Sales}(r, w, d_j) = s_p \, r_j + s w_j$$

where

$$r_j = \min(x, d_j)$$
$$= x, \text{ if } x \le d_j$$
$$= d_j, \text{ if } x \ge d_j$$
$$w_j = \max(x - d_j, 0)$$
$$= 0, \text{ if } x_i \le d_i$$
$$= x_i - d_i, \text{ if } x_i \ge d_i$$

The above information can be transformed for daily profit as follows.

$$\text{Profit} = -cx + 5/7s_p\, d_1 + 1/7s_p\, x + 1/7s_p\, x,$$
$$\text{if} \quad d_1 \le x \le d_2 \tag{5.17}$$

or

$$\text{Profit} = -cx + 5/7s_p\, d_1 + 1/7s_p\, d_2 + 1/7s_p\, x,$$
$$\text{if} \quad d_2 \le x \le d_3 \tag{5.18}$$

Notice that the problem represents two equations for the objective function, Equations (5.17) and (5.18), making the objective function a discontinuous function and is no longer an LP. Special methods such as the L-shaped decomposition or stochastic decomposition (Higle and Sen, 1991) are required to solve this problem. However, because the problem is simple, we can solve this problem as two separate LPs. The two possible solutions to the above LPs are $x = d_1 = 50$ and $x = d_2 = 100$, respectively. This provides the news vendor with an optimum profit of 1750 cents per week from Equation (5.17) and with a loss of 2750 cents per week from Equation (5.18). Obviously Equation (5.17) provides the global optimum for this problem. Earlier when we took the average value of the demand (i.e., $x = 70$) as the solution, we obtained a loss of 50 cents per week, therefore, the value of stochastic solution, VSS, is $1750 - (-50) = 1800$ cents.

Now consider the case where the vendor knows the exact demand (Table 5.7) a priori. This is the perfect information problem where we want to find the solution x_i for each day i. Let us formulate the problem in terms of x_i.

$$\underset{x_i}{\text{Maximize Profit}_i} = -cx_i + \text{Sales}(r, w, d_i)$$

subject to

$$\text{Sales}(r, w, d_i) = s_p\, r_i + s w_i$$
$$r_i = \min(x_i, d_i)$$
$$= x_i, \text{ if } \quad x_i \le d_i$$
$$= d_i, \text{ if } \quad x_i \ge d_i$$
$$w_i = \max(x_i - d_i, 0)$$
$$= 0, \text{ if } \quad x_i \le d_i$$
$$= x_i - d_i, \text{ if } \quad x_i \ge d_i$$

Here we need to solve each problem (for each i) separately, leading to the following decisions shown in Table 5.9.

Table 5.9. Supply and profit

i	Day	Supply, x_i	Profit, cents
1	Monday	50	250
2	Tuesday	50	250
3	Wednesday	50	250
4	Thursday	50	250
5	Friday	50	250
6	Saturday	100	500
7	Sunday	140	700
Average Weekly		–	2450

One can see that the difference between the two values, (1) when the news vendor has the perfect information and (2) when he does not have the perfect information but can represent it using probabilistic functions, is the expected value of perfect information, EVPI. EVPI is 700 cents per week for this problem.

The literature on optimization under uncertainties very often divides the problems into categories such as "wait and see," "here and now," and "chance constrained optimization" (Vajda, 1972; Nemhauser et al., 1989) . In wait and see we wait until an observation is made on the random elements, and then solve the deterministic problem. The first formulation, described in terms of the problem under perfect information in Examples 5.1 and 5.2, falls under this category. This is similar to the wait and see problem of Madansky (1960), originally called "Stochastic Programming" by Tintner (1955), and is not in a sense one of decision analysis. In decision-making, the decisions have to be made here and now about the activity levels. The here and now problem involves optimization over some probabilistic measure, usually the expected value. By this definition, chance constrained optimization problems can be included in this particular category of optimization under uncertainty. Chance constrained optimization involves constraints that are not expected to be always satisfied; only in a proportion of cases, or with given probabilities. These various categories require different methods for obtaining their solutions.

It should be noted that many problems have both here and now, and wait and see problems embedded in them. The trick is to divide the decisions into these two categories and use a coupled approach.

Here and Now Problems

Stochastic optimization gives us the ability to optimize systems in the face of uncertainties. The here and now problems require that the objective function and constraints be expressed in terms of some probabilistic representation (e.g., expected value, variance, fractiles, most likely values). For example, in chance constrained programming, the objective function is expressed in terms of expected value, and the constraints are expressed in terms of fractiles (probability of constraint violation), and in Taguchi's offline quality control method (Taguchi, 1986; Diwekar and Rubin, 1991), the objective is to minimize variance. These problems can be classified as here and now problems.

The here and now problem, where the decision variables and uncertain parameters are separated, can then be viewed as

$$\text{Optimize} \quad J \; = \; P_1(j(x,u)) \qquad (5.19)$$
$$x$$

subject to

$$P_2(h(x,u)) \; = \; 0 \qquad (5.20)$$

$$P_3(g(x,u) \geq 0) \; \geq \; \alpha \qquad (5.21)$$

where u is the vector of uncertain parameters and P represents the cumulative distribution functional such as the expected value, mode, variance, or fractiles. Figures 5.3 and 5.4 show the expected value, mode, variance, and fractiles for a probabilistic distribution function.

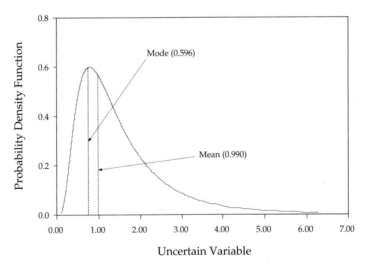

Fig. 5.3. Different probabilistic performance measures (PDF)

Unlike the deterministic optimization problem, in stochastic optimization one has to consider the probabilistic functional of the objective function and constraints. The generalized treatment of such problems is to use probabilistic or stochastic models instead of the deterministic model inside the optimization loop.

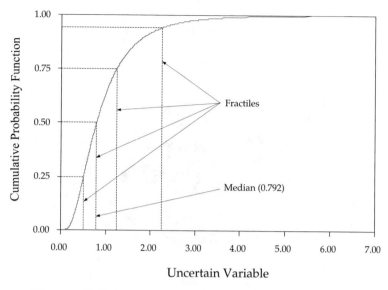

Fig. 5.4. Different probabilistic performance measures (CDF)

Figure 5.5a represents the generalized solution procedure, where the deterministic model is replaced by an iterative stochastic model with a sampling loop representing the discretized uncertainty space. The uncertainty space is represented in terms of the moments such as the mean, or the standard deviation of the output over the sample space of N_{samp} as given by the following equations (Equations (5.22) and (5.23)).

$$E(z(x, u)) = \sum_{k=1}^{N_{samp}} \frac{z(x, u_k)}{N_{samp}} \tag{5.22}$$

$$\sigma^2(z(x, u)) = \sum_{k=1}^{N_{samp}} \frac{(z(x, u_k) - \bar{z})^2}{N_{samp}} \tag{5.23}$$

where $\bar{z}$ is the average value of z. E is the expected value and σ^2 is the variance.

In chance constrained formulation, the uncertainty surface is translated into input moments, resulting in an equivalent deterministic optimization problem. This is discussed in the Section 5.2.

Wait and See

In contrast to here and now problems, which yield optimal solutions that achieve a given level of confidence, wait and see problems involve a category of formulations that shows the effect of uncertainty on optimum design. A wait and see problem involves deterministic optimal decisions at each scenario

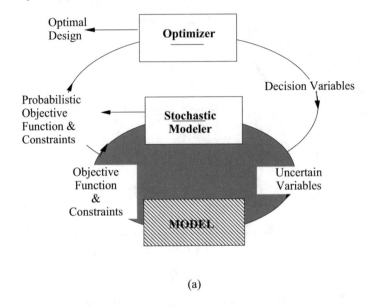

(a)

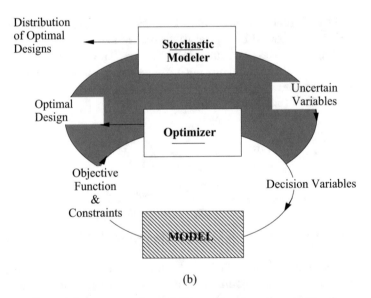

(b)

Fig. 5.5. Pictorial representation of the stochastic programming framework. (a) Here and now. (b) Wait and see

or random sample, equivalent to solving several deterministic optimization problems. The generalized representation of this problem is given below.

$$\text{Optimize } Z = z(x, u*) \tag{5.24}$$

$$x$$

subject to

$$h(x, u*) = 0 \qquad (5.25)$$

$$g(x, u*) \leq 0 \qquad (5.26)$$

where $u*$ is the vector of values of uncertain variables corresponding to each scenario or sample.

This optimization procedure is repeated for each sample of uncertain variables u and a probabilistic representation of the outcome is obtained.

Figure 5.5b represents the generalized solution procedure, where the deterministic problem forms the inner loop, and the stochastic modeling forms the outer loop. The difference between the two solutions obtained using the two frameworks is the expected value of perfect information. The concept of EVPI was first developed in the context of decision analysis and can be found in classical references such as Raiffa and Schlaifer (1961). From Figures 5.5 it is clear that by simply interchanging the position of the uncertainty analysis framework and the optimization framework, one can solve many problems in the stochastic optimization and stochastic programming domain (Diwekar, 1995). Recourse problems with multiple stages involve decisions that are taken before the uncertainty realization (here and now) and recourse actions which can be taken when information is disclosed (wait and see). These problems can be solved using decomposition methods such as the L-shaped decomposition method described in Section 5.3.

As can be seen from the above description, both here and now and wait and see problems require the representation of uncertainties in the probabilistic space and then the propagation of these uncertainties through the model to obtain the probabilistic representation of the output. This is the major difference between stochastic and deterministic optimization problems. Is it possible to propagate the uncertainty using moments (such as mean, variance) thereby obtaining a deterministic representation of the problem? This is the basis of the chance constrained programming method, developed very early in the history of optimization under uncertainty, principally by Charnes and Cooper (1959).

5.2 Chance Constrained Programming Method

In the chance constrained programing (CCP) method, some of the constraints likely need not hold as we had assumed in earlier problems. Chance constrained problems can be represented as follows.

$$\text{Optimize} \quad J = P_1(j(x, u)) = E(z(x, u)) \qquad (5.27)$$

$$x$$

subject to

$$P(g(x) \leq u) \leq \alpha \qquad (5.28)$$

In the above formulation, Equation (5.28) is the chance constraint. In the chance constraint formulation, this constraint (or constraints) is (are) converted into a deterministic equivalent under the assumption that the distribution of the uncertain variables u is a stable distribution. Stable distributions are such that the convolution of two distribution functions $F(x - m_1/v_1)$ and $F(x - m_2/v_2)$ is of the form $F(x - dmu/v)$, where m_i and v_i are two parameters of the distribution. Normal, Cauchy, uniform, and chi-square are all stable distributions that allow the conversion of probabilistic constraints into deterministic ones. The deterministic constraints are in terms of moments of the uncertain variable u (input uncertainties). For example, if constraint g in Equation (5.28) has a cumulative probability distribution F, then the deterministic equivalent of this constraint is given below.

The deterministic equivalent of the chance constraint (5.28) is

$$g(x) \leq F^{-1}(\alpha) \qquad (5.29)$$

where F^{-1} is the inverse of the cumulative distribution function F.

The major restrictions in applying the CCP formulation include that the uncertainty distributions should be stable distribution functions, the uncertain variables should appear in the linear terms in the chance constraint, and that the problem needs to satisfy the general convexity conditions. The advantage of the method is that one can apply the deterministic optimization techniques to solve the problem. The following example illustrates this method.

Example 5.3: In Example 5.1, the formulation for the here and now problem, we have allowed the manufacturer to buy more x_2 than the supplier can provide by not penalizing him. However, let us assume that the manufacturer is ready to get more x_2 from a different supplier once in a while as long as the probability of buying from another supplier is lesser than or equal to 42.714% (3/7). Formulate this problem as a chance constraint programming problem and obtain the solution using conventional deterministic optimization methods.

Solution: The problem description results in the following formulation where constraint (5.32) is the chance constraint.

$$\text{Minimize} \quad Z = 4x_1 - x_2 \qquad (5.30)$$
$$x_1, x_2$$

subject to

$$2x_1 + x_2 \leq 8 \qquad (5.31)$$

$$P(-x_2 + u \leq 0) \leq \frac{3}{7} \tag{5.32}$$

$$x_1 - x_2 \leq 4 \tag{5.33}$$

$$x_1 \geq 0; \; x_2 \geq 0$$

Earlier, Table 5.4 provided the probability distribution function for the variable u. Figure 5.6 shows the probability density function (PDF), and cumulative distribution function (CDF) F for the variable u. F^{-1} for the probability 3/7 corresponds to $u = 6$. Therefore, the deterministic equivalent of this problem results in the following problem.

$$\text{Minimize } Z = 4x_1 - x_2 \tag{5.34}$$
$$x_1, \; x_2$$

subject to

$$2x_1 + x_2 \leq 8 \tag{5.35}$$

$$x_2 \geq 6 \tag{5.36}$$

$$x_1 - x_2 \leq 4 \tag{5.37}$$

$$x_1 \geq 0; \; x_2 \geq 0$$

The solution to this problem is $x = (0, 6)$, with the average cost equal to $-5.57 per day as shown in Table 5.10.

Table 5.10. Evaluating cost under uncertainty

u	Probability, p_i	$Cost_p(0,6)$
4	1/7	−4
5	1/7	−5
6	1/7	−6
7	1/7	−6
8	1/7	−6
9	1/7	−6
10	1/7	−6
$Cost_{avg} = \sum_i p_i Cost_p$		−5.57

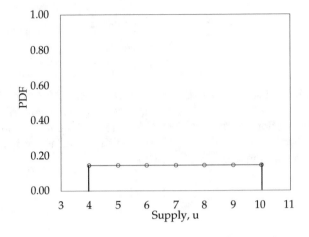

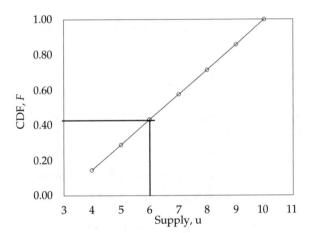

Fig. 5.6. Probability distribution functions for the uncertain variable

5.3 L-shaped Decomposition Method

In the stochastic programming problems with recourse, there is action (x), followed by observation, and then recourse r. In these problems, the objective function has the action term, and the recourse function is dependent on the uncertainties and recourse decisions. As seen earlier, the recourse function can be a discontinuous nonlinear function in x and r space. A general approach behind the L-shaped method is to use a decomposition strategy where the master problem decides x and the subproblems are solved for the recourse function (Figure 5.7). The method is essentially a Dantzig–Wolfe decomposition (Dantzig and Wolfe, 1960) (inner linearization) of the dual or Bender's decomposition of the primal. This method is due to Van Slyke and Wets

(1969), and also considers feasibility questions of particular relevance in these recourse problems. Consider the generalized representation of the recourse problem shown below, where the first term depends only on x, and R is the recourse function.

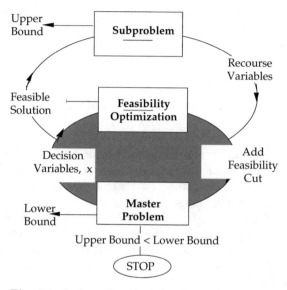

Fig. 5.7. L-shaped method decomposition strategy

$$\underset{x}{\text{Minimize}} \quad Z = f(x) + R(x, r, u) \tag{5.38}$$

subject to

$$h(x, r) = 0 \tag{5.39}$$

$$g(x, u, r) \leq 0 \tag{5.40}$$

Figure 5.7 shows the decomposition scheme used in the L-shaped method. In the figure, the master problem is the linearized representation of the nonlinear objective function (containing the recourse function) and constraints. The master problem provides the values of the action variables x (x^*) and obtains the lower bound of the objective function. In general, the multistage recourse problems involve equality constraints relating the action variables x to the recourse variables r as in the generalized representation. These constraints are included as inequalities (feasibility cuts) in terms of the dual representation (including Lagrange multipliers λ) obtained by solving the following feasibility problem for each constraint. The feasibility cut addition is continued until no constraint is violated (completely feasible solution). It should be noted that this is a very time-consuming iterative loop of the L-shaped algorithm, and variants of the L-shaped method provide improvements to this loop. The master problem then provides the values of the action variables x, and the lower bound to the objective function. At each outer iteration, for these fixed x, the subproblem is solved for r, and linearizations of the objective and recourse function (optimality cuts) are obtained along with the values of r. If the subproblem solution (upper bound) crosses or is equal to the lower bound predicted by the master problem, then the procedure stops, else iterations continue.

Feasibility Optimization

$$\text{Minimize} \quad Constraints\ Violations(x^*, r) \tag{5.41}$$
$$r, \lambda$$

The following example uses the news vendor problem described earlier to show the convergence of the L-shaped method. As indicated earlier, the inner loop of the L-shaped method consists of determining whether a first stage decision is also second stage feasible, and so on. This step is extremely computationally intensive and may involve several iterations per constraint for successive candidate first stage solutions. In some cases though (such as this news vendor problem) this step can be simplified. A first case is when the second stage is always feasible. The stochastic program is then said to have a *complete recourse*.

Example 5.4: Solve the here and now problem for the news vendor presented in Example 5.2 using the L-shaped method.

Solution: The formulation of the here and now problem is given below. News Vendor Problem (Example 5.2) Formulation:

$$\text{Maximize} \quad -Z = Profit_{avg}(u) \tag{5.42}$$
$$x$$

$$Profit_{avg}(u) = \int_0^1 [-cx + Sales_p(r, w, p(u))]dp$$
$$= \sum_j p_j Sales(r, w, d_j) - cx \tag{5.43}$$
$$Sales(r, w, d_j) = s_p r_j + s w_j \tag{5.44}$$
$$r_j = \min(x, d_j)$$
$$= x, \text{ if } x \le d_j$$
$$= d_j, \text{ if } x \ge d_j \tag{5.45}$$
$$w_j = \max(x - d_j, 0) \tag{5.46}$$

where $Sales_p$ represents the recourse function R given below. We are minimizing Z and maximizing $-Z$.

$$R = s_p\, x$$
$$\text{if} \quad 0 \le x \le d_1 \tag{5.47}$$

or

$$R = 5/7 s_p\, d_1 + 1/7 s_p\, x + 1/7 s_p\, x$$
$$\text{if} \quad d_1 \le x \le d_2 \tag{5.48}$$

or

$$R = 5/7 s_p\, d_1 + 1/7 s_p\, d_2 + 1/7 s_p\, x$$
$$\text{if} \quad d_2 \le x \le d_3 \tag{5.49}$$

As can be seen from the above formulation, this problem does not have any equality terms and is considered a problem with complete recourse. To obtain the optimal solution, we need to consider the outer loop iterations (no feasibility cut) given in Figure 5.7 for the L-shaped method. From Table 5.6, we know that the uncertain parameter u can take values 50, 100, 140, with probabilities 5/7, 1/7, and 1/7, respectively. Figure 5.8 shows the terms in the recourse function $Sales_p(50)$ and $Sales_p(100)$. Each of these functions is polyhedral. The sequence of iterations for the L-shaped method is given below.

1. Assume $x = 100$ and assume the lower bound to be $-\infty$. The recourse function that is calculated by the subproblem is calculated using Equations (5.43)–(5.46) and is equal to $Profit = -393$. To express this in the minimization term, $Z_{up} = 393$.
2. The linear cut (Equation (5.51)) for the recourse function derived from Equation (5.48) is added to the master problem, given below.

$$\text{Maximize} \quad -Z_{lo} = -20x + R \tag{5.50}$$
$$x$$

$$R \le 25(\frac{5}{7} \times 50 + \frac{2}{7} \times x) \quad \text{linear cut at x} = 100 \tag{5.51}$$

The solution to the above problem is $x = 0$ and $Z_{lo} = -892.86$. The recourse function calculated again using the Equations (5.43)–(5.46) is equal to $Z_{up} = 0$. The solution is not optimal as the upper bound (0) is greater than the lower bound (-892.86).

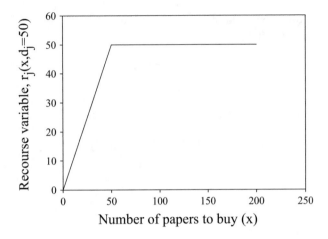

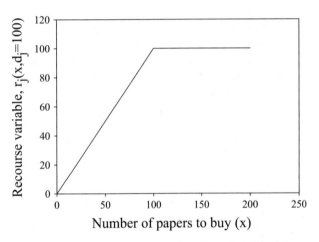

Fig. 5.8. Recourse function term as a function of the decision variable

3. Add a new cut, Equation (5.54), and solve the following problem.

$$\text{Maximize} \quad Z_{lo} = -20x + R \qquad (5.52)$$
$$x$$

$$R \leq 25(\frac{5}{7} \times 50 + \frac{2}{7} \times x) \text{ linear cut at x} = 100 \qquad (5.53)$$
$$R \leq 25x \text{ linear cut at x} = 0 \text{ from Equation (5.47)} \qquad (5.54)$$

The solution to the above problem is $x = 50$ and $Z_{lo} = -250$. The recourse function at $x = 50$ is equal to $Z_{up} = -250$, and is the optimum. So the average profit per day is 250 cents with a total weekly profit of \$1750, as found before.

The two main algorithms commonly used for stochastic linear programming with fixed recourse are the L-shaped and the stochastic decomposition methods. The L-shaped method is used when the uncertainties are described by discrete distribution. On the other hand, the stochastic decomposition method uses sampling when random variables are represented by continuous distribution functions. The chance constrained method, described earlier, uses moments to represent and propagate uncertainty in the stochastic model. Other methods use the discretized representation of uncertainty (samples or scenarios). The next section describes the uncertainty analysis and sampling for obtaining the probabilistic information necessary to solve the problems involving optimization under uncertainties.

5.4 Uncertainty Analysis and Sampling

The probabilistic or stochastic modeling (Figure 5.9) iterative procedure involves:

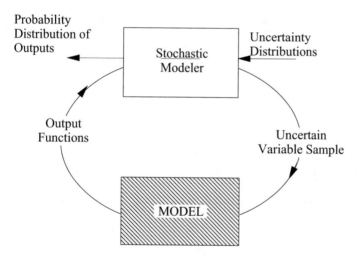

Fig. 5.9. The stochastic modeling framework

1. Specifying the uncertainties in key input parameters in terms of probability distributions.
2. Sampling the distribution of the specified parameter in an iterative fashion.
3. Propagating the effects of uncertainties through the model and applying statistical techniques to analyze the results.

5.4.1 Specifying Uncertainty Using Probability Distributions

To accommodate the diverse nature of uncertainty, different distributions can be used. Some of the representative distributions are shown in Figure 5.10. The type of distribution chosen for an uncertain variable reflects the amount of information that is available. For example, the uniform and log-uniform distributions represent an equal likelihood of a value lying anywhere within a specified range, on either a linear or logarithmic scale, respectively. Furthermore, a normal (Gaussian) distribution reflects a symmetric but varying probability of a parameter value being above or below the mean value. In contrast, log-normal and some triangular distributions are skewed such that there is a higher probability of values lying on one side of the median than the other. A beta distribution provides a wide range of shapes and is a very flexible means of representing variability over a fixed range. Modified forms of these distributions, uniform* and log-uniform*, allow several intervals of the range to be distinguished. Finally, in some special cases, user-specified distributions can be used to represent any arbitrary characterization of uncertainty, including chance distribution (i.e., fixed probabilities of discrete values).

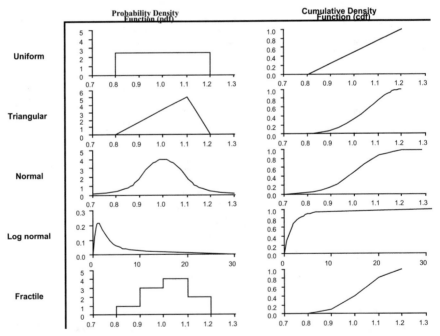

Fig. 5.10. Examples of probabilistic distribution functions for stochastic modeling

5.4.2 Sampling Techniques in Stochastic Modeling

Once probability distributions are assigned to the uncertain parameters, the next step is to perform a sampling operation from the multi-variable uncertain parameter domain. Alternatively, one can use analytical methods to obtain the effect of uncertainties on the output. These methods tend to be applicable to special kinds of uncertainty distributions and optimization surfaces only. The sampling approach provides wider applicability and is discussed below.

Crude Monte Carlo Technique

One of the most widely used techniques for sampling from a probability distribution is the Monte Carlo sampling technique, which is based on a pseudo-random generator used to approximate a uniform distribution (i.e., having equal probability in the range from 0 to 1). The specific values for each input variable are selected by inverse transformation over the cumulative probability distribution. A Monte Carlo sampling technique also has the important property that the successive points in the sample are independent. The following example illustrates how the Monte Carlo techniques can be used in probabilistic analysis to obtain the value of an output variable.

Example 5.5: Let us consider the problem of finding a maximum area circle inscribed in a square with a given area (100 square cm) as shown in Figure 5.11. We know that if one chooses any random point in the square, then the probability of that point being in the interior of a particular circle is given by

$$Pr = \frac{Area\ of\ the\ Circle}{Area\ of\ the\ Square}$$

We want to find the radius r of a circle that will maximize the area of the circle. The problem can be easily posed as a stochastic optimization problem where the objective is to maximize a probabilistic function, that is, the area that can be calculated using the Monte Carlo method.

(a) Represent this problem in probabilistic terms.
(b) Find the area of the circle using the Monte Carlo method.
(c) Find the effect of sampling on the output.

Solution: We know that if one chooses any random point in this figure, then the probability of that point being in the interior of the circle given by Pr leads to the following equation.

$$Area\ of\ the\ Circle = Pr \times Area\ of\ the\ Square$$

The optimization problem then can be represented by

$$\underset{r}{\text{Maximize}} \quad Z = Pr(r)A_{square} \tag{5.55}$$

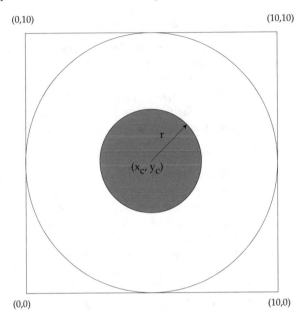

Fig. 5.11. Maximize area of a circle, sampling representation

Now, we can solve this problem using the Monte Carlo method for calculation of the probabilistic term Pr. The estimation of the area of the circle is based on the assumption that the points in the square are equally likely to occur (uniform distribution) for both sides, as shown in Figure 5.12 for the side from 0 to 10. Thus, if out of a random sample of N_{samp} points in the square, m are found to fall within the circle equation, then $Pr = m/N_{samp}$. A sample point (x^*, y^*) falls within the circle if

$$(x^* - x_c)^2 + (y^* - y_c)^2 \leq r^2$$

The problem can be written in terms of the N_{samp} as follows.

$$\text{Maximize} \quad Z = Pr(r)A_{square} \qquad (5.56)$$
$$r, Y_i$$

subject to

$$r^2 - (x_i - x_c)^2 - (y_i - y_c)^2 - UY_i \leq 0.0$$
$$i = 1, 2, \ldots, N_{samp} \qquad (5.57)$$
$$(x_i - x_c)^2 - (y_i - y_c)^2 - r^2 - U(1 - Y_i) \leq 0.0$$
$$i = 1, 2, \ldots, N_{samp} \qquad (5.58)$$
$$\frac{\sum_{i=1}^{N_{samp}} Y_i}{N_{samp}} = Pr \qquad (5.59)$$

$$x_i = (10 - 0) \times u_1 \quad i = 1, 2, \ldots, N_{samp} \tag{5.60}$$
$$y_i = (10 - 0) \times u_2 \quad i = 1, 2, \ldots, N_{samp} \tag{5.61}$$

where Y_i represents the binary decision of whether the point is inside the circle of radius r. If Y_i is 1, then the point is inside the circle; else, it is 0. U is a very large number. The first two constraints (Equations (5.57) and (5.58)) ensure this fact. Obviously, this is a mixed integer nonlinear programming problem with uncertainty (u_1 and u_2 are two random variables between 0 and 1). The solution is iterative where at each optimization iteration j, with the decision variable r_j, the area of the circle is calculated using the Monte Carlo method. Figure 5.11 shows two such iterations in terms of the two concentric circles. Figure 5.12 also shows how the samples are generated using the CDF of the uniform distribution functions for a particular circle.

It is obvious from Figure 5.11 that the solution to this problem is the larger circle with the radius $r = 5$. However, the number of samples N_{samp} to obtain the probability function Pr plays an important role in the iterative procedure. Figure 5.13 plots the area calculated using a different number of samples for $r = 5$. It can be seen that as the number of samples increases, the area of the circle approaches the exact area.

Importance Sampling

Crude Monte Carlo methods can result in large error bounds (confidence intervals) and variance. Variance reduction techniques are statistical procedures designed to reduce the variance in the Monte Carlo estimates (James and Variance, 1985). Importance sampling, Latin hypercube sampling (LHS; McKay et al. 1979; Iman and Shortencarier 1984 ; Iman and Helton 1988), descriptive sampling (Saliby, 1990), and Hammersley sequence sampling (Kalagnanam and Diwekar, 1997) are examples of variance reduction techniques. In importance Monte Carlo sampling, the goal is to replace a sample using the distribution of u with one that uses an alternative distribution that places more weight in the areas of importance. Dantzig and Infanger (1991) used such an approximate distribution function for the L-shaped method to accelerate the crude Monte Carlo method. Obviously such a distribution function is problem-dependent and is difficult to find. The following two sampling methods provide a generalized approach to improve the computational efficiency of sampling.

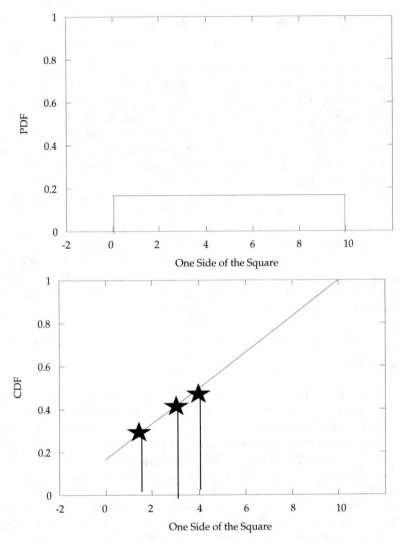

Fig. 5.12. Samples generated from the CDF of the uniform distribution

Latin Hypercube Sampling

The main advantage of the Monte Carlo method lies in the fact that the results from any Monte Carlo simulation can be treated using classical statistical methods; thus, results can be presented in the form of histograms, and methods of statistical estimation and inference are applicable. Nevertheless, in most applications, the actual relationship between successive points in a sample has no physical significance; hence the randomness/independence for approximating a uniform distribution is not critical (Knuth, 1973). Moreover,

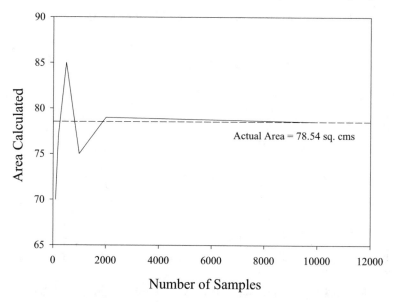

Fig. 5.13. Area Calculated Using the stochastic modeling versus number of samples

the error of approximating a distribution by a finite sample depends on the equidistribution properties of the sample used for U(0,1) rather than its randomness. Once it is apparent that the uniformity properties are central to the design of sampling techniques, constrained or stratified sampling techniques become appealing (Morgan and Henrion, 1990) .

Latin hypercube sampling is one form of stratified sampling that can yield more precise estimates of the distribution function. In Latin hypercube sampling, the range of each uncertain parameter X_i is subdivided into non-overlapping intervals of equal probability. Figure 5.14 shows the stratification scheme (intervals of equal probabilities) for a normal random variable. One value from each interval is selected at random with respect to the probability distribution in the interval. The n values thus obtained for X_1 are paired in a random manner (i.e., equally likely combinations) with n values of X_2. Figure 5.15 shows such a pairing for two uncertain variables with five samples. These n values are then combined with n values of X_3 to form n-triplets, and so on until n k-tuplets are formed. In median Latin hypercube (MLHS), this value is chosen as the midpoint of the interval. MLHS is similar to the descriptive sampling described by Saliby (1990). The main drawback of this stratification scheme is that it is uniform in one dimension (Figure 5.14) and does not provide uniformity properties in k-dimensions (Figure 5.15).

Example 5.6: The two uncertain variables in the problem have a uniform distribution with range 5–10 and a normal distribution with mean, $\mu=5$ and

Intervals used with a LHS of size n=5

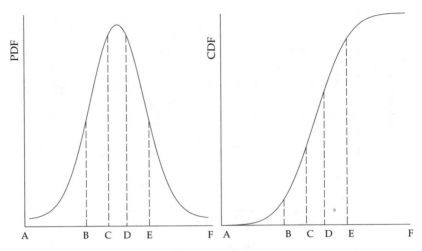

Fig. 5.14. Stratification scheme for a normal uncertain variable

Latin Hypercube Sampling (LHS)

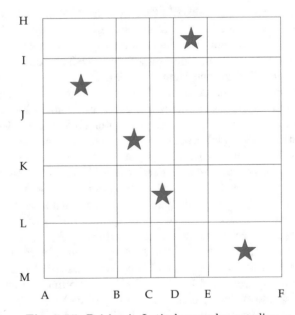

Fig. 5.15. Pairing in Latin hypercube sampling

standard deviation, $\sigma = 1.618$. Generate Monte Carlo Samples for the random numbers given below. Compare it with Latin Hypercube Sampling generated with the same random numbers.

Table 5.11. Random numbers for uncertain parameters

Random 1	Random 2
0.3370	0.0800
0.1678	0.6100
0.8419	0.5250
0.4372	0.9350
0.8127	0.6200

Solution: For Uniform and Normal distributions, the following equations provide the PDF ($f(x)$) and CDF ($F(x)$) formulas.

Uniform distribution:

$$f(x) = \frac{1}{B - A}, \qquad A \leq x \leq B \tag{5.62}$$

$$F(x) = \frac{x + A}{B - A} \tag{5.63}$$

Normal distribution:

$$f(x) = \frac{1}{\sqrt{2\pi}} e^{-\frac{(x-\mu)^2}{2\sigma^2}}, \qquad -\inf \leq x \leq \inf \tag{5.64}$$

$$F(x) = \frac{1}{2}[1 + erf(\frac{x - \mu}{\sigma\sqrt{2}})] \tag{5.65}$$

$$erf^{-1}(z) = \frac{1}{2}\sqrt{\pi}(z + \frac{\pi}{12}z^3 + \frac{7\pi^2}{480}z^5 + [\frac{127\pi^3}{40320}z^7 + \cdots) \tag{5.66}$$

For Monte Carlo, we can directly use the random numbers shown in Table 5.11 as $F(x)$, and using the inverse function we can find the sample as shown in Table 5.12.

Table 5.12. Monte Carlo samples for uncertain parameters

Variable 1	Variable 2
6.6850	2.72659
5.8390	5.4519
9.2095	5.1015
7.186	7.4498
9.0635	5.4943

For LHS, we need to first generate the values of $F(x)$ by putting the random numbers in each stratum. This involves scaling the random numbers using the following formula.

$$P_m = \frac{R_m}{N} + \frac{m - 1}{N} \tag{5.67}$$

where R_m is the m-th random number, R_m is the m-th scaled random number, and N is the total number of samples.

Using these scaled random numbers, we can then find inverse of $F(x)$ for both variables. This is shown in Table 5.13

Table 5.13. LHS samples for uncertain parameters, first step

$P(1)$	$P(2)$	Variable 1	Variable 2
0.0674	0.0160	5.3370	1.5290
0.2336	0.3220	6.1680	4.2520
0.5684	0.5050	7.8420	5.0210
0.6874	0.7870	8.4370	6.2880
0.9625	0.9240	9.8130	7.3190

By randomizing the ranks of the variables, we can generate the final sample for LHS, as shown in the last two columns of Table 5.14.

Table 5.14. LHS Samples for Uncertain Parameters, Second Step

m_1	m_2	Variable 1	Variable 2
5	1	9.8130	1.5290
3	3	7.8420	5.0210
2	4	6.1680	6.2880
1	2	5.3370	4.2520
4	5	8.4370	7.3190

Figure 5.16 shows the comparison of the two sampling techniques. Although the number of samples is very small, the LHS covers the region more uniformly than MCS.

Hammersley Sequence Sampling

Recently, an efficient sampling technique (Hammersley sequence sampling) based on Hammersley points has been developed (Kalagnanam and Diwekar, 1997), which uses an optimal design scheme for placing the n points on a k-dimensional hypercube. This scheme ensures that the sample set is more

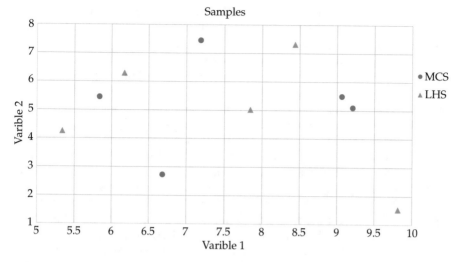

Fig. 5.16. Samples from the two sampling techniques for Example 5.6

representative of the population, showing uniformity properties in multi-dimensions, unlike Monte Carlo, Latin hypercube, and its variant, the median Latin hypercube sampling techniques. Figure 5.17 graphs the samples generated by different techniques on a unit square. This provides a qualitative picture of the uniformity properties of the different techniques. It is clear from Figure 5.17 that the Hammersley points have better uniformity properties compared to other techniques. The main reason for this is that the Hammersley points are an optimal design for placing n points on a k-dimensional hypercube. In contrast, other stratified techniques such as the Latin hypercube are designed for uniformity along a single dimension and then randomly paired for placement on a k-dimensional cube. Therefore, the likelihood of such schemes providing good uniformity properties on high-dimensional cubes is extremely small. One of the main advantages of Monte Carlo methods is that the number of samples required to obtain a given accuracy of estimates does not scale exponentially with the number of uncertain variables. HSS preserves this property of Monte Carlo. For correlated samples, the approach used, described by Kalagnanam and Diwekar (1997), uses rank correlations to preserve the stratified design along each dimension. Although this approach preserves the uniformity properties (see Figure 5.18) of the stratified schemes, the optimal location of the Hammersley points are perturbed by imposing the correlation structure. Appendix B summarizes the HSS designs. Although the original HSS technique designs start at the same initial point, they can be randomized by choosing the first prime number randomly. It has been recently found that the uniformity property of HSS for higher dimensions (more than 30 uncertain variables) gets distorted. HSS is generated based on prime numbers as bases. In order to break this distortion, leaps in prime numbers can be introduced for higher dimensions. This leaped HSS circumvents the distortion at higher dimension.

The paper by Kalagnanam and Diwekar (1997) provides a comparison of the performance of the Hammersley sampling technique to that of the Latin hypercube and Monte Carlo techniques. The comparison is performed by propagating samples derived from each of the techniques for a set of n-input variables (u_i), through various nonlinear functions $(U = f(u_1, u_2, ..., u_n))$ and measuring the number of samples required to converge to the mean and variance of the derived distribution for Y. Because there are no analytic approaches (for stratified designs) to calculate the number of samples required for convergence, a large matrix of numerical tests was conducted. It was found that the HSS technique is at least 3–100 times faster than LHS and Monte Carlo techniques and hence is a preferred technique for uncertainty analysis as well as optimization under uncertainty. For large-scale uncertainties two variants of these techniques have been proposed. These are Latin Hypercube Hammersley Sampling by Wang et al. (2004), and LHS-SOBOL recently proposed by Dige and Diwekar (2018).

5.4.3 Sampling Accuracy and the Decomposition Methods

As stated earlier, the stochastic programming formulations often include some approximations of the underlying probability distribution. The disadvantage of sampling approaches that solve the γth approximation completely is that some effort might be wasted on optimizing when approximation is not accurate (Birge, 1977). For specific structures where the L-shaped method is applicable, two approaches avoid these problems by embedding sampling within another algorithm without complete optimization. These two approaches are the method of Dantzig and Glynn (1990) which uses importance sampling to reduce variance in each cut based on a large sample, and the stochastic decomposition method proposed by Higle and Sen which utilizes a single stream to derive many cuts that eventually drop away as the iteration numbers increase (Higle and Sen, 1991). These methods require convexity conditions and dual-block angular structures, and are only applicable to continuous (decision variables) optimization. The central limit theorem is used to provide bounds for these methods.

5.4.4 Implications of Sample Size in Stochastic Modeling

In almost all stochastic optimization problems, the major bottleneck is the computational time involved in generating and evaluating probabilistic functions of the objective function and constraints. For a given number of samples (N_{samp}) of a random variable (u), the estimate for the mean or expected value ($\bar{u}$) and the unbiased estimator for standard deviation (s) can be obtained from classical statistics (Milton and Arnold, 1995). For example, the

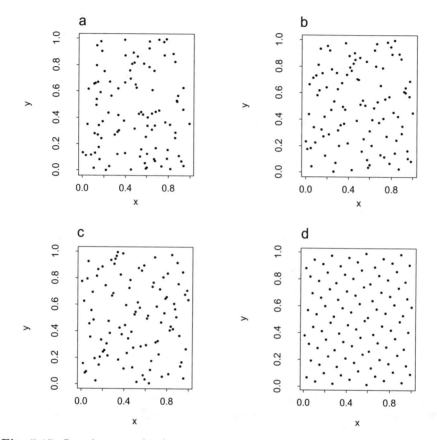

Fig. 5.17. Sample points (100) on a unit square using (**a**) Monte Carlo sampling, (**b**) Random Latin hypercube sampling, (**c**) median Latin hypercube sampling, and (**d**) Hammersley sequence sampling

error in the calculation of the expected value decreases as N_{samp} increases and is given by the central limit theorem:

$$\epsilon_\mu \propto (N_{samp})^{-0.5} \tag{5.68}$$

The accuracy of the estimates for the actual mean (μ) and the actual standard deviation (σ) is particularly important to obtain realistic estimates of any performance or economic parameter. However, as stated earlier and also shown in Example 5.5, this accuracy is dependent on the number of samples. The number of samples required for a given accuracy in a stochastic optimization problem depends upon several factors, such as the type of uncertainty

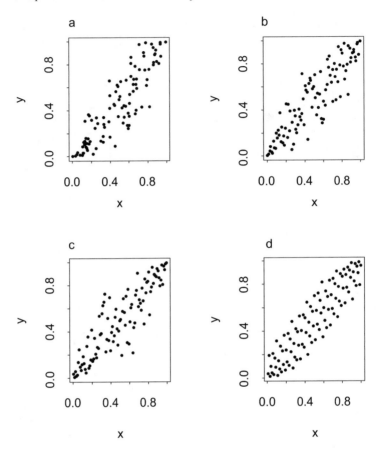

Fig. 5.18. Sample points (100) on a unit square with correlation of 0.9 using (**a**) Monte Carlo, (**b**) random Latin hypercube, (**c**) median LHS, and (**d**) HSS

and the point values of the decision variables (Painton and Diwekar, 1995). Especially for optimization problems, the number of samples required also depends on the location of the trial point solution in the optimization space. Figure 5.19 shows how the shape of the surface over a range of uncertain parameter values changes because one is at a different iteration (different values of decision variables) in the optimization loop. Therefore, the selection of the number of samples for the stochastic optimization procedure is a crucial and challenging problem. A combinatorial optimization algorithm that automatically selects the number of samples and provides the trade-off between accuracy and efficiency is presented below.

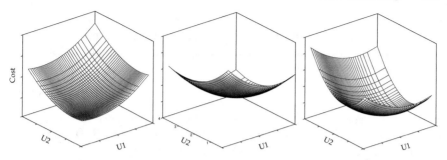

Fig. 5.19. Uncertainty space at different optimization iterations

5.5 Stochastic Annealing

The simulated annealing algorithm described in Chapter 4 is used for deterministic optimization problems. The stochastic annealing algorithm (STA)[1] is a variant of simulated annealing (Painton and Diwekar, 1995; Chaudhuri and Diwekar, 1996, 1671), and is an algorithm designed to efficiently optimize a probabilistic objective function. In the stochastic annealing algorithm, the optimizer (Figure 5.5a) not only obtains the decision variables but also the number of samples required for the stochastic model. Furthermore, it provides the trade-off between accuracy and efficiency by selecting an increased number of samples as one approaches the optimum. In stochastic annealing, the cooling schedule is used to decide the weight on the penalty term for imprecision in the probabilistic objective function. The choice of a penalty term, on the other hand, must depend on the error bandwidth of the function that is optimized, and must incorporate the effect of the number of samples.

The new objective function in stochastic annealing, therefore, consists of a probabilistic objective value P and the penalty function, which is represented as follows.

$$\text{Min} Z(cost) = P(x, u) + b(t)\, \epsilon_p \tag{5.69}$$

In the above equation, the first term represents the real objective function which is a probabilistic function in terms of the decision variables x and uncertain variables u, and all other terms following the first term signify the penalty function for error in the estimation.

The weighting function $b(t)$ can be expressed in terms of the temperature levels. At high temperatures, the sample size can be small, because the algorithm is exploring the functional topology or the configuration space to identify regions of optima. As the system gets cooler, the algorithm searches

---

[1]By "stochastic annealing" we refer to the annealing of an uncertain or stochastic function. It must be realized that the simulated annealing algorithm is a stochastic algorithm inherently, because the moves are determined probabilistically. However, for our purposes, we refer to the annealing of a deterministic objective function simply as simulated annealing.

for the global optimum; consequently it is necessary to take more samples to get more accurate and realistic objectives/costs. Thus, $b(t)$ increases as the temperature decreases. Based on these observations, an exponential function for $b(t)$ can be devised as

$$b(t) = \frac{b_o}{k^t} \qquad (5.70)$$

where b_o is small (e.g., 0.001), k is a constant which governs the rate of increase, and t is the temperature level. Remember that as the temperature level t increases the annealing temperature T decreases.

The stochastic annealing algorithm reduces the CPU time by balancing the trade-off between computational efficiency and solution accuracy by the introduction of a penalty function in the objective function. This is necessary, because at high temperatures the algorithm is mainly exploring the solution space and does not require precise estimates of any probabilistic function. The algorithm must select a greater number of samples as the solution nears the optimum. The weight of the penalty term, as mentioned before, is governed by $b(t)$, and is based on the annealing temperature.

The main steps in the stochastic annealing algorithm are given below.

1. Initialize variables: $T_{initial}$, T_{freeze}, accept and reject limits, initial configuration S.
2. If $(T > T_{freeze})$, then
   (a) Perform the following loop (i=(i)..(viii)) N (number of moves at a given temperature) times.
      i. Generate a move S' from the current configuration S as follows:
         A. Select the number of samples, N_{samp} by a random move.
            if $rand(0,1) \leq 0.5$, then

            $$N_{samp} = N_{samp} + 5 \times rand(0,1)$$

            else

            $$N_{samp} = N_{samp} - 5 \times rand(0,1)$$

         B. Select the decision variables (zero-one, integer, discrete, and continuous variables).
      ii. Generate N_{samp} samples of the uncertain parameters.
      iii. Perform the following loop (iii(A)..iii(B)) N_{samp} times.
         A. Run the model.
         B. Calculate the objective function cost(S').
      iv. Evaluate the expected value $E(cost(S'))$ and s of the cost function.
      v. Generate the weighting function $b(t) = b_o/k^t$.
      vi. Calculate the modified objective function:

      $$Obj(S') = E(Cost(S')) + b(t)\frac{1}{\sqrt{N_{samp}}}$$

      vii. Let $\Delta = Obj(S') - Obj(S)$.

viii. If $\Delta \leq 0$, then accept the move Set $S = S'$ else if $(\Delta \geq 0)$, then accept with a probability $\exp(-\Delta/T)$.

(b) Return to 2(a).

3. If $T > T_{freeze}$, set $T = \alpha T$ and return to 2(a).

4. Stop.

Note that in the above stochastic annealing algorithm, the penalty term is chosen according to the Monte Carlo simulations. For HSS sampling, recently Chaudhuri and Diwekar (1671) and Diwekar (2003) proposed a fractal dimension approach that resulted in the following error term for the stochastic annealing algorithm when Hammersley sequence sampling is used for the stochastic modeling loop in Figure 5.5a.

$$Obj(S') = E(Cost(S')) + b(t)\frac{1}{N_{samp}^{1.8}}$$

The following example illustrates the use of the stochastic annealing algorithm.

Example 5.7: In earlier chapters we have seen the maximum area problem. Now consider a different maximum area problem from the power sector. Compressors are a crucial part of any power cycle such as the Brayton cycle or the Stirling cycle where the heat energy is converted into power. Work done in the compression of a gas can be calculated using the first law of thermodynamics. From the pressure-volume diagram (Figure 5.20) it can be seen that the work done on the gas when the gas changes its state from pressure P_1, volume V_1, and temperature T_1 to a state at P_2, V_2, T_2, is essentially the area under the curve given by the following equation.

$$W = \int_{P_1, V_1}^{P_2, V_2} d\,PV \qquad (5.71)$$

For isentropic compression, this results in

$$W = c_p T_2 [(\frac{P_2}{P_1})^{(\gamma - 1)/\gamma} - 1] \qquad (5.72)$$

where W is the work done per unit mole of the gas and c_p is the molar specific heat at constant pressure. γ is the isentropic compression coefficient for ideal gas. If the required pressure ratio is large it is not practical to carry out the whole of the compression in a single cylinder because of the high temperatures that would develop. Furthermore, mechanical construction, lubrication, and the like will be difficult. In the operation of multistage compression, it not only avoids operational difficulties, but multistage compression followed by cooling results in an energy savings (more area, as shown in Figure 5.20). However, the cost and mass increase. The savings also depend on the design parameters

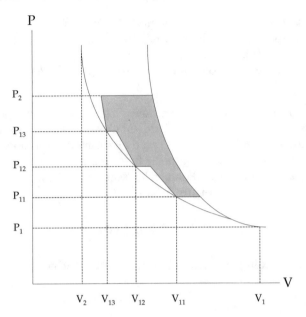

Fig. 5.20. Multistage compression cycle with interstage cooling, energy savings (shaded area) as compared with single stage compression

such as the compression ratio P_2/P_1, the amount of cooling expressed in terms of the temperature change across each heat exchanger $\Delta\,T$, and so on.

The work required in the multistage compression/expansion is given by

$$W_{N_{stages}} = \sum_{i=1}^{N_{stages}} c_p T_2^i [(\frac{P_2^i}{P_1^i})^{(\gamma-1)/\gamma} - 1)] \tag{5.73}$$

with the following associated cost for each compression stage (ASPEN Technical Reference Manual, 1982),

$$C = \sum_{i=1}^{N_{stages}} (e^{(7.7077+0.68\ \log(W_i/745.6998))} + 340 A_i^{0.68}) \tag{5.74}$$

where the first term is the cost of the compressor given in terms of the work done W in kWatts, and the second term is the cost of the heat exchanger given in terms of the heat exchanger area A in square meters. The objective is to minimize the expected value of the objective function representing the energy savings, cost, and mass (here mass is assumed to be proportional to the cost term shown in the above equation) trade-offs with the uncertainties in parameters u_1 and u_2 of the objective function given by

$$J = \frac{-u_1 W_{N_{stages}}}{0.000001 u_2 C^2} \tag{5.75}$$

Note that this objective function is representative and may be replaced by one with differing weights on the power-cost trade-offs. Here we are considering the design alternatives θ to be the number of stages N_{stages}, pressure ratios PR, and the heat exchanger capacities in terms of ΔT. Table 5.15 shows the values of the design variables for a maximum five-stage compression/cooling system. Use stochastic annealing to solve this problem and compare the solution with the fixed sampling stochastic model used in the simulated annealing framework.

Table 5.15. The decision and uncertain variables in the multistage compression synthesis problem

N_{stages}	Level i	PR^i	ΔT^i	u_1 & u_2
1	1	1.1	20	
2	2	2.2	40	
3	3	3.3	60	N(0.9, 1.1)
4	4	4.4	80	
5	5	5.5	100	

Solution: For each stage, there are N_{PR} possible pressure ratio levels, and $N_{\Delta t}$ possible heat capacities. A given number of stages i will have $N_{PR}^i \times N_{\Delta t}^i$ possible parameter combinations. Therefore, one stage will have $5 \times 5 = 25$ combinations of allowable parameters. Two stages will have $25 \times 25 = 625$, and so on. Therefore, allowing one, two, three, four, or five stages gives a state space of 10.2 million combinations, as given below.

$$N_{comb} = \sum_{i=1}^{5} N_{PR}^i \times N_{\Delta t}^i = \sum_{i=1}^{5} 5^i 5^i = 10.2 \text{ million} \qquad (5.76)$$

With the application of the simulated annealing algorithm and the stochastic annealing algorithm, it is necessary to define the analogues to the entities in physical annealing. Specifically, it is necessary to specify the following: the configuration space, the cost function, the move generator, the initial and final temperature, the temperature decrement, and the equilibrium detection method.

The cost function was defined according to the stochastic annealing criterion with the expected value of the objective function and the penalty and is given:

$$Obj = E(J) + b(t) \frac{2\sigma_j}{N_{samp}^{1/2}} \qquad (5.77)$$

If the initial temperature is too low, the search space is limited and the search becomes trapped in a local region. If the initial temperature is too high, the algorithm spends a lot of time "boiling around" and wasting CPU time.

The initial temperature is chosen to accept more than 80% of moves using the Metropolis criterion.

The final temperature was chosen so that the algorithm stopped after ten successive temperature decrements with no change in the optimal configuration. The temperature decrement was set such that the new temperature $T_{new} = \alpha T_{old}$, where $\alpha = 0.9$. Equilibrium was assumed to be reached when the accept/reject ratio, N_{acc}/N_T is 1:10.

The creation of a move generator is difficult because a move needs to be "random" yet results in a configuration that is in the vicinity of the previous configuration. An optimal move generator was created such that each move could result in one of the following permutations of the current configurations.

1. Add a random number of stages. Set the parameters of the added stages to the random possible levels.
2. Delete a random number of stages.
3. Remain at the same number of stages, but "bump" one of the parameters up or down by a random number of levels (not exceeding the maximum allowed level). When the temperature gets small enough, however, limit the move size to plus or minus one level from the current parameter level.

The above move possibilities were weighted 10:10:80. Because the objective function involves a large number of flat surfaces, the move generator had to be selected carefully.

The weighing function for the penalty term at each temperature level t was selected using the following equation.

$$b(t) = \frac{0.01}{(0.9)^t} \tag{5.78}$$

This ensures that the penalty for inaccuracy in the prediction of the expected cost function increases as one approaches optimum and also that the penalty does not outweigh the real objective function thereby defeating the purpose of optimization.

Figure 5.21 shows the progress of stochastic annealing represented in terms of the real objective function (expected cost) and the penalty function in terms of the percentage of the expected cost. It can be seen that stochastic annealing performs as expected where the penalty function increases as one approaches near optimum, accepting a few uphill moves to avoid local optima. Figure 5.22 shows the number of samples chosen at each temperature. One can see that, although the penalty function is more or less monotonic, the number of samples follows the pattern of the expected cost and not the penalty function. This is because the number of samples is correlated to the variance of the sample, which in turn is related to the expected functionals. Therefore, from Figure 5.22, one can easily infer that the stochastic optimization algorithm with the fixed samples at each optimization step may not be a right strategy to follow to obtain the given accuracy. This is because the number of samples needed to achieve a given accuracy also depends on the expected value of the objective function at that step. The performance of stochastic annealing was compared with the performance of annealing with the fixed sampled stochastic model. It was found that although both of the algorithms find the global optimum value of the expected objective function equal to -2.79, corresponding to the optimal configuration of $N_{stages} = 1$, $PR = 5.5$, and $\Delta T = 100$, stochastic annealing takes 70% less CPU time than annealing with a fixed sample stochastic model. Also, stochastic annealing automatically chooses the samples at each optimization stage, whereas the fixed sampling annealing may need extensive experimentation to come up with the right number of average samples for the given accuracy requirements.

5.6 Hazardous Waste Blending Under Uncertainty

The nuclear waste blend problem presented in the previous chapter involved consideration of discrete as well as continuous decisions and the problem was a difficult mixed integer nonlinear programming problem. However, the major problem at Hanford as Deborah Illman (1993) writes:

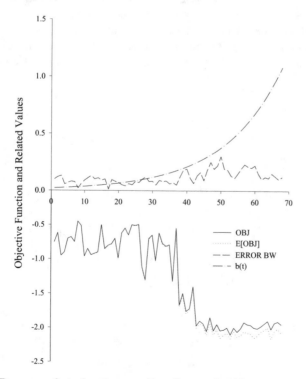

Fig. 5.21. Progress of stochastic annealing (bottom) E(j) and objective function versus T and (top) $b(t)$ and error bandwidth (penalty term) versus T

To make matters worse, wastes are often comingled on the site, unlike most hazardous waste sites. Organic wastes has co-contaminants—heavy metals, fission products, transuranics. And the mixed waste burial trenches, used from 1944–1970, may contain a mind-boggling potpourri including solid sodium, plutonium, pyrophorics, munitions, and other wastes in close proximity to one another. But no one is sure, because the records are poor.

This leads to a challenging problem of determining the optimal waste blend configuration subject to the inherent uncertainties in the waste compositions and in the glass physical property models. In this section, the two sources of uncertainty are briefly described. The characterization of the uncertainties in the model is presented in the next section.

Uncertainties in Waste Composition

The wastes in the tanks were formed as byproducts in different processes used to produce radioactive materials. Consequently, with each of these tanks a certain degree of variability is associated. Furthermore, over a period of 40–50 years, physical and chemical transformations within a tank have resulted in a nonuniform, nonhomogeneous mixture. Any experimental sample of the waste withdrawn from the tank is not representative of the tank as a whole, which contributes significantly to the uncertainty associated with the waste composition. This is supplemented, to a lesser extent, by the uncertainties associated with the analytical measurements in determining the waste compositions.

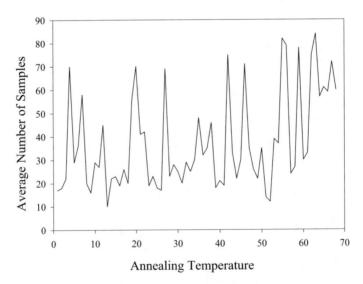

Fig. 5.22. N_{samp} versus T

Uncertainties in Glass Property Models

The glass property models are empirical equations fitted to the data (i.e., glass property values against glass compositions). Predictions made with a fitted property model are subjected to uncertainty in the fitted model coefficients. The uncertainties result from the random errors in property values introduced during the testing and measurements, as well as the minor lack-of-fit of the empirical model relative to the actual one Hopkins et al. (1994). Uncertainties in glass property models reduce the feasible range of the application of the glass property models, thereby affecting the optimal waste loading.

Characterization of Uncertainties in the Model

This section outlines the methodology adopted to characterize the uncertainties in the waste composition and the glass property models. Because this is a preliminary study, several assumptions have been made to keep the problem manageable and to focus on the key objective, namely, to develop an efficient method for solving this large-scale problem in computationally affordable time, and to illustrate how uncertainties affect the optimal blend configuration. Most of the assumptions pertain to uncertainties in the waste composition. The assumptions and simplifications used in this work are listed in the following section.

Characterization of Uncertainties in Waste Composition

As mentioned previously, the uncertainties in the waste composition arise due to many sources. The assumptions used in this study regarding waste composition uncertainties are as follows.

- For this study, "waste composition uncertainty" is a general term, covering all possible uncertainties in waste feed composition. These sources include batch-to-batch (within a waste type), sampling within a batch, and analytical uncertainties.
- The only estimate of this "lumped" uncertainty in the composition of the waste feed for high-level vitrification was based on the information available (i.e., analytical tank composition data).
- There is no bias in the composition estimates; the sample values are distributed about the true mean.
- The derived component mass fractions were assumed to follow normal distributions.
- The uncertainties of the species in the waste were assumed to be relatively independent of each other (i.e., uncorrelated).
- The relative standard deviation for each component in a particular waste tank was taken to be representative of all the tanks in the study. This assumption needs to be refined as subsequent data become available.

The procedure employed in characterizing the waste composition uncertainties is as follows.

- Based on the mean and the relative standard deviation (RSD) for each component in the tank, normal probability distributions were developed for the individual mass fractions. For a particular tank waste, the range of uncertainty is shown in Table 5.16.
- The above distributions were sampled to develop N_{samp} waste composition input sets (mass fractions). A stratified sampling technique (Latin hypercube sampling, Iman and Shortencarier, 1984), and the novel sampling technique, Hammersley sequence sampling, Diwekar and Kalagnanam (1997), were both used to generate the samples, and to observe the implication of different sampling techniques on the optimum blend configuration and the computational time.
- Given the mass fractions and the total mass of the wastes, the mass fractions were normalized to 1.0.
- The mean of the input waste mass for each component, based on N_{samp} samples of the component mass fractions, was then used in the model run.

Table 5.16. Mean mass, RSD, and the uncertainty associated with component masses for a pre-treated high-level waste in a particular tank (B-110) at the Hanford site

Components	Mass Fraction	Mass(kg)	RSD	Uncertainty(kg)
Al_2O_3	0.02002	25165.1	0.15	$25165.1(1\pm3\times0.15)$
B_2O_3	0.000856	1075.9	0.13	$1075.9(1\pm3\times0.13)$
CaO	0.011293	14195.3	0.07	$14195.3(1\pm3\times0.07)$
Fe_2O_3	0.229344	288285.2	0.04	$288285.2(1\pm3\times0.04)$
Li_2O	–	–	–	–
MgO	0.002687	3377.6	0.04	$3377.6(1\pm3\times0.04)$
Na_2O	0.080439	101111.7	0.04	$101111.7(1\pm3\times0.04)$
SiO_2	0.175263	220305.4	0.04	$220305.4(1\pm3\times0.04)$
ZrO_2	0.000041	51.4	0.12	$51.4(1\pm3\times0.12)$
Other oxides	0.480056	603429.9	0.056	$603429.9(1\pm3\times0.056)$
Cr_2O_3	0.014986	18837.4	0.03	$18837.4(1\pm3\times0.03)$
F	–	–	–	–
P_2O_5	0.248923	312895.9	0.04	$312895.9(1\pm3\times0.04)$
SO_3	–	–	–	–
Noble Metals	–	–	–	–

Characterization of Uncertainty in Physical Property Models

The uncertainty in a predicted property value for a given glass composition is defined as (Hopkins et al., 1994)

$$Uncert_{prop} = M[\mathbf{x}^\mathsf{T}\mathbf{S}\mathbf{x}]^{0.5} \tag{5.79}$$

where,

M = multiplier, which is usually the upper 95th percentile of a t-distribution $[t_{0.95}(n - ft)]$, n is the number of data points used to fit the model, and ft is the number of fitted parameters (coefficients) in the model.

$\mathbf{x}$ = glass composition vector expanded in the form of the model.

$\mathbf{S}$ = covariance matrix of the estimated parameters (coefficients) that is, b_is and b_{ij}s.

For nonlinear property models adopted in this study, the usual glass composition vector $\mathbf{x}$ is augmented by second-order terms. For example, if there are two second-order terms, x_1^2 and $x_2 x_4$, the usual composition vector $(x_1, ..., x_{10})$ becomes $(x_1, ..., x_{10}, x_1^2, x_2 x_4)$. The uncertainty expression (Equation (5.79)) corresponds to a statistical confidence statement on the property model prediction, considered a prediction of the mean property value for a glass composition $\mathbf{x}$.

The uncertainty defined in Equation (5.79) affects the glass property constraints by narrowing the feasible region determined by the glass property models. The form of the glass property constraints using this approach is given by

$$\ln(minpropval) + Uncert_{prop} \leq \sum_{i=1}^{n} b_i p^i + \sum_{i=1}^{n} \sum_{j \geq i} b_{ij} p^{(i)} p^{(j)}$$

$$\sum_{i=1}^{n} b_i f g^i + \sum_{i=1}^{n} \sum_{j \geq i} b_{ij} p^{(i)} p^{(j)} \leq \ln(maxpropval) - Uncert_{prop} \tag{5.80}$$

where $minpropval$ and $maxpropval$ are the lower and upper bounds on the glass property value. It is easily observed that if $Uncert_{prop} = 0$, this constraint formulation reduces to the deterministic equation in Chapter 3, where no uncertainties are associated with the glass property models.

5.6.1 The Stochastic Optimization Problem

The problem of determining the optimal blend configuration in the presence of uncertainties in the waste composition as well as in the physical property models is posed as a stochastic optimization problem. In the previous section, it has been shown that stochastic annealing provides an automated efficient framework for addressing such problems. The stochastic optimization problem requires that the quantities for the waste composition must be represented in terms of their expected values. Thus Equations (3.136)–(3.138) are represented as

$$g_e^{(i)} = E[w^{(i)}] + f_e^{(i)} \tag{5.81}$$

$$G_e = \sum_{i=1}^{n} g_e^{(i)} \tag{5.82}$$

$$p_e^{(i)} = \frac{g_e^{(i)}}{G_e} \tag{5.83}$$

where the subscript e signifies that the quantities are based on the expected value, and $E[w^{(i)}]$ signifies the expected value of the waste mass of the ith component in the waste.

Similarly, the individual component bounds, crystallinity constraints, solubility constraints, and the glass property constraints are formulated as

$$p_{LL}^{(i)} \leq p_e^{(i)} \leq p_{UL}^{(i)} \tag{5.84}$$

where UL and LL represent upper and lower bounds, respectively.

$$\ln(minpropval) + Uncert_{prop} \leq \sum_{i=1}^{n} b_i p_e^i + \sum_{i=1}^{n} \sum_{j \geq i} b_{ij} p_e^i p_e^j$$

$$\sum_{i=1}^{n} b_i p_e^i + \sum_{i=1}^{n} \sum_{j \geq i} b_{ij} p_e^i p_e^j \leq \ln(maxpropval) - Uncert_{prop} \tag{5.85}$$

The approach adopted for this waste blending problem is based on a coupled stochastic annealing-nonlinear programming (STA-NLP) technique, which is illustrated in Figure 5.23. The solution procedure incorporates a sequence of three loops nested within one another. The inner loop corresponds to the sampling loop, which generates the samples for the mass fractions (or masses) of the different components in the waste, and evaluates the mean of the waste mass for each tank, which is then propagated through the model that determines the glass property constraints. It must be noted that because uncertainties in the glass property models were incorporated by reducing the feasible region, as mentioned previously, a sampling exercise to account for uncertainties in the property models is not necessary. The loop above the sampling loop is the NLP optimization loop based on successive quadratic programming, a widely used technique for solving large-scale nonlinear optimization problems. The objective function for the NLP optimizer identifies the minimum amount of frit for a given blend configuration based on the expected value of the masses of the components in the waste blend.

$$\text{Min} \sum_{i=1}^{N} f_e^{(i)} \qquad \text{(NLP)} \tag{5.86}$$

subject to

Equality Constraints
Individual Component Bounds

Crystallinity Constraints
Solubility Constraints
Glass Property Constraints

where $f_e^{(i)}$ is the composition of ith component in the frit based on the expected value of the waste composition, and subject to the uncertainties in the physical property models.

Finally, the outer loop in the sequence consists of the stochastic annealing algorithm which predicts the sample size for the recursive sampling loop, and generates the blend configuration such that the total amount of frit is minimum over all the blends:

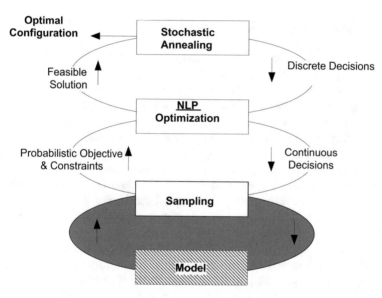

Fig. 5.23. Schematic diagram of the three-stage stochastic annealing (STA-NLP) algorithm

$$\text{Min} \quad \sum_{j=1}^{B} \sum_{i=1}^{N} f_{j_e}^{(i)} \qquad \text{(STA)} \qquad (5.87)$$

where $f_{j_e}^{(i)}$ is the mass of the ith component in the frit based on the expected values for the waste composition, and the uncertainties in the physical property models for the jth waste blend. And N and B denote the total number of components and the given number of blends that need to be formed, respectively.

The NLP problem is solved based on the expected value of the objective function, which is obtained from the runs of the model for the different samples, at each configuration predicted by the stochastic annealing algorithm. The termination of the entire procedure is governed by the stochastic annealing algorithm and is dependent on the "freezing" criterion mentioned in an earlier paper (Chaudhuri and Diwekar, 1996).

5.6.2 Results and Discussion

In order to study the effect of the uncertainties in waste composition and in the glass property models, the stochastic optimization problem of determining the optimal blend configuration was solved using two sampling techniques: namely, Latin hypercube and Hammersley sequence sampling. As mentioned previously, the presence of uncertainties in the waste composition makes this problem highly computationally intensive. In fact, a fixed sample framework for stochastic optimization using 200 samples and Hammersley sequence sampling was unable to converge on an optimal solution in 5 days (total run time was expected to be approximately 20 days), on a DEC-ALPHA 400 machine! This demanded the use of the coupled stochastic annealing-nonlinear programming (STA-NLP) approach to identify an optimal solution in a reasonable computational time.

The optimal design configuration identified by the coupled STA-NLP approach using Latin hypercube sampling and Hammersley sequence sampling are presented in Tables 5.17 and 5.18, respectively. The minimum quantity of frit required using both Latin hypercube and Hammersley sequence sampling is 11,307 kg. Nevertheless, the STA-NLP approach involving Hammersley sequence sampling, for which the error bandwidth was characterized based on a scaling relationship, was found to be computationally less intensive. For example, the STA-NLP technique using HSS and an improved formulation of the penalty term in the stochastic annealing algorithm, through accurate error bandwidth characterizations based on the scaling relationship, took 18 h, as opposed to 4 days using Latin hypercube sampling. The data, formulation, and computer code for this case study can be found online on the Springer website with the book link.

Table 5.17. The optimal waste blend configuration in the presence of uncertainties in the waste composition and glass physical property models (stochastic case)

Blends	Tank distribution
Blend-1	7,13,14,17,18,19,21
Blend-2	4,5,6,8,9,16,20
Blend-3	1,2,3,10,11,12,15

	Mass in Frit $f_e^{(i)}$ (kg)		
Component	Blend-1	Blend-2	Blend-3
SiO_2	356.49	5489.1	923.19
B_2O_3	37.997	826.70	0.6956
Na_2O	51.624	826.74	427.28
Li_2O	51.784	756.86	46.428
CaO	0.000	25.355	5.7003
MgO	0.000	0.000	43.944
Fe_2O_3	0.000	395.51	0.000
Al_2O_3	0.000	1020.0	0.000
ZrO_2	0.000	0.000	0.000
Other	0.000	21.784	0.000

The sampling exercise was performed using Latin hypercube sampling

Table 5.18. The optimal waste blend configuration in the presence of uncertainties in the waste composition and glass physical property models (stochastic case)

Blends	Tank distribution
Blend-1	7,13,14,17,18,19,21
Blend-2	4,5,6,8,9,16,20
Blend-3	1,2,3,10,11,12,15

	Mass in Frit $f_e^{(i)}$ (kg.)		
Component	Blend-1	Blend-2	Blend-3
SiO_2	356.81	5489.3	947.63
B_2O_3	38.000	828.07	1.0557
Na_2O	51.741	825.30	427.37
Li_2O	51.817	756.83	55.064
CaO	0.000	25.279	2.1108
MgO	0.000	0.000	14.208
Fe_2O_3	0.000	394.64	0.000
Al_2O_3	0.000	1020.6	0.000
ZrO_2	0.000	0.000	0.000
Other	0.000	21.590	0.000

The sampling exercise was performed using Hammersley sequence sampling

It can be observed that the presence of uncertainties significantly affects the optimal blend configuration, compared to a deterministic analysis (Chapter 4). In fact, given the uncertainties in the waste composition and the physical property models, the optimal design configuration obtained by Narayan et al. (1996) for the deterministic case (Chapter 4) estimates the total frit requirement to be 12,022 kg. The value of stochastic solution is found to be 985 kg which is significant. This study re-emphasizes the need for characterizing uncertainties in the model for the purpose of determining the optimal design configuration.

5.7 Sustainable Mercury Management: A Stochastic Optimization Problem

The previous two formulations presented in earlier chapters, for Savannah River watershed trading assumed that data is known deterministically, without any uncertainty. However, for the problem of pollutant trading, there are various possible sources of uncertainty. At the TMDL development step, the discharge from individual point sources, and fate and transportation of mercury are variable. These will affect the final bioaccumulation results, and hence the regulations and discharge allocations. Also uncertain are the efficiencies of various mercury treatment technologies implemented by the industries, affecting the actual reduction achieved.

In the first part of the case study in incorporating uncertainty in trading mechanism, the current discharge of mercury by each industry (a_i) to be uncertain, is normally distributed around a mean value. As explained before, once the TMDL has been developed, each industry is assigned a specific load reduction target based on the current discharge levels. Since the current discharge values are uncertain, load allocations and subsequent decisions are affected by the uncertainty. However, these uncertainties appear in linear constraints of the problem, and hence a chance constrained programming approach is implemented to solve this problem. The initial MILP formulation is used for this part of the stochastic problem.

In the second step, instead of considering uncertainties in the load, technological uncertainties are considered along with nonlinear models for technology. This problem is then formulated as a two-stage stochastic programming problem and solved using a decomposition strategy.

5.7.1 The Chance Constrained Programming Formulation

The formulation is an extension of MILP formulation presented in earlier chapters and given by Equations (5.88)–(5.91). In the stochastic version, the parameter red_i is uncertain. Since TMDL value is fixed, and red_i is a linear function of TMDL and a_i, red_i, and a_i have the same distribution. a_i is considered to be normally distributed with standard deviation σ_i and value used in the previous deterministic analysis as mean. This means that parameter red_i in (5.90) is also normally distributed.

Objective:

$$\text{Minimize} \quad \sum_{i=1}^{N}\sum_{j=1}^{M} TC_j . D_i . b_{ij} \tag{5.88}$$

Constraints:

$$t_{ii} = 0 \quad \forall i = 1, ..., N \tag{5.89}$$

$$red_i \le \sum_{j=1}^{M} q_j . D_i . b_{ij} + \sum_{k=1}^{N} t_{ik} - r \sum_{k=1}^{N} t_{ki} \quad \forall i = 1, ..., N \tag{5.90}$$

$$P_i \ge \sum_{j=1}^{M} b_{ij} . TC_j . D_i + F\left(\sum_{k=1}^{N} t_{ik} - \sum_{k=1}^{N} t_{ki}\right) \quad \forall i = 1, ..., N \tag{5.91}$$

The chance constrained formulation of (5.90) is, therefore, given as

$$\sum_{j=1}^{M} q_j . D_i . b_{ij} + \sum_{k=1}^{N} t_{ik} - r \sum_{k=1}^{N} t_{ki} \ge F_i^{-1}(\alpha) \quad \forall i = 1, ..., N \tag{5.92}$$

Here, F_i is the cumulative distribution function of uncertain variable red_i with mean red_i^* and standard deviation σ_i. The variations in the aggregate load will be a function of the variations in the individual loads. Also, the actual required reductions red_i for various point sources might not be correlated. However, incorporating the constraint given by (5.92) for all the point sources ensures that the worst-case scenario under the given constraint satisfaction probability (α) is accounted for. This will guarantee that there are no localized "hotspots" due to discharge uncertainties. The constraint represented by (5.92) is used in deterministic optimization techniques to solve the chance constrained programming problem, the results of which are reported in the next section.

Results and Discussions

The desired reduction red_i has a constant standard deviation of 5% for all the point sources, i.e., $\sigma_i = 0.05(red_i^*)$. This simulates ±15% uncertainty in discharge concentration. To analyze the effect of the degree of uncertainty on model solution, the chance constraint analysis is carried out for 16.67% standard deviation, i.e., $\sigma_i = 0.167(red_i^*)$, for various values of α. This simulates $\pm 50\%$ uncertainty in the current discharge concentrations. The results indicate that increase in uncertainty increases the total cost. Table 5.19 compares the solutions for the two cases of uncertainty ($\pm 15\%$ and $\pm 50\%$) for 90% constraint satisfaction ($\alpha = 0.9$) at TMDL 32 Kg/year. It can be seen that cost increase is also accompanied by higher discharge reduction and additional implementation of expensive technology (technology B). Simulations were carried out for $\sigma_i = 16.67\%$, and when point sources cannot implement more than one technology. This restriction can possibly be due to financial constraints. The simulations showed that the problem is infeasible below TMDL 30 Kg/year, even if trading is an option. These results show that the presence of uncertainty causes additional cost burden to ensure load reduction satisfaction.

Table 5.19. Solution comparison for different levels of uncertainty

	$\sigma_i = 5\%$	$\sigma_i = 16.67\%$
Total cost (Million $)	138.15	181.08
Total mercury discharge reduction (Kg)	1.032	1.359
No. of Technology A implemented	14	12
No. of Technology B implemented	4	12
No. of Technology C implemented	1	0

5.7.2 A Two-stage Stochastic Programming Formulation

As stated earlier, the data related to many mercury treatment technologies can be uncertain. This can either be due to uncertain performance characteristics of the technology (e.g., conversion efficiency, catalyst life), or due to relatively scarce data about a new treatment technology. This leads to considerable uncertainties about technology performance and cost. Under these circumstances, one has to work with the available data to arrive at optimal decisions. This means formulating the problem as a stochastic optimization (stochastic programming) problem. The idea proposed in this work is to extend the nonlinear deterministic model by considering uncertainties in nonlinear technology cost functions. The following sections discuss the problem formulation and solution methodology.

The general two-stage stochastic programming problem for pollutant trading can be represented as

Objective:

Minimize $\quad E[f_1(\zeta) + f_2(\phi, u)]$ $\qquad\qquad$ (5.93)

Constraints:

$\qquad\qquad$ Environmental constraints (regulations)

Technological constraints (technology models)

Trading constraints (5.94)

where, f_1 is the cost function corresponding to trading decisions ζ. f_2 represents the technology cost models, where f_2 depends on design variables ϕ and uncertain parameters u. E represents the expectation operator. The specific mercury trading problem formulation is given as

Objective:

$$\text{Minimize} \qquad E\Big[\sum_{i=1}^{N}\sum_{j=1}^{M} f_j(\phi_j, D_i, u_j)\, b_{ij}\Big] \qquad (5.95)$$

Constraints:

$$t_{ii} = 0 \qquad \forall i = 1, ..., N \qquad (5.96)$$

$$red_i \leq \sum_{j=1}^{M} q_j.D_i.\, b_{ij} + \sum_{k=1}^{N} t_{ik} - r\sum_{k=1}^{N} t_{ki} \qquad \forall i = 1, ..., N$$

$$(5.97)$$

$$P_i \geq \sum_{j=1}^{M} f_j(\phi_j, D_i, u_j).b_{ij} + F\Big(\sum_{k=1}^{N} t_{ik} - \sum_{k=1}^{N} t_{ki}\Big) \qquad \forall i = 1, ..., N$$

$$(5.98)$$

where, all notations have their previously assigned meanings. The nonlinear cost functions f_j are now dependent on the uncertain parameter set u_j in addition to design parameters ϕ_j and discharge volume D_i. Since the cost related to trading does not contribute to the objective function for the complete watershed, function f_1 represented in the generalized formulation is eliminated. The decision variables in the problem are b_{ij} representing the selection of a technology j by industry i, t_{ik} representing the amount of pollutant traded between industries i and k, and ϕ_j representing the design parameters for the technology. Since cost models f_j are nonlinear, this represents a stochastic nonlinear programming problem. Such problems are known to be computationally very difficult to solve.

The literature discusses various approaches to solve stochastic programming problems. One such approach is a decomposition strategy that reduces computational requirements for problem solving. In short, the given stochastic programming problem is decomposed into two or multiple stages. The first stage problem, known as the master problem, uses a linear approximation of the nonlinear recourse function to fix the first stage decision variables. The recourse function is exactly evaluated as a subproblem, referred to as the second stage problem.

For the trading model, linear approximations of the nonlinear technology models are given by TC_j, which are used to formulate a linear deterministic problem discussed earlier. Technology selection b_{ij} and trading amount t_{ik} represent the first stage decision variables. The first stage problem is represented as

Objective:

Minimize $\quad\quad\quad \theta$ $\quad\quad\quad\quad\quad\quad\quad\quad\quad\quad\quad\quad$ (5.99)

Constraints:

$$t_{ii} = 0 \quad\quad \forall i = 1, ..., N \quad\quad\quad\quad (5.100)$$

$$red_i \le \sum_{j=1}^{M} q_j . D_i . \, b_{ij} + \sum_{k=1}^{N} t_{ik} - r \sum_{k=1}^{N} t_{ki}$$
$$\forall i = 1, ..., N \quad\quad\quad\quad (5.101)$$

$$P_i \ge \sum_{j=1}^{M} TC_j . D_i . \, b_{ij} + F\left(\sum_{k=1}^{N} t_{ik} - \sum_{k=1}^{N} t_{ki}\right)$$
$$\forall i = 1, ..., N \quad\quad\quad\quad (5.102)$$

$$\theta \ge \sum_{i=1}^{N} \sum_{j=1}^{M} TC_j . D_i . \, b_{ij} \quad\quad\quad\quad (5.103)$$

$$g \le G . \zeta + \theta \quad\quad\quad\quad (5.104)$$

where, θ represents the first stage objective function. Constraints represented by Equations (5.100)–(5.102) are explained earlier in the text (5.103) puts a lower bound on the linear approximation of the nonlinear cost models, while Equation (5.104) represents the optimality cut, which is introduced after the solution of the second stage problem. This optimality cut includes the first stage decision variables b_{ij} and t_{ik} represented collectively here as ζ.

The first stage decisions are passed on to the second stage, where the recourse function is computed using nonlinear models. Here, the uncertain variables are sampled $Nsamp$ times, and the second stage subproblem is solved for each sample to calculate the expected value of the nonlinear recourse function. The second stage problem is thus given as

Objective:

Minimize $\quad\quad\quad \displaystyle\sum_{n=1}^{Nsamp} \sum_{i=1}^{N} \sum_{j=1}^{M} C_j(n) . \, b_{ij}$ $\quad\quad\quad\quad$ (5.105)

Constraints:

$$C_j(n) = f_j(n) \quad\quad\quad\quad (5.106)$$

where, $C_j(n)$ represents the exact cost computed using the nonlinear cost models f_j for a particular sample n of the uncertain parameter set u_j. The solution of the second stage problem results in a possible generation of an optimality cut, which is included in the subsequent master problem solution through the computation of G and g matrices.

These problem formulations (deterministic as well as stochastic) are quite general, applicable to any watershed and any pollutant. The next section discusses the application of the models on the case study of mercury waste management in the Savannah River basin. This exercise allows one to compare the results for the linear deterministic model with those for the nonlinear deterministic and the nonlinear stochastic models. The comparison is useful to assess the impact of nonlinearity and uncertainty on optimal decisions.

Results and Discussions

Figure 5.24 plots the annual saving due to trading implementation for the considered TMDL range (26 Kg/year–36 Kg/year) for three different models: linear deterministic, nonlinear deterministic, and nonlinear stochastic. The annual saving is computed as the difference between the total cost for technology option and total cost for trading options for a particular model under consideration. It is observed that approximate linear models underestimate the annual savings. The differences between linear deterministic and nonlinear deterministic model results are significant enough, and hence should not be ignored. The inclusion of uncertainty in the analysis predicts even higher savings for most TMDL values. It should be noted here that trends in savings do not necessarily reflect the trends in overall cost. Thus, although the nonlinear stochastic model leads to higher savings than the nonlinear deterministic model, the total cost with trading for the nonlinear stochastic model is not necessarily lower than the total cost with trading for nonlinear deterministic model. This is because saving is calculated as the difference between the total cost for the technology option and the total cost for the trading option for the same model type (linear, nonlinear deterministic, or nonlinear stochastic). Since the total cost for the technology option is different for different model types, the variations in savings do not necessarily correspond with variations in total cost for the trading option. These results highlight the importance of considering model nonlinearity and uncertainty while assessing the benefits of trading.

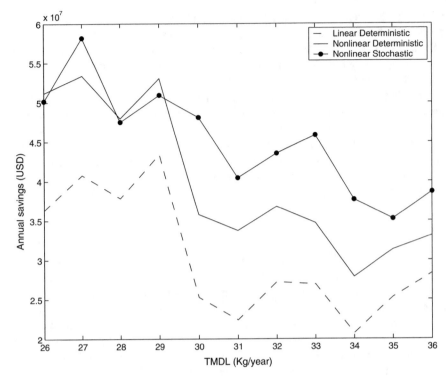

Fig. 5.24. Effect of nonlinearity and uncertainty on annual saving due to trading

Figure 5.25 shows the implications of nonlinearity and uncertainty inclusion on technology selection for the trading option. The figure shows the number of times each technology is implemented over the complete TMDL range (summation over all TMDL values). It can be seen that there are definite implications on technology selection. With linear technology models, various small industries (industries with low volumetric discharge rates) implement technologies along with large industries (industries with large volumetric discharge rates). However, when nonlinear technology models are used, large industries implement most of the technologies, and smaller industries satisfy the regulations by trading with these large industries. The distribution of technology selection is observed to be similar for both models. For both models, coagulation and filtration is the technology most commonly implemented, followed by granular activated carbon process and ion exchange process, respectively.

The inclusion of uncertainty in the model, however, has important implications on the distribution of technology selection. It can be seen from Figure 5.25 that in the presence of uncertainty, granular activated carbon treatment is implemented most often. Since this treatment is most efficient in terms of mercury removal capability, the total number of technology implementations correspondingly goes down. The trend is again similar to the nonlinear deterministic case where most of the smaller industries prefer to trade with larger industries instead of installing technologies. The amount of mercury traded is much higher than the other two cases for most TMDL values (TMDL greater than 28 Kg/year). This is because most efficient technology is getting implemented more often, and hence there is a greater scope for trading with other industries.

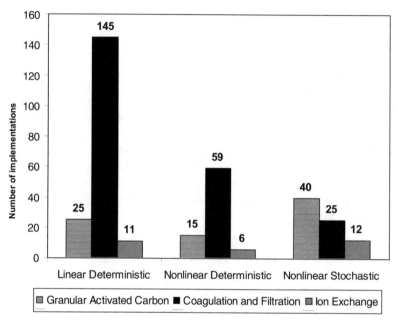

Fig. 5.25. Effect of nonlinearity and uncertainty on technology implementation decisions

5.8 Summary

The problems in optimization under uncertainty involve probabilistic objective functions and constraints. These problems can be categorized as (1) here and now problems, and (2) wait and see problems. Many problems involve both here and now, and wait and see decisions. The difference in the solution of these two formulations is the expected value of perfect information. Recourse problems normally involve both here and now, and wait and see decisions and hence are normally solved by decomposition strategies such as the L-shaped method. The major bottleneck in solving stochastic optimization (programming) problems is the propagation of uncertainties. In chance constrained programming, the uncertainties are propagated as moments, resulting in a deterministic equivalent problem. However, chance constrained programming methods are applicable to a limited number of problems. A generalized approach to uncertainty propagation involves sampling methods that are computationally intensive. New sampling techniques such as Hammersley sequence sampling reduce the computational intensity of the sampling approach. Sampling error bounds can be used to reduce the computational intensity of the stochastic optimization procedure further. This strategy is used in some of the decomposition methods and in the stochastic annealing algorithm.

Exercises

5.1 In the news vendor problem, the vendor must determine how many papers (x) to buy now at the cost of c cents for a demand which is uncertain. The selling price is s_p cents per paper. For a specific problem, whose weekly demand is shown below, the cost of each paper is $c = 20$ cents and the selling price is $s_p = 30$ cents. Assume no salvage value $s = 0$, so that any papers bought in excess of demand are simply discarded with no return. Solve the problem (1) if the news vendor knows the demand curve a priori (Table 5.20), and (2) if the vendor does not know the demand exactly and has to find an average value of x to be bought everyday. (3) Find VSS and EVPI for this problem.

Table 5.20. Weekly demand

I	Day	Demand,(u) d_i
1	Monday	50
2	Tuesday	60
3	Wednesday	60
4	Thursday	60
5	Friday	50
6	Saturday	100
7	Sunday	140

Solve (1), (2), and (3) for the following situations.
- Assume salvage value to be 5 cents $s = 5$.
- Assume $c = 25$, $s_p = 30$, and $s = 0$.
- Assume $c = 25$, $s_p = 30$, and $s = 10$.
- Compare the solutions and analyze the effect of uncertainties.

5.2 We want to evaluate the future value of an initial \$10,000 investment compounded over 30 years. The uncertainty in the percent return is summarized in Table 5.21, which is obtained from the last 50 year data of Standard and Poor's 500 Indices. Find the expected future value and its confidence interval and compare with the value based on the average percentage return.

Table 5.21. Uncertainty in percent return

Year	Return	Year	Return	Year	Return	Year	Return	Year	Return
1951	−10.50	1952	19.53	1953	26.67	1954	31.01	1955	20.26
1956	34.11	1957	−1.54	1958	7.06	1959	4.46	1960	26.31
1961	−6.56	1962	27.25	1963	12.40	1964	2.03	1965	14.62
1966	26.33	1967	1.40	1968	17.27	1969	14.76	1970	−9.73
1971	25.77	1972	12.31	1973	1.06	1974	−11.50	1975	19.15
1976	31.55	1977	−29.72	1978	−17.37	1979	15.63	1980	10.79
1981	0.10	1982	−11.36	1983	7.66	1984	20.09	1985	−13.09
1986	9.06	1987	12.97	1988	18.89	1989	−11.81	1990	23.13
1991	−2.97	1992	8.48	1993	38.06	1994	−14.31	1995	2.62
1996	26.40	1997	45.02	1998	−6.62	1999	11.78	2000	16.46

5.3 There are five beef supply vendors (v) and two distribution centers (d). We want to minimize costs associated with the production of three beef products (p) and delivery of these beef products to distribution centers while satisfying the demands of the distribution centers. The following figure (Figure 5.26) shows a conceptual diagram of this problem, where dashed arrows represent no shipment from that vendor to that distribution center.

Where

$costD(v, d, p)$ Cost of shipment from vendor to distribution center
$xD(v, d, p)$ Product shipped from vendor to distribution center
$yD(v, d, p)$ Binary variable for product shipped
$prodP(v, p)$ Beef production of p at vendor v
$costV(v)$ Cost driven by beef production
$yP(v, p)$ Binary variable of beef product
$yV(v)$ Binary variable of vendor
$dcdemand(d, p)$ Demand of product at distribution center

The input variables are given in Table 5.22.
Find the minimum cost using the here and now and wait and see methods when there is 25% uncertainty in *dcdemand*. Note that in this problem uncertainties are present only in the constraints; not in the objective function.

5.4 Introduce uncertainty in your simulated annealing cost function (Chapter 4, Exercises) as follows.

$$\text{Min} \quad Cost = \sum_{i=1}^{N_1} (N_1 - 3)^2 + (u_1 N_2(i) - 3)^2 + (u_2 N_3(i) - 3)^2$$

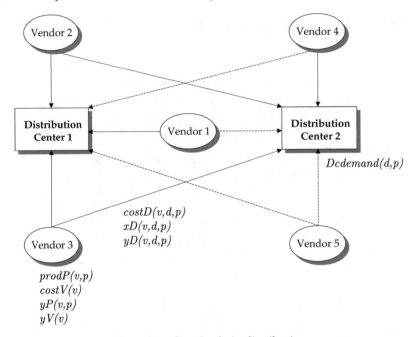

Fig. 5.26. Supply chain distribution

Table 5.22. Input variables for Problem 5.3

dcdemand, lb				costV	
$p \backslash d$	1	2		v	cost
1	1,720,000	810,000		1	0.8067
2	11,190,000	480,000		2	0.8427
3	3,570,000	0		3	0.8151
				4	0.8073
				5	0.8048

costD(v, d, p), million \$/lb/yr

$p = 1$			$p = 2$			$p = 3$		
$v \backslash d$	1	2	$v \backslash d$	1	2	$v \backslash d$	1	2
1	0.0431	0.0065	1	0.0255	0.0759	1	0.0127	0.0212
2	0.0363	0.0871	2	0.0647	0.0180	2	0.0840	0.0759
3	0.0434	0.0117	3	0.0295	0.0373	3	0.0607	0.0648
4	0.0222	0.0153	4	0.0585	0.0065	4	0.0198	0.0492
5	0.0095	0.0797	5	0.0121	0.0342	5	0.0440	0.0382

– Take uncertainties u_1 and u_2 as uniform distributions between 0 and 2 (mean 1). Plot the graphs of Cost versus u_1 and u_2 for two configurations ($N_1 = 1, N_2(1) = 2, N_3(1) = 3$ and $N_1 = 2, N_2(1) =$

$1, N_2(2) = 2; N_3(1) = 1, N_2(2) = 3)$. Which configuration will require more samples to evaluate the moments correctly?
- Modify the simulated annealing algorithm to become the stochastic annealing algorithm and plot the graph of temperature versus average expected cost.

Bibliography

ASPEN (1982), *ASPEN Technical Reference Manual*, Cambridge, MA.

Beale E.M. L. (1955), On minimizing a convex function subject to linear inequalities, *Journal of the Royal Statistical Society* **17B**, 173.

Birge J. R. (1997), Stochastic programming computation and applications, *INFORMS Journal on Computing*, **9(2)**,111.

Birge J. R. and F. Louveaux (1997), *Introduction to Stochastic Programming*, Springer Series in Operations Research, Springer, New York, NY.

Charnes A. and W. W. Cooper (1959), Chance-constrained programming, *Management Science* **5**, 73.

Chaudhuri P. (1996), *Process synthesis under uncertainty, Ph.D. Thesis*, Department of Environmental Engineering, Carnegie Mellon University, Pittsburgh, PA.

Chaudhuri P. and U. M. Diwekar (1996), Synthesis under uncertainty: A penalty function approach, *AIChE Journal* **42**, 742.

Chaudhuri P. and U. Diwekar (1999), Synthesis approach to optimal waste blend under uncertainty, *AIChE Journal* **45**, 1671.

Dantzig G. B. (1955), Linear programming under uncertainty, *Management Science* **1**, 197.

Dantzig G. B. and P. Glynn (1990), Parallel processors for planning under uncertainty, *Annals of Operations Research* **22**, 1.

Dantzig G. B. and G. Infanger (1991), Large scale stochastic linear programs–Importance sampling and bender decomposition, *Computational and Applied Mathematics*, Brezinski and U. Kulisch (ed.), 111.

Dantzig G. B. and P. Wolfe (1960), The decomposition principle for linear programs, *Operations Research* **8**, 101.

Dige N. and U. Diwekar (2018), Efficient sampling algorithm for large-scale optimization under uncertainty problems, *Computers and Chemical Engineering*, **115**, 431.

Diwekar U. M. (1995), A process analysis approach to pollution prevention, *AIChE Symposium Series on Pollution Prevention Through Process and Product Modifications*, **90**, 168.

Diwekar U. (2003), A novel sampling approach to combinatorial optimization under uncertainty, *Computational Optimization and Applications*,**24**, 335.

Diwekar U. M. and J. R. Kalagnanam (1997), An efficient sampling technique for optimization under uncertainty, *AIChE Journal*, **43**, 440.

Diwekar U. M. and E. S. Rubin (1994), Parameter design method using Stochastic Optimization with ASPEN, *Industrial Engineering Chemistry Research*, **33**, 292.

Diwekar U. M. and E.S. Rubin (1991), Stochastic modeling of chemical Processes, *Computers and Chemical Engineering*, **15**, 105.

Diwekar U. and Y. Shastri, Green process design, green energy, and sustainability: a systems analysis perspective, (2010),*Computers and chemical Engineering*, **34**, 1348.

Edgeworth E. (1888), The mathematical theory of banking, *J. Royal Statistical Society*, **51**, 113.

Higle J. and S. Sen (1991), Stochastic decomposition: An algorithm for two stage linear programs with recourse, *Mathematics of Operations Research*, **16**, 650.

Hopkins, D. F., M. Hoza, and C. A. Lo Presti (1994), *FY94 Optimal Waste Loading Models Development*, Report prepared for U.S. Department of Energy under contract DE-AC06-76RLO 1830.

Illman D. L. (1993), Researchers take up environmental challenge at Hanford, *Chemical and Engineering News*, **9**, July 21.

Iman R. L. and W. J. Conover (1982), Small sample sensitivity analysis techniques for computer models, with an application to risk assessment, *Communications in Statistics*, **A17**, 1749.

Iman R. L. and J. C. Helton (1988), An investigation of uncertainty and sensitivity analysis techniques for computer models, *Risk Analysis*, **8(1)**, 71.

Iman, R. L. and M. J. Shortencarier(1984), A FORTRAN77 Program and User's Guide for Generation of Latin Hypercube and Random Samples for Use with Computer Models, *NUREG/CR-3624, SAND83-2365*, Sandia National Laboratories, Albuquerque, N.M.

James B. A. P., Variance reduction techniques (1985), *Journal of Operations Research Society*, **36(6)**, 525.

Luckacs E. (1960),*Characteristic Functions*, Charles Griffin, London.

Kalagnanam J. R. and U. M. Diwekar (1997), An efficient sampling technique for off-line quality control, *Technometrics*, **39(3)**,308.

Knuth D. E. (1973), *The Art of Computer Programming, Volume 1: Fundamental Algorithms*, Addison-Wesley, Reading, MA.

Madansky A.(1960), Inequalities for stochastic linear programming problems, *Management Science*, **6**, 197.

McKay M. D., R. J. Beckman, and W. J. Conover (1979), A comparison of three methods of selecting values of input variables in the analysis of output from a computer code, *Technometrics*, **21(2)** 239.

Milton J. S. and J. C. Arnold (1995), *Introduction to Probability and Statistics : Principles and Applications for Engineering and the Computing Sciences*, McGraw-Hill, New York.

Morgan G. and M. Henrion (1990), *Uncertainty: A Guide to Dealing with Uncertainty in Quantitative Risk and Policy Analysis*, Cambridge University Press, Cambridge, UK.

Narayan, V., U. Diwekar and M. Hoza (1996), Synthesizing optimal waste blends, *Industrial and Engineering Chemistry Research*, **35**, 3519.

Nemhauser, G. L., A. H. G. Ronnooy Kan, and M. J. Todd (1989), *Optimization: Handbooks in operations research and management science*, Vol. 1. North-Holland Press, New York.

Niederreiter, H. (1992), *Random Number Generation and Quasi-Monte Carlo methods*, SIAM, Philadelphia.

Painton, L. A. and U. M. Diwekar (1995), Stochastic annealing under uncertainty, *European Journal of Operations Research*, **83**, 489.

Petruzzi N. C. and M. Dada (1999), Pricing and the newsvendor problem: A review with extensions, *Operations Research*, **47(2)**, 183.

Prékopa, A. (1980), Logarithmic concave measures and related topics, in *Stochastic Programming*, M. A. H. Dempster (ed.), Academic Press, New York.

Prékopa A. (1995), *Stochastic Programming*, Kluwer Academic, Dordrecht, Netherlands.

Raiffa H. and R. Schlaifer (1961), *Applied Statistical Decision Theory*, Harvard University, Boston.

Saliby E. (1990), Descriptive sampling: A better approach to Monte Carlo simulations, *Journal of Operations Research Society*, **41(12)**, 1133.

Shastri Y., U. Diwekar and S. Mehrotra.,(2011),An innovative trading approach for mercury waste management,*International Journal of Innovation Science*,**3**, 9.

Taguchi G. (1986), *Introduction to Quality Engineering*, Asian Productivity Center, Tokyo.

Tintner G. (1955), Stochastic linear programming with applications to agricultural economics, *Proc. 2nd Symp. Lin. Progr.* , Washington, 197.

Vajda S. (1972), *Probabilistic Programming*, Academic Press, New York.

Van Slyke R. and R. J. B. Wets (1969), L-shaped linear programs with application to optimal control and Stochastic Programming, *SIAM Journal on Applied Mathematics*, **17**, 638.

Wang R., U. Diwekar, and C. Gregoire-Padro, Latin hypercube Hammersley sampling for risk and uncertainty analysis, (2004), *Environmental Progress.*, **23**, 141.

Wets R. J. B. (1990), Stochastic programming, in *Optimization Handbooks in Operations Research and Management Science, Volume 1*, G. L. Nemhauser, A. H.G. Rinooy Kan, and M. J. Todd, (ed.), North-Holland, Amsterdam (1990).

Wets R. J. B (1996), Challenges in stochastic programming, *Math. Progr.*, **75**, 115.

6

Multiobjective Optimization

Life is a compromise, often involving more than one objective. Even Noah at the time of the great flood faced the same dilemma. Noah's problem was to build an ark to accommodate a maximum number of animals and to store the maximum amount of food on the ark.

Noah had to satisfy at least two objectives (as stated above) while satisfying constraints: a multiobjective optimization problem (MOP). MOP is a cousin of (and subset of) multiple criteria decision making (MCDM). MCDM deals with problems in which alternatives are known and perspectives are sought. The theory behind MOP has been around for almost 50 years. Kuhn and Tucker actually dealt with it, in passing, in their seminal paper on conditions of optimality (Kuhn and Tucker, 1951). MOP deals with problems in which the alternatives are represented implicitly with decision variables and constraints. Obviously, keeping in line with the focus of this book, we talk about MOP in this chapter.

Multiobjective problems appear in virtually every field and in a wide variety of contexts. The importance of multiobjective optimization can be seen from the large number of applications presented in the literature. The problems solved vary from designing spacecraft (Sobol, 1992), aircraft control systems (Schy and Giesy, 1988), bridges (Ohkubo et al., 1998), vehicles (Starkey et al., 1988), and highly accurate focusing systems (Eschenauer, 1988) to fore-

---

Electronic Supplementary Material: The online version of this chapter (https://doi.org/10.1007/978-3-030-55404-0_6) contains supplementary material, which is available to authorized users.

This chapter is based on his class notes from Jared Cohon, President, Carnegie Mellon University.

© Springer Nature Switzerland AG 2020
U. M. Diwekar, *Introduction to Applied Optimization*, Springer Optimization and Its Applications 22,
https://doi.org/10.1007/978-3-030-55404-0_6

casting manpower supplies (Silverman etal., 1988), selecting portfolios (Tamiz and Jones, 1996), blending sausages (Olson and Tchebycheff, 1993), planning manufacturing systems (Kumar et al., 1991), and solving pollution control and management problems (Collins et al., 1988).

An MOP problem is any decision problem that can be stated in the following format.[1]

$$\text{Minimize (or Maximize)} \quad \text{Set of objectives}$$

subject to

$$\text{Set of constraints}$$

Therefore, a generalized MOP can be represented as follows.

$$\text{Optimize} \quad \bar{Z} \;=\; (Z_1, Z_2, \ldots, Z_k)$$

subject to

$$h(x) \;=\; 0 \tag{6.1}$$

$$g(x) \;\leq\; 0 \tag{6.2}$$

The objective function and constraints are mathematical functions of a set of decision variables and parameters. The form (LP, NLP, MIP, etc.) of the equations determines the particular type of the MOP, such as MOLP for linear programming problems, MONLP for nonlinear programming, and so on.

MOP can be thought of as a set of methodologies for generating a preferred solution or range of efficient solutions to a decision problem (Cohon, 1978). For example, consider the graduate school selection problem given below.

Example 6.1: Shivani wanted to select a graduate school on the basis of the *US News and World Report* rankings for engineering schools. She selected the seven schools given in Table 6.1 from the list of top schools published in *US News* in 1998. The criteria she decided to base her decision on included consideration of academic rank, recruiting, and research, as shown in Tables 6.2 and 6.3. What schools should she prefer given the different criteria at which she is looking?

Solution: From Tables 6.2 and 6.3, it can be seen that a college is better if the rank is lower for each of the criteria, except the doctoral student-to-faculty ratio where the higher the ratio is, the better the college. In short, Shivani wants to minimize rankings and maximize the ratio R. To have her selection consistent with rankings, she converted the last criterion as $1/R$ to be minimized. Also, every criterion is normalized as shown in Table 6.4.

---

[1] Note that because an MOP involves a set of (a vector) objectives, instead of a single objective, it is also referred to as vector optimization. The difference between optimal control problems described in the next chapter and MOP is that the vector optimization in optimal control is in the decision domain where the decision variable is a trajectory but in MOP the vector is in the objective space.

Figure 6.1 shows the plot of the different normalized criteria (normalized using the maximum value in each criteria column) versus the college for Shivani. From the graph and from Table 6.1, it is easier to see that MIT (School 1) is better than Georgia Tech (School 4) as the slopes of all the lines joining

Table 6.1. Schools of engineering

School	Index
Massachusetts Institute of Technology	1
Stanford University	2
Carnegie Mellon University	3
Georgia Institute of Technology	4
University of Michigan- Ann Arbor	5
California Institute of Technology	6
Cornell University	7

Table 6.2. Different criteria

Criteria	Index
Academic Rank	1
Engineering Recruiters	2
Student Selectivity	3
Research Activity	4
Doctoral Student-to-Faculty Ratio	5

Table 6.3. *US News* criteria and ranks

Schools	Criteria				
	1	2	3	4	5
1	1	1	11	1	3.21
2	1	8	31	7	4.71
3	8	12	4	6	3.36
4	8	2	20	2	2.72
5	5	3	31	3	3.18
6	3	7	1	26	3.88
7	7	10	6	13	2.87

Schools 1–4 for each criterion are positive. Similarly, MIT is also better than the University of Michigan in all the criteria she considered, whereas other schools such as Stanford, Cal. Tech., Cornell, and Carnegie Mellon are better or worse than MIT in at least one criterion. At this stage, Shivani can look at the five colleges shown in Figure 6.2 as the preferred set for further selection.

Table 6.4. Normalized objectives

Schools	Criteria				
	1	2	3	4	5
1	0.1250	0.0833	0.3584	0.0385	0.8465
2	0.1250	0.6667	1.0000	0.2692	0.5769
3	1.0000	1.0000	0.1290	0.2308	0.8088
4	1.0000	0.1667	0.6452	0.0769	0.9990
5	0.6250	0.2500	1.0000	0.1154	0.8545
6	0.3750	0.5833	0.0322	1.0000	0.7004
7	0.8750	0.8333	0.1936	0.5000	0.9468

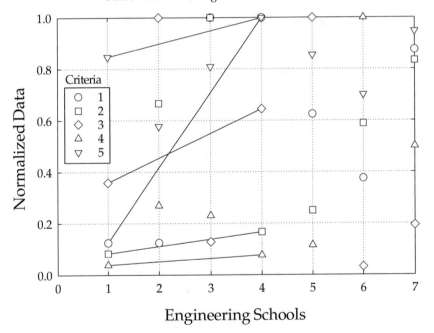

Fig. 6.1. The idea of nondominance

6.1 Nondominated Set

The preferred set in the above example is also known as the *nondominated set*, a most important concept in the MOP solution method. In fact, the solution to the MOP is not a single solution, but rather is the nondominated set, also known as the Pareto set after the French–Italian economist and sociologist Vilfredo Pareto (1964; 1971). This set is a collection of alternatives that

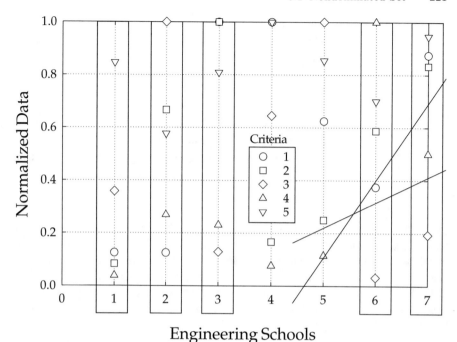

Engineering Schools

Fig. 6.2. The preferred set for further selection

represent potential compromise solutions among the objectives. This concept is illustrated using the MOLP problem derived from the chemical manufacturer's problem described in Chapter 2.

Example 6.2: Consider Example 2.1 from Chapter 2. In this example, the chemical manufacturer was using chemicals X_1 and X_2 to obtain a minimum cost solvent, given that there are constraints related to storage, safety, and the availability of materials. Let us add another dimension to the problem: the manufacturer not only wants to minimize the cost of solvents, but also desires to reduce the environmental impacts from the solvents as given by the following equation.

$$\text{Environmental Impacts} \quad \propto \quad -0.5x_1 + x_2$$

Furthermore, he found out that a minimum amount of solvent X_1 is necessary to increase the durability of the process equipment, a constraint given below.

$$x_1 \geq 1$$

Find the nondominated set of alternatives for this problem.

Solution: This problem can be formulated as the following MOLP.

$$\text{Minimize} \quad Z_1 = 4x_1 - x_2 \tag{6.3}$$
$$\text{Minimize} \quad Z_2 = -0.5x_1 + x_2 \tag{6.4}$$
$$x_1, \; x_2$$

subject to

$$x_1 \geq 1 \quad \text{Durability Constraint} \tag{6.5}$$
$$2x_1 + x_2 \leq 8 \quad \text{Storage Constraint} \tag{6.6}$$
$$x_2 \leq 5 \quad \text{Availability Constraint} \tag{6.7}$$
$$x_1 - x_2 \leq 4 \quad \text{Safety Constraint} \tag{6.8}$$
$$x_1 \geq 0; x_2 \geq 0$$

Figure 6.3 defines the feasible set of decision variables x_1 and x_2. The shaded region with extreme points ABCD provides the feasible region for this problem in the decision space. Because we only have two objectives, we can graph these extreme points (Table 6.5) in the objective space as shown in Figure 6.4. The shaded region in this figure represents the feasible objective value combinations in the objective space, corresponding to the feasible solutions in the feasible decision space.

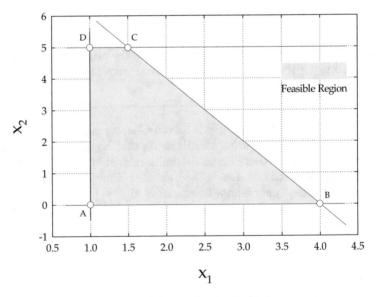

Fig. 6.3. Feasibility region in decision space

To understand the concept of a nondominated set, consider the points M_1 and M_2 in Figure 6.5. The solution corresponding to point A gives a lower level of both objectives than the solution corresponding to the points M_1 and

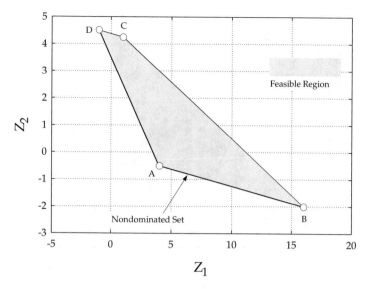

Fig. 6.4. Feasible region in the objective space

Table 6.5. Decision variables and objective values for the extreme points

Extreme points	x_1	x_2	Z_1	Z_2
A	1	0	4	-0.5
B	4	0	16	-2.0
C	1.5	5	1	4.25
D	1	5	-1	4.0

M_2. Point A is said to dominate these points. Considering the point M_2, all points within the area indicated by the dashed lines are said to dominate M_2 because they all yield lower levels of both objectives. Using similar logic, it is possible to show that for all points inside the boundaries of the feasible region, there is at least one point along the BAD boundary of the feasible region that dominates each of the inside points. Also, for points on the boundary BAD, there are no points that dominate them. Optimal trade-offs lie along BAD. The collection of these points is the nondominated or the Pareto set. A Pareto optimal is also known as an Edgeworth–Pareto optimal, an efficient solution, a nondominated solution, a noninferior, or a functional efficient solution.

Mathematically, the nondominated solution can be defined if $\bar{x}$ is a particular set of feasible values for the decision variables x. A solution $\bar{x}^*$ is

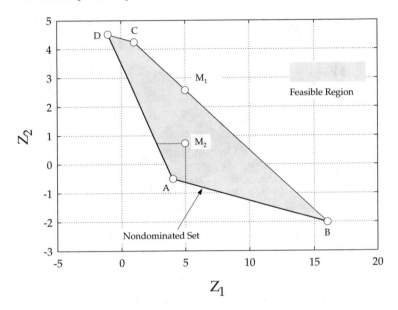

Fig. 6.5. Concept of nondominated set

nondominated if it is feasible and if there is no other feasible solution $\bar{x}$ such that

$$Z_p(\bar{x}) \leq Z_p(\bar{x}^*) \quad p = 1, 2, \ldots, k$$

where p is the number of objectives, and with at least one of these inequalities being a strict inequality (assuming all objectives are to be minimized).

It should be noted that each point along the nondominated set in objective space (Figure 6.4) has an equivalent point in the decision space (Figure 6.3) but the graphical interpretation of nondominance applies only in objective space. The corresponding decision variables can be found by using the objective values. All the solutions in the nondominated set (an infinite number for the case of continuous variable optimization such as MOLP) are candidates for selection, and are selected depending on the decision-maker's preference. As you move along the nondominated set, you are essentially trading off one objective for another. "Perfect is the enemy of good," is the basis of all MOP solution methods.

6.2 Solution Methods

There is a large array of analytical techniques for multiobjective optimization problems. Cohon (1978) reviewed many of the methods. Zeleny et al. (1982) provided a comprehensive treatment of the entire multicriteria endeavor. Hwang and Masud (1979) illustrated a large number of methods by solving numerical examples in detail. Stadler (1988) offered broad coverage of the field with many examples from engineering and the sciences. Chankong and Haimes (1983a) included a rigorous development of most multicriteria techniques. Steuer (1986) provided an especially useful review of multicriteria linear programming theories and algorithms. Miettinen (1999) gave a thorough review of nonlinear multiobjective optimization theories and methods. The large number of multiobjective optimization methods can be classified in many ways according to different criteria. Hwang and Masud (1979), followed by Buchanan (1986), Lieberman (1991), and Miettinen (1999), classified the methods according to the participation of the decision-maker in the solution process: no preference methods, a priori methods, interactive methods, and a posteriori methods. Rosenthal (1985) suggested three classes of solution methods: partial generation of the Pareto set, explicit value function maximization, and interactive implicit value function maximization. In Carmichael (1981), methods were classified according to whether a composite single objective function, a single objective function with constraints, or many single objective functions were the basis for the approach. Here we apply the classification presented by Cohon (1985), but extend its content, as shown in Figure 6.6.

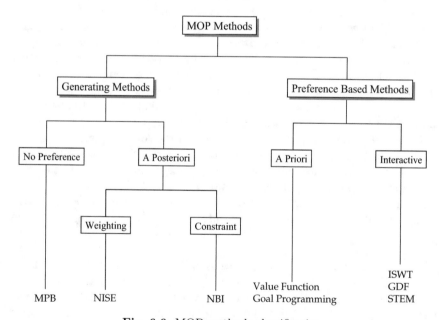

Fig. 6.6. MOP methods classification

In general, the multiobjective optimization methods are divided into two basic types: preference-based methods and generating methods. Preference-based methods attempt to quantify the decision-maker's preference, and with this information, the solution that best satisfies the decision-maker's preference is then identified. Generating methods have been developed to find the exact Pareto set or an approximation of it, and one of the generated Pareto optimal solutions is chosen for implementation. The two sets of methods imply very different things for the respective roles of the decision-maker and the analyst/designer. Preference-based methods require the decision-maker to articulate his or her preferences in a formal structured way. The analyst becomes a counselor, in effect. Generating techniques put the analyst/designer in the role of information provider, and the decision-maker is expected to make the necessary value judgments by selecting from among the Pareto optimal solutions.

Preference-based methods and generating methods exhibit both strengths and weaknesses. Even though preference-based techniques have advantages, such as reducing the computational burden to generate many solution alternatives to approximate the whole Pareto set, the demand of a decision-maker's time, knowledge, and experience, which provide consistent preference, is sometimes rather difficult. Furthermore, the decision-maker may not be able to state her preference exactly or may simply not want to reveal her preferences to the analysts. Many of the preference-based methods suffer from an information inadequacy; they require the decision-maker to state preferences before she knows what the choices are, thereby stripping the analysis of that which is of most interest to her. It is sometimes difficult for decision-makers to give consistent preference during the process of finding one best-compromise solution. The more desirable scenario would be to present the decision-maker with the set of Pareto optimal solutions determined independent of a priori or interactive preferences. Then the decision-makers could consider their relative preferences for the objectives and select the final solution with the benefit of knowing their choices, which are represented by the Pareto set.

Generating methods provide a great deal of information, emphasizing the Pareto optimal set or the range of choice available to decision-makers, and providing the trade-off information of one objective versus another. Generating techniques also do not require explicit value judgments from decision-makers, allowing them instead to express their values implicitly through their selection of an alternative. There are, however, problems with generating techniques that are not observed with most of the preference-based techniques. First, the generating algorithms are often complex and difficult for decision-makers to understand. Second, the number of Pareto optimal solutions is often too large for the decision-maker to analyze effectively. Third, the computational cost of the existing generating methods increases rapidly with the number of objectives, and it is difficult to solve high-dimensional problems. Overall, selecting an appropriate multiobjective optimization method itself is a problem with multiple objectives, as a large variety of methods exists for these problems and none can claim to be superior to the others in every aspect.

As described in Figure 6.6, generating techniques can be further divided into two subclasses: no-preference methods and a posteriori methods. No-preference methods, including compromise programming (Zeleny 1974), multiobjective proximal bundle (MPB); (Miettinen, 1999), and feasibility-based methods, such as the parameter space investigation (PSI) methods (Osyczka, 1984; Sobol and Statnikov, 1982); , focus on generating a feasible solution (e.g., all the points in the feasible region ABCD in Figure 6.4) or all the feasible solutions instead of the Pareto set (the best feasible solutions, e.g., the boundary BAD in Figure 6.4). In compromise programming and MPB, a single solution is obtained and presented to the decision-maker. The decision-maker may either accept or reject the solution, and it is unlikely that the best-compromised solution can be obtained by these methods. In PSI methods the continuous decision space is first uniformly discretized using the Monte Carlo sampling technique; next a solution is checked with the constraints. If one of the constraints is not satisfied, the solution is eliminated and the objective values are finally calculated, but only for those feasible solutions. Therefore, a "discretized approximation" of the feasible objective region, instead of the Pareto set, is retained by the PSI method. The solutions of this feasibility-based method cover the whole feasible objective region rather than covering only the optimal solutions in the Pareto set. Because most of the feasible solutions are not Pareto optimal, a relatively small number of the nondominated (relatively better, but not necessary to be Pareto optimal) solutions must be extracted from the whole feasible solution set to formulate an approximate representation of the Pareto set for feasibility-based methods. A large number of runs must be used to obtain maximum feasible solutions to ensure that a certain number of nondominated solutions can be extracted from them to ensure an accurate representation of the Pareto set. Therefore, the computational efficiency is low for this feasibility-based method.

Steuer and Sun (1995) used multiobjective linear problems to test the PSI method and they found that this method is difficult to apply to problems with more than about ten decision variables even though it has the advantage of being insensitive to the number of objectives. On the other hand, a posteriori methods, such as weighting methods and constraint methods, can obtain each point of the Pareto set. It is believed that these methods are more efficient than the feasibility-based methods such as PSI as long as there are no numerical difficulties for a particular application.

In this chapter, I present the most commonly used and generalized techniques, namely (1) the weighting method and (2) the constraint method, and the goal programming method as one of the preference-based techniques. For other methods, please refer to Miettinen (1999).

6.2.1 Weighting Method

The weighting method is used to approximate the nondominated set through the identification of extreme points along the nondominated surface. An approximation of the nondominated set is formed by "connecting" the extreme points identified. The idea of the weighting methods (Gass and Saaty, 1955; Zadeh, 1963) is to associate each objective function with a weighting coefficient and minimize the weighted sum of the objectives. In this way, the multiobjective optimization problem is transformed into a series of single objective optimization problems. The problem takes the following form.

$$\text{Optimize} \quad Z_{mult} = \sum_{i=1}^{k} w_i Z_i \tag{6.9}$$

subject to

$$h(x) = 0 \tag{6.10}$$

$$g(x) \leq 0 \tag{6.11}$$

Theory (Kuhn–Tucker conditions) tells us that as long as all the weights are greater than zero then the optimal solution of the weighted problem is a nondominated solution of the original MOP. The Pareto set can be derived by solving the number of single-objective problems of the form shown above by modifying the weighing factors w_i.

To explain this method, we return to our two-objective example.

Example 6.3: Solve the MOLP described in Example 6.1 using the weighting method.

Solution: The single-objective representation of the MOLP in Example 6.1 is given below.

$$\text{Minimize} \quad Z_{mult} = w_1 Z_1 + w_2 Z_2 \tag{6.12}$$
$$x_1, \, x_2$$

subject to

$$w_i \geq 0 \tag{6.13}$$
$$x_1 \geq 1 \tag{6.14}$$
$$2x_1 + x_2 \leq 8 \tag{6.15}$$
$$x_2 \leq 5 \tag{6.16}$$
$$x_1 - x_2 \leq 4 \tag{6.17}$$
$$w_1 \geq 0; \, w_2 \geq 0 \, x_1 \geq 0; \, x_2 \geq 0$$

where w_1, w_2 represent the weights on Z_1 and Z_2, respectively. The solution to this single-objective problem would be the optimal solution for a decision-maker whose preference for these objectives was represented accurately by these weights. Rewriting this equation in the standard form of a line gives us:

$$Z_2 = -\frac{w_1}{w_2} Z_1 + \frac{1}{w_2} Z_{mult} \tag{6.18}$$

This objective can be graphed as a line in an objective space where the slope of the line is $-w_1/w_2$, and the intercept is given by $Z = Z_{mult}/w_2$. Figure 6.7

shows the objective space representation of our two-objective problem where contours of the line are drawn for $w_1/w_2 = 0.5$ and Z is varied from 1.5 to 5.0.

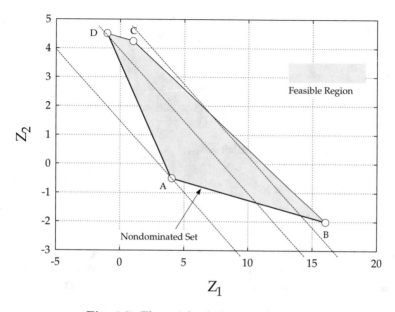

Fig. 6.7. The weighted objective function

The solution to minimization problem (6.12) can be found graphically by pushing the line given by Equation (6.18) as far to the southwest boundary as possible until the line touches the boundary of the feasible region. In this example, that solution occurs at extreme point A. Mathematically, the problem can be solved as a single-objective LP.

In this two-dimensional form it is possible to visualize that for decision problems with strictly linear equations, the solution to the weighting problem will always occur at the extreme points. Furthermore, as long as the ratio $w = w_1/w_2$ is greater than zero, the solution for minimization would be on the southwest boundary of the feasible region. Consider the two extremes $w = 0$ and $w = \infty$, which produce solutions D and B, respectively, in Figure 6.7. All other nonnegative values of w will produce solutions between these two points. The approximation of the nondominated surface would be just the straight lines that connect these extreme points. In this case, lines AD and AB form the nondominated surface.

The steps involved in this method are given below.

1. Find the individual optima for each objective. These represent the "ends" of the nondominated set.

$$\text{Optimize} \quad Z_1 \tag{6.19}$$
$$\text{Optimize} \quad Z_2 \tag{6.20}$$
$$\text{Optimize} \quad \ldots$$
$$\text{Optimize} \quad Z_k \tag{6.21}$$

2. Choose the set of nonnegative weights and solve the weighted problem.

$$\text{Optimize} \quad Z_{mult} = \sum_{i=1}^{k} w_i Z_i \tag{6.22}$$

subject to
$$h(x) = 0 \tag{6.23}$$
$$g(x) \leq 0 \tag{6.24}$$

Observe where this point is in objective space and repeat with new weights chosen to move towards regions of the nondominated set that you would like to explore (Figure 6.8). Repeat until the approximation is good enough.

Note that:

- It is important to have comparable scales for the objectives. If not, then the weighting process can be difficult as only the relative weights matter.
- It was claimed that you always get nondominated solutions from the weighted problem as long as the weights are positive. There is an important exception: when $w_i = 0$ for one or more i. In this case, you may get a dominated solution. Consider the two-objective case shown in Figure 6.9, for $w_1 = 0$, where A and B are alternate optima for Z_2 but only A is nondominated. So when applying the weighting method, if (in Step 1) you obtain an alternate optimum (multiple solutions), be sure to resolve that problem to get a nondominated one.

The noninferior set estimation (NISE) method (Cohon, 1978; Chankong and Haimes, 1983a,b) is one of the most referred weighting methods. However, there are several major disadvantages of using the weighting method.

1. Its inefficiency arising from the linear combination of objectives.

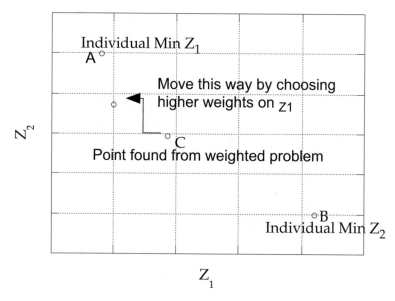

Fig. 6.8. Weighting scheme to obtain the nondominated set

2. Its difficulty to control the region of the nondominated surface on which the decision-maker is heavily favored. For example, a small change in the weighting coefficients may cause big changes in the objective vectors, and dramatically differing weighting coefficients may produce nearly similar objective vectors.
3. In addition, an evenly distributed set of weighting vectors does not necessarily produce an evenly distributed representation of the Pareto set, even if the problem is convex (Das and Dennis, 1997). This shows a lack of robustness. Furthermore, all of the Pareto optimal points cannot be found if the problem is nonconvex (Miettinen, 1999).

6.2.2 Constraint Method

The constraint methods (Haimes et al., 1971; Cohon, 1978; Zeleny et al., 1982) belong to another type of posterior methods for generating the Pareto set. The normal boundary intersection (NBI) method (Das and Dennis, 1998) and the minimization of single-objective optimization problems (MINSOOP) method (Fu and Diwekar, 2003) are examples of the constraint methods. The basic strategy is also to transform the multiobjective optimization problem into a series of single-objective optimization problems. The idea is to pick one of the objectives to minimize (say Z_1), whereas each of the others ($Z_i, i = 2, \ldots, k$) is turned into an inequality constraint with parametric right-hand sides ($\epsilon_i, i = 1, 2, \ldots, k$). In the MINSOOP method the values of $\epsilon_i, i = 1, 2, \ldots, k$ are generated using the Hammersley sequence sampling. The problem takes the

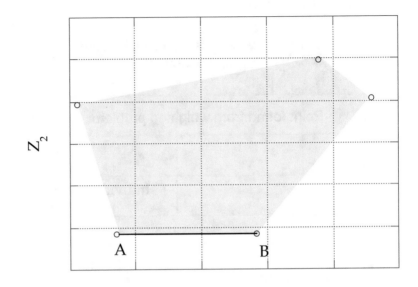

Fig. 6.9. An exception to the rule

following form.

$$\text{Optimize} \quad Z_{mult} = Z_i \tag{6.25}$$

subject to

$$\text{For Minimization} \quad Z_j \leq \epsilon_j \quad j = 1, 2, \ldots, k; j \neq i \tag{6.26}$$

or

$$\text{For Maximization} \quad Z_j \geq \epsilon_j \quad j = 1, 2, \ldots, k; j \neq i \tag{6.27}$$

$$h(x) = 0 \tag{6.28}$$

$$g(x) \leq 0 \tag{6.29}$$

Again, theory tells us that the original solution of this constrained problem is a nondominated solution of the MOP. Solving repeatedly for different values of ϵ_i the Pareto set is generated.

Example 6.4: Solve the MOLP described in Example 6.1 using the constraint method.

Solution: The single-objective representation of the MOLP in Example 6.1 is given below.

$$\text{Minimize} \quad Z_1 \ = \ 4x_1 - x_2 \qquad (6.30)$$
$$x_1, \ x_2$$

subject to

$$Z_2 \ = \ -0.5x_1 + x_2 \ \le \ \epsilon_2 \ (\text{e.g., } \epsilon_2 = 1.0) \qquad (6.31)$$
$$x_1 \ \ge \ 1 \qquad (6.32)$$
$$2x_1 + x_2 \ \le \ 8 \qquad (6.33)$$
$$x_2 \ \le \ 5 \qquad (6.34)$$
$$x_1 - x_2 \ \le \ 4 \qquad (6.35)$$
$$x_1 \ \ge \ 0; \ x_2 \ \ge \ 0$$

It can be seen from Figure 6.10 that the new constraint (6.31) reduced the feasible region. The above minimization problem gives the solution N_1. Notice that this point N_1 lies on the nondominated set of the original problem. To find the other points on the nondominated surface, the right-hand side of constraint (6.31) is changed and the problem is resolved. By connecting these points, an approximation to the Pareto set is obtained. Table 6.6 shows the points generated on the surface by changing the right-hand side of the constraint values for this problem. The approximate Pareto surface generated by this method is shown in Figure 6.11.

Table 6.6. RHS constraint values used to estimate the nondominated set

Point	ϵ_1	ϵ_2	Z_1	Z_2
B	∞	–	16	−2
D	–	∞	−1	4
N_1	–	1	2.5	1
N_2	–	−1	8.0	−1

The steps of the constraint method are given below.

1. Solve k individual optimization problems to find the optimal solutions for each of the individual objectives.
2. Compute the value of each of the k objectives for each of the individual optimal solutions. In this way, the potential range of values for each of the objectives is determined. The minimum possible value is the individual-minimization solution.

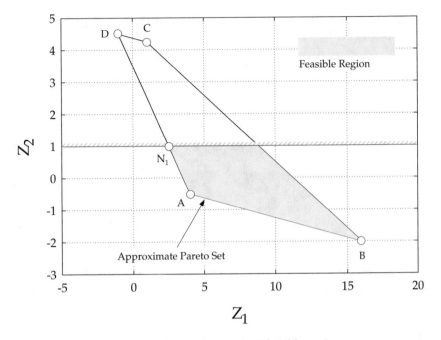

Fig. 6.10. Constraint method feasible region

3. For each objective and its range of potential values, select a desired level of resolution and divide the range into the number of intervals determined by this level of resolution. These intervals will be used as the RHS values for the constraints that will be formed for each objective.

4. Select a single objective to be optimized. Transform the remaining objectives into constraints of the form:

$$\text{For Minimization} \quad Z_j \leq \epsilon_j \quad j = 1, 2, \ldots, j \neq i, k \tag{6.36}$$

or

$$\text{For Maximization} \quad Z_j \geq \epsilon_j \quad j = 1, 2, \ldots, j \neq i, k \tag{6.37}$$

and add these new $k - 1$ constraints to the original set of constraints, where ϵ_j represents the RHS values that will be varied.

5. Solve the constrained problem setup in Step 4 for every combination of RHS values determined in Step 3. These solutions form the approximation for the nondominated surface.

There is a mapping between the weighting method and the constraint method. For details, please see Chankong and Haimes (1983a,b).

The strength of the constraint method is its ability to have better control over the exploration of the nondominated set. However, in general, this method has difficulty locating the extreme points.

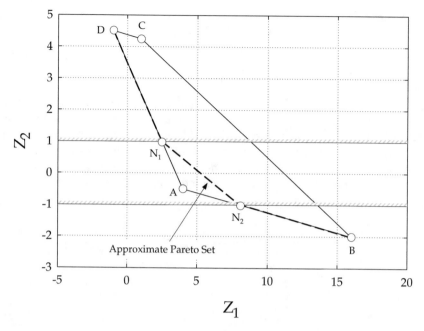

Fig. 6.11. The approximate Pareto surface using the constraint method

6.2.3 Goal Programming Method

In preference-based methods, the commonly used approaches are the value function approach and goal programming. In the value function approach, the decision-maker provides an exact representation of the value function which shows her preferences globally. Then the value function problem is readily solved using any single-objective optimization method described in earlier chapters. In goal programming, the decision-maker decides a goal for each objective and the optimization is used to minimize the total deviations from goals. Goal programming is one of the oldest (Charnes and Cooper, 1961) and most widely known methods in the preference-based category. The single-objective optimization problem in goal programming then takes the following form.

Minimize   Total deviations from the goals

$$\text{Minimize} \quad Z_{goal} = \sum_{i=1}^{k} |(Z_i - G_i)| \qquad (6.38)$$

subject to

Original constraints

$$h(x) = 0 \tag{6.39}$$

$$g(x) \leq 0 \tag{6.40}$$

This formulation involves defining negative (δ^-) and positive deviation (δ^+) from the goals (G_i) and solving the optimization problem for the original decision variables, and also for deviation variables as shown below. The advantage of using this formulation for MOLP is that the resultant goal programming problem is an LP, as it does not include the nonlinear absolute value function. Goal programming was originally developed for MOLP problems, as can be evident from this formulation.

$$\text{Minimize} \quad Z_{goal} = \sum_{i=1}^{k} \delta_i^+ + \delta_i^- \tag{6.41}$$

$$x, \delta_i^+, \delta_i^-$$

subject to

$$Z_i - G_i = \delta_i^+ - \delta_i^- \quad i = 1, 2, \ldots, k \tag{6.42}$$

$$h(x) = 0 \tag{6.43}$$

$$g(x) \leq 0 \tag{6.44}$$

$$\delta_i^+ \geq 0; \delta_i^- \geq 0$$

The following two objective problem explains this concept:

Example 6.5: Solve the MOLP described in Example 6.1 using the goal programming method. The goal is to reduce the cost to -5 and emission function Z_2 to -5.

Solution: The single-objective goal programming representation of the MOLP in Example 6.1 is given below.

$$\text{Minimize} \quad Z_{goal} = \sum_{i=1}^{2} \delta_i^+ + \delta_i^- \tag{6.45}$$

$$x_1, \ x_2$$
$$\delta_1^+, \delta_2^+, \delta_1^-, \delta_2^-$$

subject to

$$Z_1 - (-5) = \delta_1^+ - \delta_1^- \tag{6.46}$$

$$Z_2 - (-5) = \delta_2^+ - \delta_2^- \tag{6.47}$$

$$x_1 \geq 1 \tag{6.48}$$

$$2x_1 + x_2 \leq 8 \qquad (6.49)$$
$$x_2 \leq 5 \qquad (6.50)$$
$$x_1 - x_2 \leq 4 \qquad (6.51)$$
$$x_1 \geq 0; \quad x_2 \geq 0$$

$$\delta_1^+ \geq 0; \ \delta_1^- \geq 0$$
$$\delta_2^+ \geq 0; \ \delta_2^- \geq 0$$

The objective space for the above problem is shown in Figure 6.12 as the decision space remained no longer two-dimensional. The figure also shows the compromise solution obtained using the above formulation. The solution to the above LP is found to be $x = (-1.0, 4.0)$ where the deviational variables are $\delta^+ = (4.0, 9.0)$ and $\delta^- = (0.0, 0.0)$. The goal was to reach $Z = (-5, -5)$.

There are a number of variations of goal programming used in the literature. For example, depending on the decision-maker's preference and priorities one can assign different weights to the deviations to take the weighted average deviation as the objective function. Lexicographic ordering uses weights different by an order of magnitude thus driving the high-priority objective to its goal at the expense of other objectives. One-sided goal programming does not care about either positive or negative deviations and sets the appropriate priority weights to zero.

Goal programming is a popular method due to its age. Goal setting is also an understandable concept. However, this is also a major drawback of this method as it is not a trivial task to set goals. If the goals are not set properly, the solution may not be in the Pareto set. Furthermore, this method is not an appropriate method if it is desired to obtain a trade-off.

6.3 Hazardous Waste Blending and Value of Research

In the earlier chapters, we have looked at the nuclear waste blending problem formulated as LP, NLP, and MINLP by progressively including more information about the model and/or adding more decision variables. In the last chapter, we considered uncertainties associated with the models as well as data in terms of probabilistic distribution functions. Sources of uncertainty have important technical implications and reflect significant aspects of the decision-making process. In this chapter, the policy dimension of the problem is added to the problem through progressive extensions to the objective functions to include implications of uncertainty. The details of this case study can

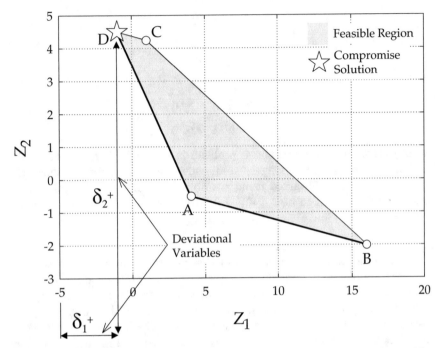

Fig. 6.12. The goal programming compromise solution and deviational variables

be found in Johnson and Diwekar (1999, 2001). This analysis also introduces a new criterion called value of research and illustrates the usefulness of using the multiobjective framework.

Previous efforts to address the blending problem (Narayan et al., 1996; earlier chapters), for instance, have focused solely on the cost of vitrification (i.e., minimization of frit, which is equivalent to minimizing glass volume and, hence, disposal costs). Although these efforts have included a representation of the different sources of uncertainty inherent in the blending problem, they have not recognized reduction of this uncertainty as an important objective in itself. Significant policy dimensions related to the vitrification process have thus been ignored. The augmented framework described in the next section facilitates a comparative analysis of the resulting trade-offs. Although the case study illustrates the concepts on MOP, more emphasis is placed on the implications of uncertainty and less on the accuracy of the Pareto surface.

For this illustrative analysis, we have chosen a subset of 12 tanks divided evenly into three blends. Initial remediation efforts at the Hanford site focus on a limited number of storage tanks; the criticality of a tank's condition (its position on a "watch list") and the compatibility of its contents with the demands of vitrification govern the selection process (Gephart and Lundgren, 1995).

6.3.1 Variance as an Attribute: The Analysis of Uncertainty

Sources of uncertainty in the blending problem have important technical implications and reflect significant aspects of the policy-making process surround-
ing Hanford's remediation efforts. Expansion of the objective from minimization of frit to include different sources of variation represents an important
methodological development, one that capitalizes on the STA-NLP framework
to make stochastic optimization a more robust mathematical technique and a
more useful tool. This section illustrates the multiobjective STA-NLP framework's advantages through progressive extensions to the blending problem's
objective function. The base analysis is presented first, and results accom-
pany the description of each extension. The following section discusses the
corresponding implications.

6.3.2 Base Objective: Minimization of Frit Mass

In Chapter 4, we have used the SA-NLP framework for the deterministic
analysis of 21 tanks. Similar deterministic analysis of this 12 tank blending
problem yields a basis for comparison.

Table 6.7 presents the frit requirements from these preliminary solution
schemes, based on the base case objective: minimization of frit mass.

Table 6.7. Frit requirements as determined by basic solution techniques

Solution method	Required frit mass (kg)
Worst case (no blending)	13410
Best case (one blend of all tanks)	9839
Deterministic solution (SA-NLP)	11,161
Single objective stochastic solution (STA-NLP)	10,060

Note that the difference between the deterministic and stochastic solutions,
the value of the stochastic solution (VSS) is 1101 kg.

Figure 6.13 presents histograms (generated using LHS) of the frit mass
requirements and the corresponding proportion of constraint violations when
the individual waste mass fraction sample values are used with the tank-blend
configuration derived from their expected value.

6.3.3 Robustness: Minimizing Variance

It can be shown that the variance in objective (e.g., var_{frit}) is a measure of
the STA-NLP algorithm's robustness and can be used as another objective
in the exercise. The magnitude of var_{frit}, for instance, directly affects the
probability that the NLP/glass property constraints are met when actual (i.e.,

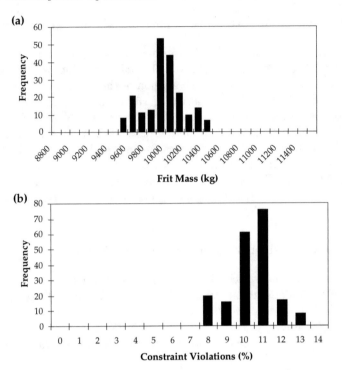

Fig. 6.13. Distribution of (**a**) frit masses and (**b**) constraint violations for the base Case objective: minimize frit mass

sample) values of the waste component mass fractions are used in place of their sample mean. Hence, there is a desire to keep this source of variation as low as possible. Including variance as an attribute produces the following objective.

$$\text{Minimize} \quad Z = \sum_{i=1}^{n} f^{(i)} + w_1 var_{frit} \tag{6.52}$$

Note that "frit mass" ($\sum_{i=1}^{n} f^{(i)}$) in Equation (6.52) is an expected value, and that var_{frit} has been scaled so that both terms have the same order of magnitude. The variance of frit mass is used instead of its standard deviation. Although portfolio theory optimization frameworks employ the latter, quality control models like the loss function—which, such as the blending problem, are characterized by a nonlinear domain—feature variance. This approach is derived from Taguchi's robust design methodology (Kacker, 1985). The decrease in NLP constraint violations (produced by using the waste component mass fraction sample values rather than their means, for which the constraints are always met) can be examined as a function of increasing frit mass (Table 6.8).

Table 6.8. The balance between expected value and variance minimization

w_1	Frit Mass (kg)	$\sqrt{var_{frit}}$(kg)	Constraints violated
0.50	10,255	293	11
1.0	10,075	190	6
2.0	10,647	138	2
4.0	11,558	118	0

The optimization framework illustrated here and extended in the following section, unlike formal multiattribute decision analysis, is qualitative in nature. A specific meaning, for instance, cannot be attached to w_1. The highly nonconvex, nonlinear, and discrete character of the blending problem precludes the assessment of "weights" customary with multiobjective optimization algorithms. The parsimonious choice of an additive objective function in Equation (6.52), as well as the selection of units and scaling factors for its terms, determines the trade-offs produced by variation of w_1. Attention therefore should focus not on the w_1 term, but on the relative changes in frit mass, its variance, and the number of constraint violations that parametric adjustments of w_1 produce. The scale of w_1 values explored was selected iteratively in order to observe the complete range of the criteria of interest, in this case, constraint violations (which decreased from a maximum of 11% to 0; see Table 6.8). An illustration clarifies this caveat. The decrease in NLP constraint violations (produced by using waste component mass fraction sample values rather than their means, for which the constraints are always met) can be examined as a function of increasing frit mass (Table 6.8).

As shown in Table 6.8, an increase in expected frit mass of approximately 4% yields an 80% reduction in constraint violations, an important factor for decision making at Hanford. Again, the corresponding increase in w_1 (from 0.5 to 2.0) that produces this result does not have a meaningful interpretation. Nor is the trade-off between frit mass and its variance constant over the range of w_1. Instead, the value of this framework lies in its ability to explore trade-offs in terms of relative changes between different factors relevant to the blending problem (the compromise between increasing frit mass and decreasing constraint violations). The following section builds on this flexibility.

6.3.4 Reducing Uncertainty: Minimizing the Time Devoted to Research

We have known that better characterization of the Hanford tank wastes and glass property models would result in lower frit requirements. The decrease in frit mass that a reduction in uncertainty yields, however, must be weighed against the opportunity costs of pursuing the objective. The extensions introduced here facilitate this analysis: an examination of the trade-offs inherent in allocating scarce resources to reduce uncertainty. Such extensions are generalizable to similar situations, which are ubiquitous, especially in nuclear waste management where the long-lived nature of the waste creates large uncertainties.

The analysis rests on a key assumption: that time spent on research increases understanding, and therefore decreases variation in quantitative estimates derived from this knowledge. Research activities introduce their own costs and risks; hence, time spent learning and experimenting need to be minimized. Reducing uncertainty is profitable, however the time required to achieve a reduction tempers the benefit. An augmented objective captures this trade-off:

$$\text{Minimize} \qquad \text{processing and disposal costs}$$
$$\text{and time devoted to reducing uncertainty} \qquad (6.53)$$

As before, processing and disposal costs are represented by the expected frit mass and its associated variance. As illustrated below, the sampling variance of the tank waste component mass fractions and the uncertainty in the empirical glass property models (through its effect on constraint width) serve as proxies for resources devoted to reducing uncertainty. The expanded blending objective therefore attempts to minimize frit mass, but—beyond finding an optimal tank-blend assignment—limits the extent to which improved waste characterization and more accurate glass property models contribute to this goal. Research efforts, for instance, could aim at easing the constraint bounds via improvements in the glass property models' prediction error; as the constraints govern frit requirements, less conservative limits in an optimization framework translate into the need for a smaller safety margin and therefore less frit. Proportional relaxation of the constraints, however, carries an increasing penalty: the time and opportunity costs of related research activities.

To understand how the augmented blending objective captures this trade-off in mathematical terms, note that the type of investigation relevant to the problem will exhibit diminishing marginal returns as uncertainty declines nonlinearly with time spent on research. For characterization of the tank waste components, an exponential relationship between sampling variance and time provides an adequate first-order functional approximation of this nonlinear dependence:

$$uncertainty\ in\ waste\ composition\ <=>\ var_{samp}\ \propto\ \exp(-time)$$

or

$$time\ \propto\ -\ln(var_{samp}) \qquad (6.54)$$

A similar relationship holds for the constraint width term. Note, however, that the width of the constraint bounds varies inversely with the prediction error of the empirical glass property models.

$$time\ \propto\ -\ln(prediction\ error)\ \propto\ -\ln(Constraint\ width)^{-1}$$
$$-\ln(Constraint\ width)^{-1}\ =\ \ln(Constraint\ width) \qquad (6.55)$$

Once again, minimization of resources devoted to reducing uncertainty, taken by itself, is captured in this model by seeking tank-blend combinations with larger input sampling variances and prediction errors (i.e., narrower constraint bounds). Excessive values, however, are simultaneously penalized through their detrimental effect on the expected frit mass and its associated sample variance. The optimum reflects a balance in this trade-off: a low frit mass and var_{frit}, with moderate values of var_{samp} and the constraint widths. Combining these arguments, the augmented blending objective (multiobjective) becomes

$$Minimize \quad Z_{mult} = \sum_{i=1}^{n} f^{(i)} + w_1 var_{frit} - w_2 \ln\left(\sum var_{samp}\right)$$
$$+ w_3 \ln\left(\sum constraint\ width\right) \tag{6.56}$$

Table 6.9 presents results of a parametric analysis of changes in the weights w_i, similar to those presented earlier in Table 6.8. Table 6.10 presents a qualitative summary of these results, the implications of which are discussed in the following section.

6.3.5 Discussion: The Implications of Uncertainty

The results from the previous section have implications for optimization under uncertainty in general, and the blending problem in particular. The importance of attending to matters of robustness, for instance, is apparent in Table 6.9; as reduction in frit variance is emphasized (i.e., as w_1 increases), the proportion of constraint violations decreases to zero and the frit masses become clustered more tightly around their mean. The expected frit mass, however, is uniformly higher with fewer constraint violations, a compromise that illustrates the balance between reducing the volume of immobilized waste and increasing the probability that vitrification succeeds. The multiobjective STA-NLP framework facilitates such an analysis.

Table 6.9. Parametric results of the trade-off in reducing sources of variation

w_1 (var_{frit})	w_2 (var_{samp})	w_3 (c.width)	$E[fritmass]$ (kg)	$\sqrt{var_{frit}}$ (kg)	% const. violated
0	0	0	10,255	293	11
1	1	1	10,932	214	8
1	1	3	11,061	192	9
1	3	1	9931	478	5
1	3	3	9971	337	5
3	1	1	10,815	184	2
3	1	3	10,050	175	3
3	3	1	12,008	245	2
3	3	3	11,217	230	3

Table 6.10. A qualitative summary of the trade-off in reducing sources of variation

Focus of research	E[frit mass]	var_{frit}	% constraint violations
Robustness/ minimization of frit variance	Increases	Decreases	Decreases
Minimize time for tank characterization	Increases	Increases	Increases
Minimize time for improving property models	No change	Decreases	Increases

Beyond providing a framework in which similar trade-offs may be assessed, however, policy-makers desire answers to specific questions. Note that the most important question concerning the blending problem is not minimization of frit mass, per se; indeed, consideration of the entire context of Hanford's remediation effort and the politics of radioactive waste disposal may decrease the priority of reducing frit mass, especially on the order of the savings seen above (compare the values in Tables 6.8 and 6.9). Expanding the problem scale by including a larger subset of tanks, however, would increase the importance of lowering the frit mass; a greater number of tanks would also take better advantage of blending, and result in more impressive reductions of frit. More important are questions concerning uncertainty. To what extent is imperfect information acceptable, and where should scarce resources be allocated to leverage the impact of this narrow part of Hanford's waste remediation effort on the whole of its strategy? Not all sources of uncertainty, after all, are significant. In pursuing answers to such questions, multiobjective optimization works more as an exploratory tool than as a means of providing a "one best" solution. The preceding analysis illustrates this capacity. An examination of the constraints, for instance, reveals that the crystallinity requirements are most consistently violated, with the P_2O_5 solubility limit and the component bound on Al_2O_3 frequently exceeded as well (see also Hopkins et al. (1994)). Resources would be profitably allocated to reducing the error in the corresponding glass property models ahead of additional waste pretreatment efforts designed to mitigate the effects of these limiting components. Perhaps more significant is the ability to determine what sources of uncertainty need to be reduced and which, in contrast, may be tolerated. The relationship, however, among the required frit mass, its variance, and constraint violations is complicated. As described, the constraint width terms enter the objective function as penalties; considered in isolation on their effects on frit mass, larger values are desired (i.e., the devotion of resources to reducing uncertainty is minimized). The "benefit" of greater uncertainty in the tank waste distributions and glass property models, however, is balanced by its detrimental effect on the expected frit mass and its variance.

Results from the preceding section illustrate this relationship. As the sampling variance term increases (i.e., characterization of the tank wastes is less complete), variation in frit mass increases and constraint violations become more numerous. This effect is not surprising: a change in the variance of the waste component sampling distributions leads to a proportionate shift in the frit variance and a similar impact on both the average frit mass and extent of constraint violations. Compared to these changes, however, the variance in frit mass decreases and the percentage of constraint violations increases with the constraint width uncertainty (compare parts (a) and (b) of Figures 6.14 and 6.15 which illustrate the effect of increasing w_3); greater uncertainty in the glass property models translates into narrower constraint bounds, and a smaller range across which frit requirements may vary without consequence. This impact on process robustness leads to the conclusion that improvements in the glass property models should come before efforts to reduce uncertainty in the tank waste composition. The presence of nonlinearities in the glass property models (constraints)—which inflate the effects of variance—provides one explanation for the pattern of these results.

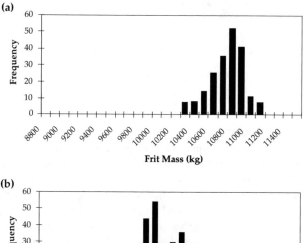

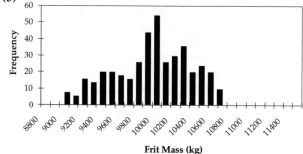

Fig. 6.14. Distribution of frit masses for the different objectives (**a**) $w_3 = 1.0$, (**b**) $w_3 = 3.0$

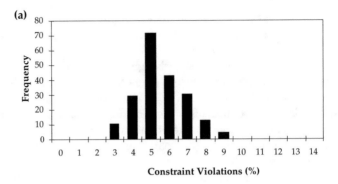

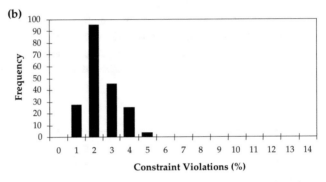

Fig. 6.15. Distribution of constraint violations for different objectives: (**a**) $w_3 = 1.0$; (**b**) $w_3 = 3.0$

The case study illustrated the usefulness of multiobjective optimization analysis. The data, formulation, and computer code for this case study can be found online on Springer website with the book link. A new paradigm called value of research is introduced to provide a policy dimension to the traditional optimization problem. This new paradigm is based on a key assumption: that time spent on research increases understanding, and therefore decreases variation in quantitative estimates derived from this knowledge. Research activities, however, introduce their own costs and risks; hence, time spent learning and experimenting needs to be minimized. Although reducing uncertainty is profitable, the time required to achieve a reduction tempers the benefit. The qualitative nature of the value of research objective in contrast to the quantitative benefits of environmental policy programs demands the multiobjective framework with generating methods such as the weighting method used here.

6.4 Sustainable Mercury Management: A Multiobjective Optimization Problem

The discharge of mercury to the watershed, although below TMDL limit, is still harmful to humans, and consequently associated with some health care cost. From a social perspective, it is preferable to reduce the discharge as much as possible to avoid any harmful health effects. The inclusion of health care cost results in a multiobjective optimization problem and is discussed below.

6.4.1 Health Care Cost

The bioaccumulative nature of mercury and its slow dynamics make the long term effects of mercury exposure important. Hence, it is essential to account for such effects while quantifying health care costs. Majority of mercury accumulates in the food chain as methyl mercury. Therefore, quantification of the health care costs based on methyl mercury concentration is most appropriate. IRIS (Integrated Risk Information System) database reports methyl mercury reference dose for chronic oral exposure (RfD), which is the highest dose of methyl mercury without any harmful effects. However, since the TMDL for the Savannah River is developed on the discharge of total mercury to the watershed, not just methyl mercury, the model needs a quantifying measure based on total mercury. IRIS database does not report the RfD value for mercury (elemental). Given these considerations, this case study quantifies the health care costs through LC50 (Lethal Concentration 50%) value for mercury. LC50 is defined as the concentration of a toxic substance (mercury) at which 50% of the population exposed to it dies within a certain time. LC50 value of a substance is often used to quantify its harmful exposure effects.

The health care cost is a function of the final overall mercury discharge. This discharge value is used to calculate average mercury consumption by humans. This value is then compared with the mercury consumption rate, at which 50% of the human population will die (calculated using the LC50 value). The comparison gives an approximate estimate of the human mortality rate due to the discharge of mercury in the watershed.

The presence of mercury in water is dangerous to humans primarily through fish consumption. During the development of TMDL for Savannah watershed, it has been established that a WQS of 2.8 ng/liter leads to safe mercury concentration in fishes (Hg_{safe}). Assuming a linear relationship between WQS and average mercury concentration in fishes, knowing the actual WQS after compliance (WQS_{final}) gives the average fish tissue mercury concentration after compliance. Then, knowing the average fish consumption per person per day in the watershed (F_{avg}), the total mercury intake by an individual is computed. Hg_{safe} is given as 0.4 mg/Kg, while F_{avg} is 17.5 g per person per day. Thus, mercury intake per person per day in grams is given as:

$$\text{Fish mercury intake} = \frac{WQS_{final}}{WQS}.Hg_{safe}.F_{avg}.10^{-6} \qquad (6.57)$$

This mercury intake rate is to be compared with the rate resulting in 50% mortality. Since the LC50 value for humans is not available, this work assumes that the value for the fish being consumed by humans in the watershed is close to the value for humans. The most commonly consumed fish in Savannah watershed is largemouth bass. According to the Pesticide Action Network (PAN) database, mercury LC50 value for largemouth bass is $50\,\mu g/liter$. This value is, therefore, used in this analysis. It is assumed that chronic exposure to mercury is through the consumption of contaminated water by humans. If W_{avg} is the average water consumption per person per day, then the consumption rate leading to 50% mortality can be computed. For the Savannah River watershed, W_{avg} is 2 liters per person per day. The 50% leather consumption rate per person per day in grams is given as:

$$\text{Water mercury intake} = LC50.W_{avg}.10^{-6} \qquad (6.58)$$

Let P be the population in the watershed affected by mercury pollution, and C_{health} be the health care compensation per mortality. Then the total health care cost for the watershed is given as:

$$\text{Health Cost} = \left(\frac{WQS_{final}}{WQS}.Hg_{safe}.F_{avg}.\frac{1}{LC50.W_{avg}} \right)\left(\frac{P}{2} \right).C_{health} \qquad (6.59)$$

For the Savannah River watershed case,
Population affected by the consumption $= P = 10000$, and
Compensation for the health cost $= C_{health} = \$ 3$ Million per person

The affected population in the watershed is based on the data reported by the US Census Bureau. The compensation amount is estimated from USEPA. Using various published studies to quantify mortality risk to humans due to pollution, this technical report by the USEPA proposes $\$ 6.1$ Million as the value of statistical life (1999 Dollars). However, it is mentioned that for slow bioaccumulative chemicals like mercury showing long term effects, the perceived value is less. Hence, for this analysis, the compensation value per human mortality is taken as $\$ 3$ Million.

The data presented here helps in the quantification of health care costs for the Savannah River watershed. There are various simplifying assumptions due to the lack of sufficient data. It should, however, give an approximate estimate that can later be refined with the availability of new information. The next section presents model results for the Savannah River watershed.

6.4.2 The Multiobjective Optimization Formulation

The multiobjective optimization problem formulation using the weighting method, with health care cost as a part of the objective, is given by (6.60)–

(6.67). The objective function of MINLP formulation has been appended to include the health care cost in (6.60). This health care cost is calculated as a function of the mortality rate, as mentioned before. Accordingly, additional equations are included in the problem to calculate the mortality rate. red_i^{final} is the final reduction achieved by point source i. All other symbols have their previously assigned meanings.

Objective:

$$\text{Minimize} \quad \sum_{i=1}^{N}\sum_{j=1}^{M} f_j(\phi_j, D_i).\, b_{ij} + W_{health}.\text{Mortality}.C_{health} \tag{6.60}$$

Constraints:

$$t_{ii} = 0 \qquad \forall i = 1, ..., N \tag{6.61}$$

$$red_i \leq \sum_{j=1}^{M} q_j.D_i.\, b_{ij} + \sum_{k=1}^{N} t_{ik} - r\sum_{k=1}^{N} t_{ki} \qquad \forall i = 1, ..., N \tag{6.62}$$

$$P_i \geq \sum_{j=1}^{M} b_{ij}.TC_j.D_i + F\Big(\sum_{k=1}^{N} t_{ik} - \sum_{k=1}^{N} t_{ki}\Big) \qquad \forall i = 1, ..., N \tag{6.63}$$

$$red_i^{final} = \sum_{j=1}^{M} q_j.D_i.\, b_{ij} + \sum_{k=1}^{N} t_{ik} - r\sum_{k=1}^{N} t_{ki} \qquad \forall i = 1, ..., N \tag{6.64}$$

$$WQS_i = \frac{(red_i - red_i^{final})}{D_i} \qquad \forall i = 1, ..., N \tag{6.65}$$

$$WQS_{final} = \frac{\sum_{i=1}^{N} WQS_i.D_i}{\sum_{i=1}^{N} D_i} \tag{6.66}$$

$$\text{Mortality} = \Big(\frac{WQS_{final}}{WQS}.Hg_{safe}.F_{avg}.\frac{1}{LC50.W_{avg}}\Big)\Big(\frac{P}{2}\Big) \tag{6.67}$$

where W_{health} is the weight given to the health care cost. Tables 6.11 and 6.12 present the solution of this multiobjective problem for different values of weight for TMDL 32 using the MINLP solution technique presented in Chapter 4. It can be seen that as the weight of the health care cost increased, total trading decreases, and the solution approaches to technology, the only solution at the highest value of the weight. Health care cost is minimum for this weight. If one looks at the optimal technology distributions shown in Table 6.12, the distribution changes significantly for large difference in weights. As the weights increase, the most expensive technology (B) is preferred for both technology only and trading options.

Table 6.11. Savannah River watershed Trading: solution for the Multiobjective MINLP

Weight	Cost Technology	Trading	Saving	Total trading	Health care cost Technology	Trading
50	130452235.1	93673310.48	36778924.57	0.115079227	869496.3247	1034449.433
90	140261054	89699583.8	50561470.25	0.234616535	754686.413	1023061.677
100	140261054	89699583.8	50561470.25	0.234616535	754686.413	1023061.677
110	140261054	92058567.86	48202486.19	0.199095074	754686.413	1001193.029
125	140508365.4	93392704.4	47115661.05	0.190978968	752644.2849	989746.5313
150	148064638.6	113525743.3	34538895.25	0.095947367	701511.352	846255.7694
175	148064638.6	116469511.9	31595126.67	0.076056035	701511.352	829416.9824
200	153621723.9	127452952.6	26168771.21	0.073393855	671650.1444	770648.4206
250	162706049.6	147612341.1	15093708.51	0.029483739	628064.9002	677403.85
350	171659525.4	169734139.7	1925385.724	0.002873991	596619.7411	601391.8552
500	171659525.4	171585412.8	74112.6215	5.18E-05	596619.7411	596690.702

Table 6.12. Savannah river watershed trading: solution technology distribution for the multiobjective MINLP

	Technology distribution					
	Technology			Trading		
Weight	A	B	C	A	B	C
50	9	12	8	6	1	1
90	8	13	8	3	2	0
100	8	13	8	3	2	0
110	8	13	8	3	2	0
125	7	14	8	3	2	0
150	5	16	8	4	3	0
175	5	16	8	5	3	0
200	4	17	8	3	5	0
250	2	19	8	3	9	0
350	0	28	1	0	24	0
500	0	28	1	0	27	0

6.5 Summary

Multiobjective optimization is also referred to as a vector optimization because it deals with a vector of objectives. Multiobjective programming can be thought of as a set of methodologies for generating a preferred solution or a range of optimum solutions to a decision problem. The form of equations determines the particular type of MOP, such as MOLP for linear programming, MONLP for nonlinear programming, and so on. In 1950, Kuhn-and-Tucker presented the theory for MOP. Since then the field has grown tremendously and there are a large number of solution methods available to solve problems. The idea of nondominance forms the basis for most of the methods. There are generating methods where the complete nondominated solutions (the Pareto set) or all feasible solutions are generated. The most widely used methods among these categories are weighting methods and constraint methods. Goal programming is one of the oldest and most commonly used preference-based techniques. In problems involving uncertainties, the MOP framework, with a new paradigm called value of research, can help in identifying crucial sources of uncertainties.

Exercises

6.1 A dietitian is planning a menu that consists of three main foods: A, B, and C. Each ounce of food A contains 3 units of fat, 1 unit of carbohydrates, 4 units of protein, 2 units of cholesterol, 1 unit of Vitamin B, and 6 units of Vitamin D. Each ounce of food B contains 6 units of fat, 2 units of carbohydrates, 8 units of protein, 3 units of cholesterol, 3 units of Vitamin B, and 3 units of Vitamin D. Each ounce of food C contains 2 units of fat, 5 units of carbohydrates, 2 units of protein, 1 unit of cholesterol, 2 units of Vitamin B, and 5 units of vitamin D. The dietitian wants the meal to provide at least 36 units of carbohydrate, 48 units of protein, 18 units of Vitamin B, and 30 units of Vitamin D. If the prices for foods A, B, and C are 15, 25, and 20 cents per ounce, respectively, then how many ounces of each food should be served to minimize the cost of the meal, minimize the fat and cholesterol contained in the meal, and satisfy the dietitian's requirements?

6.2 Consider a refinery that produces three types of motor oil: Standard, Extra, and Super. The selling prices are $9.00, $13.00, and $19.00 per barrel, respectively. These oils can be made from three basic ingredients; crude oil, paraffin, and filler. The costs of the ingredients are $19.00, $9.00, and $11.00 per barrel, respectively. Company engineers have developed the following specifications for each oil.

–   Standard—60% paraffin, 40% filler
–   Extra—at least 25% crude oil and no more than 45% paraffin
–   Super—at least 50% crude oil and no more than 25% paraffin

The CO_2 emissions of Standard, Extra, and Super oils are 13.0, 11.8, and 8.0 units per barrel. With a supply capacity of 110, 90, and 70 thousand barrels per week for crude oil, paraffin, and filler, what should be blended in order to maximize profits as well as satisfy the requirements of the EPA to minimize the CO_2 emissions from all the products of this industry? Solve the problem using the goal programming approach, if the goals are to have a profit greater than or equal to $5 per barrel and CO_2 emissions less than or equal to 10 units per barrel.

6.3 Solve the following two-objective NLP problem using (a) a weighting method and (b) a constraint method with 20 parameters (w_i and ε_i) [from Das and Dennis (1998).

$$\min_{x} \begin{bmatrix} Z_1(x) = x_1^2 + x_2^2 + x_3^2 + x_4^2 + x_5^2 \\ Z_2(x) = 3x_1 + 2x_2 - \frac{x_3}{3} + 0.01 \times (x_4 - x_5)^3 \end{bmatrix} \qquad (6.68)$$

subject to

$$x_1 + 2x_2 - x_3 - 0.5x_4 + x_5 = 2 \qquad (6.69)$$

$$4x_1 - 2x_2 + 0.8x_3 + 0.6x_4 + 0.5x_5^2 = 0 \qquad (6.70)$$

$$x_1^2 + x_2^2 + x_3^2 + x_4^2 + x_5^2 \leq 10 \qquad (6.71)$$

6.4 There is an isothermal batch reactor in which the following series reaction occurs.

$$A \to \quad k_1 \quad B \to \quad k_2 \quad S \qquad (6.72)$$

where A, B, and S are reactant, desired product, and side product, respectively, and k_1 and k_2 are reaction constants. The differential equations for each chemical are expressed as follows.

$$\frac{dC_A}{dt} = -k_1 C_A \qquad (6.73)$$

$$\frac{dC_B}{dt} = k_1 C_A - k_2 C_B \qquad (6.74)$$

$$\frac{dC_S}{dt} = k_2 C_B \qquad (6.75)$$

where C is the concentration in moles per liter and t is reaction time in hours. In this reaction we want to maximize the yield and selectivity of product B. Yield (ξ) and selectivity (y) are defined as follows.

$$Z_1 = \xi = \frac{C_B}{C_{A,\ initial}} \quad \sim \text{Production rate} \qquad (6.76)$$

$$Z_2 = y = \frac{C_B}{C_B + C_S} \quad \sim \text{Purity} \qquad (6.77)$$

Formulate the two-dimensional multiobjective optimization problem and solve this problem when k_1 is 0.1/h, k_2 is 0.01/h, $C_{A,\ initial}$ is 100 moles per liter, and the reaction time is 50 h.

6.5 The Reynolds Manufacturing Company manufactures rings and bracelets. The production of a ring requires 1 unit of cutting, 2 units of grinding, 2 units of polishing, 1 unit of packaging, and an initial capital investment of $100 per ring. The units of workers exposure time to the hazards are given by

$$t_{ring} = \frac{200}{(x + 1000)^{\frac{2}{3}}} \qquad (6.78)$$

where x is the number of rings produced per day. A bracelet requires 2 units of cutting, 2 units of grinding, 3 units of polishing, 2 units of packaging, an initial investment of $90 per bracelet, and workers' exposure time to the hazards can be obtained by

$$t_{bracelet} = \frac{1000}{(y + 8000)^{0.6}} \qquad (6.79)$$

where y is the number of bracelets produced per day. The unit cost of cutting is $2, grinding $3, polishing $3.5, and packaging $1. The selling price of rings is given by the following equation according to the production rate of the company.

$$P_{ring} = \frac{1500}{(x + 1000)^{\frac{1}{3}}} \qquad (6.80)$$

where x is the number of rings produced per day. While the selling price of bracelets is given by

$$P_{bracelet} = \frac{5000}{(y + 9000)^{0.4}} \qquad (6.81)$$

If the availability of units of cutting is limited to 6000, units of grinding 3800, units of polishing 4900, and units of packaging 3400 per day and a positive daily profit of $8000 is needed to keep the company running, how many rings and bracelets should be manufactured in order to maximize profits and minimize workers' exposure time to the hazards?

Bibliography

Benson H. P. (1991), Complete efficiency and the initialization of algorithms for multiple objective programming, *Operational Research Letters*, **10(4)**, 481.

Birge J. R. (1997), Stochastic programming computation and applications, *INFORMS journal on computing*, **9(2)**,111.

Buchanan J. T. (1986), Multiple objective mathematical programming: A Review, *New Zealand Operational Research*, **14(1)**, 1.

Carmichael D. G. (1981), *Structural Modeling and Optimization*, Ellis Horwood, Chichester, UK.

Chankong V., and Y. Y. Haimes(1983a), Optimization-based methods for multiobjective decision-making: An overview, *Large Scale Systems*, **5(1)**, 1.

Chankong V., and Y. Y. Haimes (1983b), *Multiobjective decision making theory and methodology*, Elsevier Science, New York.

Chankong V., Y.Y. Haimes, J. Thadathil, and S. Zionts (1985), Multiple criteria optimization: A state of the art review in *Decision Making with Multiple Objectives*, Lecture Notes in Economics and Mathematical Systems 242, Edited by Y.Y. Haimes, V. Chankong, Springer-Verlag, New York, 36.

Charnes A. and W. Cooper (1961), *Management Models and Industrial Application of Linear Programming*, Wiley, New York.

Cohon J. L. (1978), *Multiobjective Programming and Planning*, Academic Press, New York.

Cohon J. L. (1985), Multicriteria programming: Brief review and application, in *Design Optimization*, J. S. Gero (ed.), Academic Press, 163.

Cohon J., G. Scavone, and R. Solanki (1988), Multicriterion optimization in resources planning in *Multicriteria Optimization in Engineering and in the Sciences*, W. Stadler (ed.), Plenum Press, New York, 117.

Das, J. Dennis A closer look at drawbacks of minimizing weighted sums of oobjective for pareto set generation in multicriteria optimization problems Structural Optimization, 14 (1997), pp. 63–69

Das I. and J. E. Dennis (1998), Normal-boundary intersection: A new method for generating the Pareto surface in nonlinear multicriteria optimization problems, *SIAM Journal on Optimization*, **8(3)**, 631.

Eschenauer H. A. (1988), Multicriteria optimization techniques for highly accurate focusing systems, in *Multicriteria Optimization in Engineering and in the Sciences*, W. Stadler (ed.), Plenum Press, New York, 309.

Evans G. W. (1984), An overview of techniques for solving multiobjective mathematical programs, *Management Science*, **30 (11)**, 1268.

Fu Y. and U. Diwekar (2003), An efficient sampling approach to multiobjective optimization, *Annals of Operations Research*, **132**, 109.

Gass S. and T. Saaty (1955), The computational algorithm for the parametric objective function, *Naval Research Logistics Quarterly*, **2**, 39.

Gephart R. E. and Lundgren, R.E. (1995), *Hanford tank clean up: A guide to understanding the technical issues*, Report BNWL-645, Richland, Pacific Northwest Laboratory, WA.

Gerber M. S. (1998), *Historical generation of Hanford site wastes*, Report WHC-SA-1224, Richland, Pacific Northwest Laboratory, WA.

Haimes Y. Y., L. S. Lasdon and D. A. Wismer (1971), On a bicriterion formulation of the problems of integrated system identification and system optimization, *IEEE Transactions on Systems, Man, and Cybernetics*, **1**, 296.

Hopkins D. F., M. Hoza, and C. A. Lo Presti (1994), *FY94 Optimal Waste Loading Models Development*, Report prepared for U.S. Department of Energy under contract DE-AC06-76RLO 1830.

Hoza M. (1994), Multipurpose optimization models for high-level waste vitrification, *Proceedings of the International Topical Meeting on Nuclear and Hazardous Waste Management—SPECTRUM 94*, La Grange Park, IL: American Nuclear Society, 1072–1077.

Hrma P. R. and A. W. Bailey (1995), *High level waste at Hanford: Potential for waste loading maximization*, Report PNL-SA-26441, Richland, Pacific Northwest Laboratory, WA.

Hwang C. L. and A. S. M. Masud (1979), Multiple objective decision making—methods and applications: A state-of-the-art survey, *Lecture Notes in Economics and Mathematical Systems*, 164, Springer-Verlag, Berlin, Heidelberg.

Jantzen C. M. and K. G. Brown (1993), Statistical process control of glass manufactured for nuclear waste disposal, *American Ceramic Society Bulletin* **72**, 55.

Johnson T. and U. Diwekar (1999), The value of design research: Stochastic optimization as a policy tool, *Foundations of Computer-Aided Design*, AIChE Symposium Series, Vol. 96, Malone et al. Editors, 454.

Johnson T. and U. Diwekar (2001), Hanford nuclear waste disposal and the value of research, *Journal of Multi-Criteria Decision Analysis*, **10**, 87.

Kacker R. S. (1985), Off-line quality control, parameter design, and the Taguchi method, *Journal of Quality Technology*, **17**, 176.

Kumar P., N. Singh, and N. K. Tewari (1991), A Nonlinear goal programming model for multistage, multiobjective decision problems with application to grouping and loading problem in a flexible manufacturing system, *European Journal of Operational Research*, **53(2)**, 166.

Kuhn H. W. and A. W. Tucker (1951), Nonlinear programming in *Proceedings of Second Berkeley Symposium on Mathematical Statistics and Probability*, J. Neyman (ed.), University of California Press, Berkeley, LA., 481.

Lieberman E. R. (1991), Soviet multiobjective mathematical programming methods: An overview, *Management Science*, **37(9)**, 1147.

Mendel J. E., W. A. Rawest, R. P. Turcotte, and L. McElroy (1980), Physical properties of glass for immobilization of high-level radioactive waste, *Nuclear and Chemical Waste Management*, **1**, 17.

Miettinen K. M. (1999), *Nonlinear Multiobjective Optimization*, Kluwer Academic, Norwell, MA.

Narayan V., U. Diwekar and Hoza M. (1996), Synthesizing optimal waste blends, *Industrial and Engineering Chemistry Research*, **35**, 3519.

Ohkubo S., P. B. R. Dissanayake, and K. Taniwaki (1998), An approach to multicriteria fuzzy optimization of a prestressed concrete bridge system considering cost and aesthetic feeling, *Structural Optimization*, **15(2)**, 132.

Olson D. L.(1993), Tchebycheff Norms in multiobjective linear programming, *Mathematical and Computer Modeling*, **17(1)**, 113.

Osyczka A. (1984), *Multicriterion Optimization in Engineering with FORTRAN programs*, Ellis Horwood, London.

Pareto V. (1964), *Cours d'Economie Politique*, Libraire Droz, Genéve.

Pareto V. (1971), *Manual of Political Economy*, MacMillan Press, New York.

Painton L. A. and U. M. Diwekar (1995), Stochastic annealing under uncertainty, *European Journal of Operations Research*, **83**, 489.

Ravindran A., D. T. Phillips, and J. J. Solberg (1987), *Operations Research: Principles and Practice*, (Second Edition), John Wiley and Sons. New York.

Rosenthal R. E. (1985), Principles of multiobjective optimization, *Decision Sciences*, **16(2)**, 133.

Schulz W. W. and N. Lombado (Eds.) (1998), Science and technology for disposal of radioactive tank wastes. *Proceedings of the American Chemical Society Symposium on Science and Technology for Disposal of Radioactive Tank Waste* (1997: Las Vegas, Nevada). Plenum Press, New York.

Schy A. A. and D. P. Giesy (1988), Multicriteria optimization methods for design of aircraft control systems, *Multicriteria Optimization in Engineering and in the Sciences*, W. Stadler (ed.), Plenum Press, New York, 225.

Shastri Y., U. Diwekar and S. Mehrotra.,(2011),An innovative trading approach for mercury waste management,*International Journal of Innovation Science*,**3**, 9.

Silverman J., R. E. Steuer, and A. W. Whisman (1988), A multi-period, multiple criteria optimization system for manpower planning, *European Journal of Operational Research*, **34(2)**, 160.

Sobol I. M., and R. B. Statnikov (1982), *Nailuchshie Resheniyagde Ikh Iskat' (The Best Decisions—Where to Seek Them)*, Mathematics/Cybernetics Series, No. 1, Znanie, Moscow.

Sobol I. M. (1992), A global search for multicriterial problems, in *Multiple Criteria Decision Making: Proceedings of the Ninth International Conference*, A. Goicoechea, L. Duckstein and S. Zionts (eds.), Springer-Verlag, New York, 401.

Stadler W. (ed.) (1988), *Multicriteria Optimization in Engineering and in the Sciences*, Plenum Press, New York.

Starkey J. M., S. Gray, and D. Watts (1988), Vehicle performance simulation and optimization including tire slid, *ASME 881733*.

Steuer R. E. (1986), *Multiple Criteria Optimization: Theory, Computation, and Applications*, John Wiley & Sons, New York.

Steuer R. E., and M. Sun (1995), The parameter space investigation method of multiple objective nonlinear programming: A computational investigation, *Operations Research*, **43(4)**, 641.

Stewart T. J. (1992), A critical survey on the status of multiple criteria decision making theory and practice, *OMEGA*, **20(5–6)**, 569.

Tamiz M. and D. D. Jones (1996), An overview of current solution methods and modeling practices in goal programming, *multiobjective Programming and goal programming: Theories and Applications*, M. Tamiz (ed.), Lecture Notes in Economics and Mathematical Systems 432, Springer-Verlag, Berlin, 198.

VanLaarhoven P. J. M. and E. H. Aarts (1987), *Simulated Annealing Theory and Applications*, D. Reidel, Holland.

Wood E., N. P. Greis, and R. E. Steuer (1982), Linear and nonlinear applications of the Tchebycheff metric to the multi criteria water allocation problem, in *Environmental Systems Analysis and Management*, S. Rinaldi (ed.), North-Holland, 363.

Yu P. L. (1985), *Multiple-Criteria Decision Making Concepts, Techniques, and Extensions*, Plenum Press, New York.

Zadeh L. (1963), Optimality and non-scalar-valued performance criteria, *IEEE Transactions on Automatic Control*, **8**, 59.

Zeleny M. (1974), Linear multiobjective programming,in *Lecture Notes in Economics and Mathematical Systems 95*, Springer-Verlag, Berlin.

Zeleny M. (1982), *Multiple Criteria Decision Making*, McGraw-Hill, New York.

Zionts S. (1989), Multiple criteria mathematical programming: An updated overview and several approaches, *Multiple Criteria Decision Making and Risk Analysis Using Microcomputers*, B. Karpak, S. Zionts (ed.), Springer-Verlag, Berlin 7.

7

Optimal Control and Dynamic Optimization

Optimal control problems involve vector decision variables. These problems are one of the most mathematically challenging problems in optimization theory.

Consider the historic isoperimetric problem in its original form below.

Example 7.1: Formulate the isoperimetric problem faced by Queen Dido.

Solution: Queen Dido's problem was to find the maximum area that could be covered by a rope (curve) whose length (perimeter) was fixed. This problem is equivalent to tracking the path of the point "P" shown in Figure 7.1 so as to maximize the area covered by the path, given that the path length is fixed.

The area of kinematics deals with geometry of motion. Suppose object P is traveling in the $x - y$ plane. Then the area covered by this object is given by

$$A = \int_0^X y(x)dx \qquad (7.1)$$

where y represents the displacement and A is the area covered by the curve. The perimeter of the curve can be expressed in terms of the following equations.

Electronic Supplementary Material The online version of this chapter (https://doi.org/10.1007/978-3-030-55404-0_7) contains supplementary material, which is available to authorized users.

One section of this chapter is written by Professor Benoit Morel, Engineering and Public Policy, Carnegie Mellon University.

© Springer Nature Switzerland AG 2020
U. M. Diwekar, *Introduction to Applied Optimization*, Springer Optimization and Its Applications 22,
https://doi.org/10.1007/978-3-030-55404-0_7

$$Le = \int \sqrt{dy^2 + dx^2} \qquad (7.2)$$

$$Le = \int_0^X \sqrt{1 + (\frac{dy}{dx})^2} dx \qquad (7.3)$$

We want to find the maximum area covered when the perimeter is fixed at Le. The velocity of P is defined in terms of the path characteristics such as

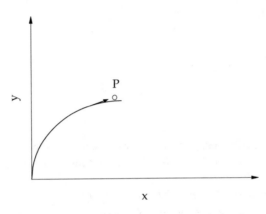

Fig. 7.1. Isoperimetric problem as a path optimization problem

the length of the arc and a unit vector tangent to the path. By introducing kinematic terms such as displacement and velocity, we can identify the decision variable vector as the velocity vector u_x, so the problem that needs to be solved is then given by

$$\underset{u_x}{\text{Maximize}} \quad A = \int_0^X y(x) \, dx \qquad (7.4)$$

subject to

$$\frac{dy(x)}{dx} = u_x \qquad \text{Kinematic Constraint} \qquad (7.5)$$

$$Le = \int_0^X \sqrt{1 + (\frac{dy}{dx})^2} \, dx \qquad \text{Perimeter Constraint} \qquad (7.6)$$

As can be seen above, the problem involves path optimization where the vector u_x is the decision variable. The constraints constitute differential equations in terms of the path-dependent state variables. Surprisingly enough, these types of problems gave rise to the first systematic theory of optimization.

Groningen,
January 1, 1697

AN ANNOUNCEMENT

I, Johann Bernoulli, greet the most clever mathematicians in the world. Nothing is more attractive to intelligent people than an honest, challenging problem whose possible solution will bestow fame and remains as a lasting monument. Following the example set by Pascal, Fermat, etc., I hope to earn the gratitude of the entire scientific community by placing before the finest mathematicians of our time a problem which will test their methods and the strength of their intellect. If someone communicates to me the solution of the proposed problem, I shall then publicly declare him worthy of praise.

Calculus of variations defines the first systematic theory of optimization as a solution to the famous *Brachistochrone* (Greek for the "shortest time") problem presented by John Bernoulli to challenge the whole world (please see the announcement above). In 1696 John Bernoulli challenged the mathematicians to find the Brachistochrone, that is, the planar curve that would provide the shortest transit time. The Brachistochrone problem is as follows. What is the slide down which a frictionless object would slip in the least possible time? Thus it was natural for Galileo in 1637 to propose that the solution is a circular arc. In falling under gravity, an object accelerates quickly so that a wire bent in the shape of the circular arc shown in Figure 7.2a would offer a faster time of transit to a bead sliding down it under the action of gravity than a straight line joining the two points. However, the correct solution to this problem was a cycloid (Figure 7.2b) derived using various physical and mathematical analogies. The satisfactory solution was, however, based on the method of calculus of variations. The name of the method is derived from the fact that it is based on the vanishing of the first variation of a functional. A functional is defined as a quantity or function that depends upon the entire course or path (path optimization) of one or more functions rather than on a number of scalar variables.

These problems, also known as optimal control problems, are a subset of problems called differential algebraic optimization problems (DAOPs), as the underlying model for these problems is a dynamic model consisting of differential and algebraic equations.

A differential algebraic optimization problem in general can be stated as follows.

$$\text{Optimize} \quad J \;=\; j(\overline{x}_T) \;+\; \int_0^T k(\overline{x}_t, \theta_t, x_s)\, dt \qquad (7.7)$$
$$\theta_t, x_s$$

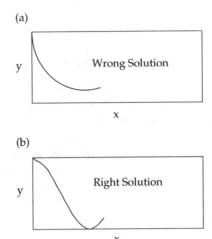

Fig. 7.2. The Brachistochrone problem solutions. (**a**) Circular arc: Galileo's solution. (**b**) Cycloid: Bernoulli's solution

subject to

$$\frac{d\overline{x}_t}{dt} = f(\overline{x}_t, \theta_t, x_s) \tag{7.8}$$

$$h(\overline{x}_t, \theta_t, x_s) = 0 \tag{7.9}$$

$$g(\overline{x}_t, \theta_t, x_s) \leq 0 \tag{7.10}$$

$$\overline{x}_0 = \overline{x}_{initial}$$

$$\theta(L) \leq \theta_t \leq \theta(U)$$

$$x_s(L) \leq x_s \leq x_s(U)$$

where J is the objective function given by Equation (7.7), $\overline{x}_t$ is the state variable vector ($nx \times 1$ dimensional) at any time t, θ_t is the control vector, and x_s represents the scalar variables. It is obvious that the objective function can only be calculated at the end of operation T. Equations (7.9) and (7.10) represent the equality (m_1 constraints) and inequality constraints (m_2 constraints, including bounds on the state variables), respectively (constituting a total of m constraints). $\theta(L)$ and $x_s(L)$ represent the lower bounds on the set of control variables θ_t and the scalar variable x_s, respectively, and $\theta(U)$, $x_s(U)$ are the corresponding upper bounds. In the absence of the scalar decision variables x_s, a DAOP is equivalent to an optimal control problem and is the focus of this chapter. As most of the solution methods to optimal control problems, in their original form, did not consider bounds on the control variables ($\theta(L)$ and $\theta(U)$), initially we neglect the bounds on the control variables.

Calculus of variations had its origin in the belief that God had constructed the universe to operate in the most efficient manner, and to understand the principles of the universe "something" needs to be minimized. For example, in 1957 Fermat invoked such a principle in declaring that light travels through a medium along the path of least time of transit. Engineering efforts to design an efficiently self-correcting electromechanical apparatus, relative to some target object, gave rise to the discipline of optimal control. Systematic methods to solve these problems involve the maximum principle due to Pontryagin (Boltyanskii et al., 1956; Pontryagin, 1956, 1957), a Russian mathematician, and Bellman's principle of optimality (Bellman, 1957), leading to the dynamic programming technique.

Calculus of variations considers the entire path of the function and optimizes the integral by minimizing the functional by making the first derivative vanish (first-order condition for nonlinear systems, Chapter 3), resulting in second-order differential equations that can be difficult to solve. Other approaches keep the first-order differential system as is but transform

- The integral objective function into a Hamiltonian H_t, a nonlinear objective function for each time step that can be optimized using a (discretized) variable θ_t for that step. This results in n NLP optimization problems corresponding to n time steps. However, this maximum principle transformation needs to include additional variables and corresponding first-order differential equations, referred to as adjoint variables and adjoint equations, respectively.
- The problem into an equivalent first-order system involving partial differential equations based on the principle of optimality. This results in the Hamilton–Jacobi–Bellman equations that may not be easy to solve. However, this dynamic programming method provides the basis for stochastic optimal control problems.

In short, the general mathematical techniques used to solve optimal control problems include the calculus of variations, Pontryagin's maximum principle, and dynamic programming. Nonlinear programming optimization methods can also be applied to optimal control problems provided that the complete system of differential equations is transformed into nonlinear algebraic equations. The first three methods treat the decision variables as vectors, whereas the NLP approach requires the variables to be transformed into scalars and then the nonlinear programming techniques defined in Chapter 3 can be used.

7.1 Calculus of Variations

As seen earlier, the theory of optimization began with the calculus of variations, which is based on the theorem of minimum potential energy (because energy is a path-dependent quantity), leading to the Euler equations and natural boundary conditions. A functional is defined as a quantity or function that depends upon the entire course or path of one or more functions rather than on a number of scalar variables. Application of the minimum-energy principle involves the definition of stationary values for a function, or a set of functionals. In the above optimal control definition, the objective function J is a functional that depends upon the entire path from time equal to zero to time equal to T. Remember that we are neglecting the bounds $\theta(L)$ and $\theta(U)$ on the control variables and are assuming that the scalar variables x_s are fixed. Also, at the first part of the derivation, the constraints are not included. To obtain the extremum value of J, the total differential of Equation (7.7) is equated to zero, as follows.

$$\int_0^T dJ = \int_0^T \left[\frac{\partial J}{\partial \theta} \delta\theta + \frac{\partial J}{\partial \theta'} \delta\theta' \right] dt = 0 \qquad (7.11)$$

The left-hand side is called the first variation of the integral J. In order to eliminate the variations with respect to $\delta\theta'$, where $\theta' = d\theta/dt$, the second term of the above equation is integrated by parts.

$$\int_0^T \frac{\partial J}{\partial \theta'} \delta\theta' dt = \left[\frac{\partial J}{\partial \theta'} \right]_0^T [d\theta]_0^T - \int_0^T \frac{d\left(\frac{\partial J}{\partial \theta'}\right)}{dt} \delta\theta dt \qquad (7.12)$$

By substituting Equation (7.12) in Equation (7.11) and imposing the boundary condition that $d\theta = 0$ at $t = 0$ and $t = T$, the following equation results.

$$\int_0^T dJ = \int_0^T \left[\frac{\partial J}{\partial \theta} - \frac{d\left(\frac{\partial J}{\partial \theta'}\right)}{dt} \right] \delta\theta dt \qquad (7.13)$$

The above integral must vanish for all admissible values of $\partial\theta$, which requires that the expression inside the brackets in Equation (7.13) be zero (the first-order necessary condition for minimization and maximization of a nonlinear programming problem); that is,

$$\frac{\partial J}{\partial \theta} - \frac{d\left(\frac{\partial J}{\partial \theta'}\right)}{dt} = 0 \qquad (7.14)$$

The above differential equation is known as the Euler differential equation, corresponding to the functional given in Equation (7.7). This, together with the boundary conditions, determines the function θ.

If the functional J is also constrained by equality constraints, then the application of the calculus of variations leads to Euler-Lagrangian equations. In the Euler-Lagrangian formulation, the objective function is augmented to include constraints through the use of Lagrangian multipliers μ_i and λ_j, as given below.

$$\text{Optimize} \quad L = j(\overline{x}_T) \int_0^T k(\overline{x}_t, \theta_t, x_s) \, dt$$

$$\theta_t, \mu_i, \lambda_{jj}, \lambda_{kk} \quad + \sum_i \mu_i^T \left(\frac{dx_i}{dt} - f(\overline{x}_t, \theta_t, x_s) \right) +$$

$$+ \sum_{jj=1}^{JJ} \lambda_{jj} h(\overline{x}_t, \theta_t, x_s)$$

$$+ \sum_{kk=JJ+1}^{KK} \lambda_{kk} (g(\overline{x}_t, \theta_t, x_s) \qquad (7.15)$$

By applying the first-order condition for optimization, that is, the first derivative with respect to the control variable, and the Lagrange multipliers (for equality constraints) should cause results in Euler-Lagrangian differential equations to disappear.

$$\frac{\partial L}{\partial \theta} - \frac{d(\frac{\partial L}{\partial \theta'})}{dt} = 0 \qquad (7.16)$$

$$\frac{\partial L}{\partial x_i} - \frac{d(\frac{\partial L}{\partial x_i'})}{dt} = 0 \qquad (7.17)$$

$$h(\overline{x}_t, \theta_t, x_s) = 0 \qquad (7.18)$$

$$g_l(\overline{x}_t, \theta_t, x_s) = 0 \; \lambda_l \geq 0 \qquad (7.19)$$

$$g_m(\overline{x}_t, \theta_t, x_s) \leq 0 \; \lambda_m = 0 \qquad (7.20)$$

The following example demonstrates the application of the calculus of variations to the isoperimetric problem.

Example 7.2: Example 7.1 formulated the isoperimetric problem in terms of the differential equations derived using kinematics. Solve this problem using the calculus of variations.

Solution: Example 7.1 resulted in the following formulation. For simplicity, we are replacing the displacement variable y by x_1, the variable x by t, and the velocity vector u_x by u_t.

$$\text{Maximize} \quad A = \int_0^T x_1(t) \, dt \qquad (7.21)$$
$$u_t$$

subject to

$$\frac{dx_1}{dt} = u_t \; x_1(0) = 0.0 \quad \text{Kinematic Constraint} \qquad (7.22)$$

$$Le = \int_0^T \sqrt{1 + (\frac{dx_1}{dt})^2} \, dt \quad \text{Perimeter Constraint} \qquad (7.23)$$

To include the perimeter constraint in the formulation, let us introduce a new state variable x_2 that relates to the perimeter constraint as follows.

$$\frac{dx_2}{dt} = \sqrt{1 + (\frac{dx_1}{dt})^2} \quad x_2(0) = 0.0; \quad x_2(T) = Le \qquad (7.24)$$

$$\frac{dx_2}{dt} = \sqrt{1 + u_t^2} \quad x_2(0) = 0.0; \quad x_2(T) = Le \qquad (7.25)$$

Now, the problem is reduced to solving the maximization problem given by Equation (7.21) subject to the two constraints given by the differential equations (7.22) and (7.25). Combining these two constraints with the objective function, the problem results in the following Euler-Lagrangian formulation.

$$\underset{u_t, \mu_1, \mu_2}{\text{Minimize}} \int_0^T L = \int_0^T (-x_1(t) + \mu_1(\frac{dx_1}{dt} - u_t)$$
$$+ \mu_2(\frac{dx_2}{dt} - \sqrt{1 + u_t^2}))dt \qquad (7.26)$$

where μ_1 and μ_2 are t-dependent Lagrange multipliers for the two constraints.

Introducing $x_1' = dx_1/dt$ and $x_2' = dx_2/dt$ results in

$$\underset{u_t, \mu_1, \mu_2}{\text{Minimize}} \int_0^T L = \int_0^T (-x_1(t) + \mu_1(x_1' - u_t)$$
$$+ \mu_2(x_2' - \sqrt{1 + u_t^2}))dt \qquad (7.27)$$

Application of the calculus of variations to the above problem results in the following Euler-Lagrangian formulation.

Taking the partial derivative of the functional given in Equation (7.27) with respect to x_1 results in

$$\frac{\partial L}{\partial x_1} - \frac{d(\frac{\partial L}{\partial x_1'})}{dt} = 0 \implies \frac{d\mu_1}{dt} = -1 \qquad (7.28)$$

The partial derivative with respect to x_2 and u_t leads to the following equations.

$$\frac{\partial L}{\partial x_2} - \frac{d(\frac{\partial L}{\partial x_2'})}{dt} = 0 \implies \frac{d\mu_2}{dt} = 0 \qquad (7.29)$$

$$\frac{\partial L}{\partial u_t} - \frac{d(\frac{\partial L}{\partial u_t'})}{dt} = 0 \implies \mu_1 + \mu_2 \frac{u_t}{\sqrt{(1 + u_t^2)}} = 0 \qquad (7.30)$$

Equations (7.28) and (7.29) integrate into

$$\mu_1 = -t + c_1 \qquad (7.31)$$
$$\mu_2 = c_2 \qquad (7.32)$$

where c_1 and c_2 are integration constants.

Substituting in Equation (7.30) results in the following expression for the u_t.

$$0 = -t + c_1 + \frac{c_2 u_t}{\sqrt{1 + u_t^2}} \tag{7.33}$$

This leads to

$$u_t = \frac{dx_1}{dt} = \pm \frac{t - c_1}{\sqrt{c_2^2 - (t - c_1)^2}} \tag{7.34}$$

This in turn leads to

$$x_1(t) = \pm \sqrt{c_2^2 - (t - c_1)^2} \tag{7.35}$$

The boundary condition of x_1 ($x_1(0) = 0.0$) leads to $c_2 = c_1$. This is the equation for a semicircle with its center at c_1 and a radius of c_2 as shown in Figure 7.3.

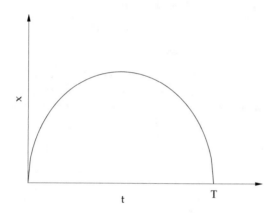

Fig. 7.3. Solution to the isoperimetric problem

$$x_2(t) = \int_0^t \frac{c_2}{\sqrt{c_2^2 - (t - c_1)^2}} dt \tag{7.36}$$

$$x_2(t) = c_2 \arcsin \frac{t - c_1}{c_2} + c_3 \tag{7.37}$$

From the boundary conditions for x_2, the values of the integration constant can be determined as follows.

The boundary condition $x_2(0) = 0$ leads to $c_3 = c_2 \pi/2$, and $x_2(T) = Le = \pi c_2$ results in $c_1 = T/2$ and $c_2 = T/2$. This implies that the curve is the semicircle with a radius equal to Le/π, the same solution as given by Queen Dido.

7.2 Maximum Principle

In the maximum principle formulation (the right-hand side of Equation (7.38)), the objective function is represented as a linear function in terms of the final values of $\bar{x}$ and the values of $\bar{c}$, where $\bar{c}$ represents the vector of constants. The maximum principle formulation for the above-mentioned DAOP is given below.

$$\text{Maximize } J = j(\bar{x}_T) + \int_0^T k(\bar{x}_t, \theta_t, x_s)dt = \bar{c}^T \bar{x}_T = \sum_{i=1}^{nx} c_i x_i(T)$$

$$\theta_t \tag{7.38}$$

subject to

$$\frac{d\bar{x}_t}{dt} = f(\bar{x}_t, \theta_t, x_s) \tag{7.39}$$

$$h(\bar{x}_t, \theta_t, x_s) = 0 \tag{7.40}$$

$$g(\bar{x}_t, \theta_t, x_s) \leq 0 \tag{7.41}$$

$$\bar{x}_0 = \bar{x}_{initial}$$

where nx refers to the number of state variables x_t in the problem ($\bar{x}_t$ is an $nx \times 1$-dimensional vector). By using the Lagrangian formulation for the above problem, fixing scalar variables x_s, and removing the bounds $\theta(L)$ and $\theta(U)$ on the control variable vector θ_t, one obtains

$$\text{Maximize } J^* = \bar{c}^T \bar{x}_T + \overline{\lambda_1}(h(\bar{x}_t, \theta_t, x_s)) + \overline{\lambda_2}(g(\bar{x}_t, \theta_t, x_s))$$

$$\theta_t \tag{7.42}$$

subject to

$$\frac{d\bar{x}_t}{dt} = f(\bar{x}_t, \theta_t, x_s) \tag{7.43}$$

$$\bar{x}_0 = \bar{x}_{initial}$$

where

$$\bar{\lambda} = [\overline{\lambda_1}, \overline{\lambda_2}]$$

Application of the maximum principle to the above problem involves the addition of nx adjoint variables z_t (one adjoint variable per state variable), nx adjoint equations, and a Hamiltonian, which satisfies the following relations.

$$H(\bar{z}_t, \bar{x}_t, \theta_t) = \bar{z}_t^T f(\bar{x}_t, \theta_t, x_s) = \sum_{i=1}^{nx} z_i f_i(\bar{x}_t, \theta_t) \tag{7.44}$$

$$\frac{dz_i}{dt} = -\sum_{j=1}^{n} z_j \frac{\partial f_j}{\partial x_i} \tag{7.45}$$

$$\bar{z}_T = \bar{c} \tag{7.46}$$

The boundary conditions given above (Equation (7.46)) are often true, but not always. When present, they play an important role in the final stages of the solution. Therefore, it is important to keep track of the boundary conditions. As stated earlier, we have one objective H for each time step. The optimal decision vector θ_t can be obtained by extremizing the Hamiltonian given by Equation (7.44) for each time step. θ_t can then be expressed as:

$$\theta_t = H^*(\bar{x}_t, \bar{z}_t, \bar{\lambda}) \tag{7.47}$$

where H* denotes the function obtained by using the stationary condition $(dH_t/d\theta_t)$ for the Hamiltonian. It should be noted that this principle does not apply in all situations. It applies only if the functions in the maximization problem are convex. It is possible to derive the necessary condition for optimality in the calculus of variations from the maximum principle when the decision vector is not constrained. Conversely, by using the technique of the calculus of variations, the weakened form of the maximum principle can be derived. These derivations are presented in Fan (1966), and the interested reader is referred to this book on the maximum principle for further details.

Example 7.3: Formulate the isoperimetric problem using the maximum principle.

Solution: The isoperimetric problem formulated earlier (see Examples 7.1 and 7.2) written in terms of two state variables x_1 and x_2 is given below.

$$\underset{u_t}{\text{Maximize}} \quad A = \int_0^T x_1(t)\, dt \tag{7.48}$$

subject to

$$\frac{dx_1}{dt} = u_t \quad x_1(0) = 0.0; \tag{7.49}$$

$$\frac{dx_2}{dt} = \sqrt{1 + u_t^2} \quad x_2(0) = 0.0; \quad x_2(T) = Le \tag{7.50}$$

Remember that in the maximum principle, we need to express the objective function in terms of linear combinations of the final state variables $\bar{x}(T)$ as shown in Equation (7.38). To solve this problem, an additional state variable $x_3(t)$ is introduced which is given by

$$x_3(t) = \int_0^t x_1(t)\, dt \tag{7.51}$$

The problem can then be written as:

$$\text{Maximize } x_3(T), \tag{7.52}$$
$$u_t$$

subject to the following differential equations for the three state variables.

$$\frac{dx_1}{dt} = u_t \quad x_1(0) = 0.0; \ x_1(T) = 0.0 \tag{7.53}$$

$$\frac{dx_2}{dt} = \sqrt{1 + u_t^2} \quad x_2(0) = 0.0; \ x_2(T) = Le \tag{7.54}$$

$$\frac{dx_3}{dt} = x_1(t) \tag{7.55}$$

The Hamiltonian function, which should be maximized, is

$$H_t = z_1 u_t + z_2 \sqrt{1 + u_t^2} + z_3 x_1(t) \tag{7.56}$$

And the adjoint equations are

$$\frac{dz_1}{dt} = -z_3, \tag{7.57}$$

$$\frac{dz_2}{dt} = 0, \tag{7.58}$$

Note that we are not imposing any final boundary condition on the above equations, as we know both boundary conditions for the variables x_1 and x_2. However, for z_3, the final boundary condition (derived from Equation (7.46)) is active and is given by

$$\frac{dz_3}{dt} = 0, \ z_3(T) = 1 \implies z_3(t) = 1 \tag{7.59}$$

From the above equations, we have

$$\frac{dz_1}{dt} = -1 \implies z_1(t) = -t + c_1 \tag{7.60}$$

$$z_2(t) = c_2, \tag{7.61}$$

The Hamiltonian function in Equation (7.56) can be written as:

$$H_t = z_1 u_t + z_2 \sqrt{1 + u_t^2} + x_1(t) \tag{7.62}$$

From the optimality condition $\partial H / \partial u_t = 0$, it follows that

$$u_t = \frac{dx_1}{dt} = \pm \frac{t - c_1}{\sqrt{c_2^2 - (t - T)^2}} \tag{7.63}$$

$$x_1(t) = \pm \sqrt{c_2^2 - (t - c_1)^2} \tag{7.64}$$

$$x_2(t) = c_2 \arcsin \frac{t - c_2}{c_2} + c_3 \tag{7.65}$$

It can be easily seen from Equations (7.63)–(7.65) and from Equations (7.34)–(7.37) in Example 7.2 that the formulations lead to the same results, where in the case of the calculus of variations the t-dependent Lagrange multipliers μ_i are equivalent to the adjoint variables z_i in the maximum principle formulation.

7.3 Dynamic Programming

Dynamic programming is based on Bellman's principle of optimality, as described below.

An optimal policy has the property that whatever the initial state and initial decision are the remaining decisions must constitute an optimal policy with regard to the state resulting from the first decision.

In short, the principle of optimality states that the minimum or maximum value (of a function) is a function of the initial state and the initial time.

In the calculus of variations, we locate a curve as a locus of points as shown in Figure 7.4a, whereas dynamic programming considers a curve to be an envelope of tangents (Figure 7.4b). In that sense, the two theories are dual to each other. However, the duality and equivalence remain valid only for deterministic processes.

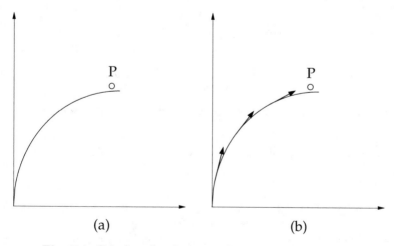

Fig. 7.4. Calculus of variations and dynamic programming

Dynamic programming is best suited for multistage processes, where these processes can be decomposed into N stages as shown in Figure 7.4b. However, application of the dynamic programming technique to a continuously operating system leads to nonlinear partial differential equations, the Hamilton–Jacobi–Bellman (H-J-B) equation that can be tedious to solve. A brief derivation of the H-J-B equation is given below. For details, please refer to Bellman (1957), Aris (1961) , and Kirk (1970).

The optimal control problem described earlier involves the process described by the state equations:

$$\frac{d\overline{x}_t}{dt} \;=\; f(\overline{x}_t, \theta_t, x_s) \tag{7.66}$$

which are to be controlled so as to minimize the performance measure given by J:

$$\underset{\theta_t}{\text{Optimize}} \quad J \;=\; j(\overline{x}_T) \;+\; \int_0^T k(\overline{x}_t, \theta_t, x_s)\, dt \tag{7.67}$$

Introducing a dummy variable of integration τ, where $t \leq \tau \leq T$, the performance measure in the interval $[t, T]$ is

$$\underset{\theta_\tau}{\text{Optimize}} \quad j(\overline{x}_T) \;+\; \int_t^T k(\overline{x}_\tau, \theta_\tau, x_s)\, d\tau \tag{7.68}$$

By subdividing into $ns + 1$ intervals, we obtain

$$\underset{\theta_\tau}{\text{Optimize}} \;\; j(\overline{x}_T) + \sum_{j=1}^{ns} \int_{t_j}^{t_j + \triangle t} k(\overline{x}_\tau, \theta_\tau, x_s)\, d\tau \;\ldots$$

$$+ \int_{t_{ns} + \triangle t}^{T} k(\overline{x}_\tau, \theta_\tau, x_s)\, d\tau \tag{7.69}$$

The principle of optimality requires that $J_{opt}(\overline{x}(t),\ t)$ is equal to

$$\underset{\theta_\tau}{\text{Optimize}} \quad J_{opt}(\overline{x}(t + \triangle t),\ t + \triangle t) \;+\; \int_t^{t + \triangle t} k(\overline{x}_\tau, \theta_\tau, x_s)\, d\tau \tag{7.70}$$

where $J_{opt}(\overline{x}(t + \triangle t),\ t + \triangle t)$ is the optimum objective for the time interval $t + \triangle t \leq \tau \leq T$ with the initial state $\overline{x}(t + \triangle t)$.

Assuming that the second partial derivative of the function J exists and is bounded, we can expand $J_{opt}(\overline{x}(t + \triangle t),\ t + \triangle t)$ as a Taylor series (neglecting the higher derivatives) about the point $(\overline{x}(t), t)$ to obtain

$$J_{opt}(\bar{x}(t),\ t)\ =\ \underset{\theta_\tau}{\text{Optimize}}\ \int_t^{t+\triangle t} k(\bar{x}_\tau, \theta_\tau, x_s)\ d\tau\ +\ J(\bar{x}(t),\ t)$$

$$+\quad \left[\frac{\partial J}{\partial t}\right]\triangle t\ +\ \left[\frac{\partial J}{\partial \bar{x}}\right]^T [x(t+\triangle t)\ -\ x(t)]$$

$$(7.71)$$

For a small $\triangle t$, the above equation reduces to

$$J_{opt}(\bar{x}(t),\ t)\ =\ \underset{\theta_t}{\text{Optimize}}\ k(\bar{x}_t, \theta_t)\ \triangle t\ +\ J(\bar{x}(t),\ t))$$

$$+\quad \left[\frac{\partial J}{\partial t}\right]\triangle t\ +\ \left[\frac{\partial J}{\partial \bar{x}}\right]^T [dx(t)] \quad (7.72)$$

Dividing Equation (7.72) by $\triangle t$ and substituting the value of dx/dt from the differential equation (7.66) and further by virtue of the fact that the left-hand side is not a function of θ_t, the following equation results.

$$0\ =\ \underset{\theta_t}{\text{Optimize}}\ k(\bar{x}_t, \theta_t)\ +\ \left[\frac{\partial J}{\partial t}\right]\ +\ \left[\frac{\partial J}{\partial \bar{x}}\right]^T [f(\bar{x}_t, \theta_t, x_s)]$$

$$(7.73)$$

Defining the Hamiltonian as a function of $\bar{x}(t)$, $\partial J/\partial \bar{x}$, t, the above equation results in what is referred to as the Hamilton–Jacobi–Bellman equation:

$$0\ =\ \left[\frac{\partial J}{\partial t}\right]\ +\ H\left(\bar{x}(t),\ \frac{\partial J}{\partial \bar{x}},\ t\right) \quad (7.74)$$

where

$$H\ =\ \underset{\theta_T}{\text{Optimize}}\ k(\bar{x}_T, \theta_T)\ +\ \left[\frac{\partial J}{\partial \bar{x}}\right]^T [f(\bar{x}_T, \theta_T, x_s)]$$

$$(7.75)$$

As can be seen, the dynamic programming optimality conditions lead to the H-J-B equation, a first-order partial differential equation as compared to the second-order differential equations of the calculus of variations, but it can also be tedious to solve.

Although the mathematics of dynamic programming look different from the calculus of variations or the maximum principle, in most cases it leads to the same results, as can be seen from the following isoperimetric problem.

Example 7.4: Solve the isoperimetric problem using dynamic programming.

Solution: Once again the isoperimetric problem formulated earlier can be written as follows.

$$\text{Maximize}_{u_t} \quad A = \int_0^T x_1(t)\, dt \qquad (7.76)$$

subject to

$$\frac{dx_1}{dt} = u_t \quad x_1(0) = 0.0; \qquad (7.77)$$

$$\frac{dx_2}{dt} = \sqrt{1 + u_t^2} \quad x_2(0) = 0.0; \quad x_2(T) = Le \qquad (7.78)$$

Introducing a new variable $I(t) = \int_t^T x_1(\tau)d\tau$ leads to

$$\text{Maximize}_{u_t} \quad A = \int_0^t x_1(t)\, dt \; + I(t) \qquad (7.79)$$

The H-J-B equation derived for (7.79) is

$$\frac{\partial I}{\partial t} + \text{Maximize}_{u_t} \; x_1(t) \; + u_t\frac{\partial I}{\partial x_1(t)} \; + \sqrt{1 + u_t^2}\,\frac{\partial I}{\partial x_2(t)} = 0 \qquad (7.80)$$

or

$$\text{Maximize}_{u_t} \; x_1(t) + I_t \; + \; I_{x_1}u_t \; + \; I_{x_2}\sqrt{1 + u_t^2} = 0 \qquad (7.81)$$

where

$$I_t = \frac{\partial I}{\partial t}; \; I_{x_1} = \frac{\partial I}{\partial x_1(t)}; \; I_{x_2} = \frac{\partial I}{\partial x_2(t)}$$

$$H = \text{Maximize}_{u_t} \; I_{x_1}u_t \; + \; I_{x_2}\sqrt{1 + u_t^2} \qquad (7.82)$$

Maximizing with respect to u_t leads to

$$I_{x_1} + \frac{u_t}{\sqrt{1 + u_t^2}}I_{x_2} = 0 \qquad (7.83)$$

Differentiating Equation (7.81) with respect to x_1 and x_2 leads to

$$\frac{dI_{x_1}}{dt} = -1 \implies I_{x_1} = -t + c_1 \qquad (7.84)$$

$$\frac{dI_{x_2}}{dt} = 0 \implies I_{x_2} = c_2 \tag{7.85}$$

Substituting the values in Equation (7.83) results in

$$u_t = \frac{dx_1}{dt} = \pm \frac{t - c_1}{\sqrt{c_2^2 - (t - T)^2}} \tag{7.86}$$

$$x_1(t) = \pm \sqrt{c_2^2 - (t - c_1)^2} \tag{7.87}$$

$$x_2(t) = c_2 \arcsin \frac{t - c_1}{c_2} + c_3 \tag{7.88}$$

It can be seen from the earlier Examples 7.2 and 7.3 and the above equations that the formulation using the calculus of variations and the maximum principle, dynamic programming, leads to the same results, where in the case of the calculus of variations the t-dependent Lagrange multipliers μ_i are equivalent to adjoint variables z_i in the maximum principle formulation. These are equivalent to the partial derivatives of the function with respect to the state variables in dynamic programming, I_{x_1} and I_{x_2}. Thus it leads to the same solution as before.

The advantage of dynamic programming over the other methods is that it is possible to use dynamic programming when the constraints are stochastic, as is discussed in the next section. However, dynamic programming formulation leads to a solution of partial differential equations that can be tedious to solve. Recently, a first version of the stochastic maximum principle was presented using the analogy between dynamic programming and the maximum principle. Interested readers are referred to Rico-Ramirez and Diwekar (2004). In the last section of this chapter, we present a real-world case study where we show that a problem solution can be simplified when one uses a combination of these methods.

7.4 Stochastic Processes and Stochastic Optimal Control

A stochastic process is a variable that evolves over time in an uncertain way. A stochastic process in which the time index t is a continuous variable is called a continuous-time stochastic process. Otherwise, it is called a discrete-time stochastic process. Similarly, according to the conceivable values for x_t (called the states), a stochastic process can be classified as being continuous state or discrete state.

Stochastic processes do not have time derivatives in the conventional sense and, as a result, they cannot always be manipulated using the ordinary rules of calculus. This is because, in general, the solution to a stochastic differential equation is not a single value for the function, but rather is a probability distribution. As a result, the typical mathematical techniques used to solve optimal control problems namely calculus of variations, Pontryagin's maximum principle, and nonlinear programming algorithms cannot be directly applied. To work with stochastic processes, one must make use of Ito's lemma and the dynamic programming formulation. This lemma, called the fundamental theorem of stochastic calculus, allows us to differentiate and to integrate functions of stochastic processes.

One of the simplest examples of a stochastic process is the random walk process. The Wiener process, also called a Brownian motion, is a continuous limit of the random walk and is a continuous-time stochastic process. The Wiener process can be used as a building block to model an extremely broad range of variables that vary continuously and stochastically through time. For example, consider the price of a technology stock. It fluctuates randomly but over a long time period has had a positive expected rate of growth that compensated investors for risk in holding the stock. Can the stock price be represented as a Wiener process? The following paragraphs establish that stock prices can be represented as a Wiener process, as it has the following important properties.

1. It satisfies the Markov property. The probability distribution for all future values of the process depends only on its current value. Stock prices can be modeled as Markov processes, on the grounds that public information is quickly incorporated in the current price of the stock and past patterns have no forecasting values.
2. It has independent increments. The probability distribution for the change in the process over any time interval is independent of any other time interval (nonoverlapping).
3. Changes in the process over any finite interval of time are normally distributed, with a variance that is linearly dependent on the length of time interval dt.

From the example of the technology stock above, it is easier to show that the variance of the change distribution can increase linearly. However, given that stock prices can never fall below zero, price changes cannot be represented as a normal distribution. However, it is reasonable to assume that changes in the logarithm of prices are normally distributed. Thus, stock prices can be represented by the logarithm of a Wiener process.

As stated earlier, stochastic processes do not have time derivatives in the conventional sense and, as a result, they cannot be manipulated using the ordinary rules of calculus as needed to solve the stochastic optimal control problems. Ito provided a way around this by defining a particular kind of uncertainty representation based on the Wiener process.

An Ito process is a stochastic process $x(t)$ on which its increment dx is represented by the equation:

$$dx = a(x,t)dt + b(x,t)dz \qquad (7.89)$$

where dz is the increment of a Wiener process, and $a(x,t)$ and $b(x,t)$ are known functions. By definition, $E[(dz)] = 0$ and $(dz)^2 = dt$, where E is the expectation operator and $E[dz]$ is interpreted as the expected value of dz.

The simplest generalization of Equation (7.89) is the equation for Brownian motion with drift given by

$$dx = \alpha dt + \sigma dz \qquad \text{Brownian motion with drift} \qquad (7.90)$$

where α is called the drift parameter and σ is the variance parameter. The discretized form of Equation (7.90) is the following:

$$x_t = x_{t-1} + \alpha \Delta t + \sigma \epsilon_t \sqrt{\Delta t} \qquad (7.91)$$

where ϵ_t is normally distributed with a mean of 0 and a standard deviation of 1.0. Figure 7.5 shows the sample paths of Equation (7.90). For details, please refer to Dixit and Pindyck (1994). Over any time interval Δt, the change in x, denoted by Δx, is normally distributed and has an expected value variance:

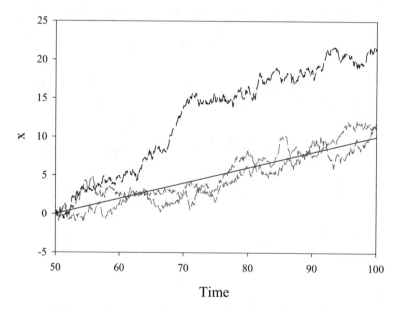

Fig. 7.5. Sample paths for a Brownian motion with drift

$$E[\Delta t] = \alpha \Delta t \qquad (7.92)$$
$$\nu[\Delta t] = \sigma^2 \Delta t \qquad (7.93)$$

For calculation of α, the average value of the differences in x $(E[x_t - x_{t-1}])$ is computed. Then this value is divided by the time interval Δt to obtain α.

However, for σ, the variance of the differences in x is found and divided by the time interval Δt. Then the square root of this value is computed.

Other examples of Ito processes are the geometric Brownian motion with drift (Equation (7.94) given below) and the mean reverting process (Equation (7.98), Figure 7.6).

$$dx = \alpha x dt + \sigma x dz \quad \text{geometric Brownian motion with drift} \quad (7.94)$$

In geometric Brownian motion, the percentage changes in x and $\Delta x / x$ are normally distributed (absolute changes are lognormally distributed). We can write Equation (7.94) in the following form if we write $F(x) = \log x$.

$$dF = (\alpha - \frac{\sigma^2}{2})dt + \sigma dz \quad (7.95)$$

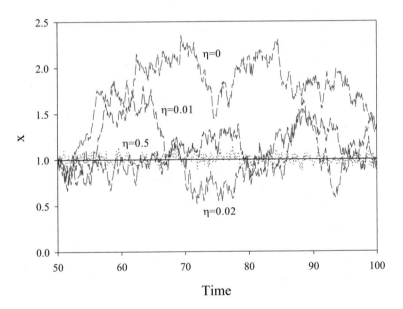

Fig. 7.6. Sample paths of a mean reverting process

Over the time interval t, the change in the logarithm of x is normally distributed with mean $(\alpha - (\sigma^2/2))t$ and variance $\sigma^2 t$. We can estimate the parameters of this Ito process following this procedure. First we find the variance of the changes in the logarithm of x, $(\ln x_t - \ln x_{t-1})$. When we divide this value by Δt, we can obtain σ^2. Once we know the value of σ, we can then calculate the mean value of the changes in logarithm of x, $(\ln x_t - \ln x_{t-1})$, which is equal to $(\alpha - (\sigma^2/2))t$. From that value, we can calculate α. It was shown that for the absolute value of x, Equations (7.96) (expected value) and (7.97) (variance) hold true:

$$E[x(t)] = x_0 . \exp(\alpha t) \tag{7.96}$$
$$\nu[x(t)] = x_0^2 . \exp(2\alpha t(\exp \sigma^2 t - 1)) \tag{7.97}$$

In mean reverting processes, the variable may fluctuate randomly in the short run, but in the longer run it will be drawn back toward the marginal value of the variable:

$$dx = \eta(x_{avg} - x)dt + \sigma dz \quad \text{mean reverting process} \tag{7.98}$$

where η is the speed of reversion and x_{avg} is the nominal level to which x reverts. The expected value of change in x depends on the difference between x and x_{avg}. If the current value of x is x_0, then the expected value of x at any future time and the variance of $x_t - x_{avg}$ is given by the following equations.

$$E[x(t)] = x_{avg} + (x_0 - x_{avg}) \exp(-\eta t) \tag{7.99}$$
$$\nu[x_t - x_{avg}] = \frac{\sigma^2}{2\eta}(1 - \exp(-2\eta t)) \tag{7.100}$$

From these equations, it could be observed that the expected value of x_t converges to x_{avg} as t becomes large and the variance converges to $\sigma^2/2\eta$. We can write Equation (7.98) in the following form.

$$x_t - x_{t-1} = \eta x_{avg} \Delta t - \eta x_{t-1} \Delta t + \sigma \epsilon_t \sqrt{\Delta t} \tag{7.101}$$
$$x_t - x_{t-1} = C_1 + C_2 x_{t-1} + e_t \tag{7.102}$$

In order to estimate the parameters, we can run the regression with the available discrete-time data (Equation (7.102)). In this equation, $C_1 = \eta x_{avg} \Delta t$, $C_2 = -\eta \Delta t$, and $e_t = \sigma \epsilon_t \sqrt{\Delta t}$. From the standard error of regression e_t, one can calculate the standard deviation σ.

In Equation (7.98), if the variance rate grows with x, we obtain the geometric mean reverting process:

$$dx = \eta(x_{avg} - x)dt + \sigma x dz \quad \text{geometric mean reverting process} \tag{7.103}$$

The procedure for parameter estimation for this process is the following. We can write Equation (7.103) in the following form.

$$x_t - x_{t-1} = \eta x_{avg} \Delta t - \eta x_{t-1} \Delta t + \sigma x_{t-1} \epsilon_t \sqrt{\Delta t} \tag{7.104}$$

If we divide both sides by x_{t-1}, Equation (7.105) is Obtained.

$$\frac{x_t - x_{t-1}}{x_{t-1}} = \frac{C_1}{x_{t-1}} + C_2 + e_t \tag{7.105}$$

In this equation, $C_1 = \eta x_{avg} \Delta t$, $C_2 = -\eta \Delta t$, and $e_t = \sigma \epsilon_t \sqrt{\Delta t}$. By running this regression using the available discrete-time data, we can find the values of C1 and C2, which enable us to predict the parameters in Equation (7.103). Again, from the standard error of regression, one can calculate the standard deviation σ.

7.4.1 Ito's Lemma

Ito's lemma is easier to understand as a Taylor series expansion. Suppose that $x(t)$ follows the process of Equation (7.89) and consider a function F that is at least twice differentiable in x and once in t. We would like to find the total differential of this function dF. The usual rules of calculus define this differential in terms of first-order changes in x and t:

$$dF = \frac{\partial F}{\partial t}\, dt + \frac{\partial F}{\partial x}\, dx \qquad (7.106)$$

But suppose that we also include higher-order terms for changes in x:

$$dF = \frac{\partial F}{\partial t}\, dt + \frac{\partial F}{\partial x}\, dx + \frac{1}{2}\frac{\partial^2 F}{\partial x^2}\,(dx)^2 + \frac{1}{6}\frac{\partial^3 F}{\partial x^3}\,(dx)^3 + \ \dots \ (7.107)$$

In ordinary calculus, these higher-order terms all vanish in the limit. For an Ito process following Equation (7.89), it can be shown that the differential dF is given in terms of first-order changes in t and second-order changes in x. Hence, Ito's lemma gives the differential dF as

$$dF = \frac{\partial F}{\partial t}\, dt + \frac{\partial F}{\partial x}\, dx + \frac{1}{2}\frac{\partial^2 F}{\partial x^2}\,(dx)^2 \qquad (7.108)$$

By substituting Equation (7.89) and $(dz)^2 = dt$ in Equation (7.108) and neglecting terms containing $(dt)^2$ and $dtdz$, an equivalent expression is obtained.

$$dF = \left[\frac{\partial F}{\partial t} + a(x,t)\frac{\partial F}{\partial x} + \frac{1}{2}b^2(x,t)\frac{\partial^2 F}{\partial x^2}\right] dt + b(x,t)\frac{\partial F}{\partial x}\, dz \quad (7.109)$$

Compared to the chain rule for differentiation in ordinary calculus (Equations (7.106) and (7.108)) has one extra term that captures the effect of convexity or concavity of F.

7.4.2 Dynamic Programming Optimality Conditions

We have seen that for the deterministic case when no uncertainty is present, the principle of optimality states that the minimum value is a function of the initial state and the initial time, resulting in the Hamilton–Jacobi–Bellman equation. The H-J-B equation states that, for the optimal control problem:

$$\text{Maximize } J = j(\bar{x}_t) + \int_0^T k\,(\bar{x}_t, \theta_t)\ dt \qquad (7.110)$$
$$\theta_t$$

subject to

$$\frac{d\bar{x}_t}{dt} = f(\bar{x}_t, \theta_t) \tag{7.111}$$

The optimality conditions are given by

$$0 = \frac{\partial J}{\partial t} + \underset{\theta_t}{\text{Maximize}} \left[k(\bar{x}_t, \theta_t) + \sum_i \frac{\partial J}{\partial x_i} \frac{dx_i}{dt} \right] \tag{7.112}$$

$$0 = \frac{\partial J}{\partial t} + \underset{\theta_t}{\text{Maximize}} \left[k(\bar{x}_t, \theta_t) + \sum_i \frac{\partial J}{\partial x_i} f(\bar{x}_t, \theta_t) \right] \tag{7.113}$$

where i represents the state variables in the problem.

However, when uncertainty is present in the calculation, the H-J-B equations are modified to obtain the following objective function.

$$\underset{\theta_t}{\text{Maximize}} \, J = E \left[j(\bar{x}_t) + \int_0^T k(\bar{x}_t, \theta_t) \, dt \right]$$

where E is the expectation operator. If the state variable i can be represented as an Ito process given by Equation (7.114), then Merton and Samuelson (1990) found that the optimality conditions are given by Equation (7.115).

$$dx_i = f_i(\bar{x}_t, \theta_t) \, dt + g_i(\bar{x}_t, \theta_t) \, dz \tag{7.114}$$

$$0 = \underset{\theta_t}{\text{Maximize}} \left[k(\bar{x}_t, \theta_t) + \frac{1}{dt} E(dJ) \right] \tag{7.115}$$

Following Ito's lemma Equation (7.115) results in

$$0 = k(\bar{x}_t, \theta_t^*) + \frac{\partial J}{\partial t} + \sum_i \frac{\partial J}{\partial x_i} f_i(\bar{x}_t, \theta_t^*)$$
$$+ \sum_i \frac{g_i^2}{2} \frac{\partial^2 J}{(\partial x_i)^2} + \sum_{i \neq j} g_i g_j \frac{\partial^2 J}{\partial x_i \, \partial x_j} \tag{7.116}$$

where θ^* represents the optimal solution to the maximization problem.

In Equation (7.114), σ_i is the variance parameter of the state variable x_i. Note that this definition implicitly restricts our analysis for the cases in which the behavior of the state variables can be represented as an Ito process. Also, the extra terms in Equation (7.116) come from the fact that second-order contributions of stochastic state variables are not negligible (see Equation (7.108) and Ito's lemma).

As stated earlier, the solution of a stochastic differential equation is not a value for the function, but it is a probability distribution that varies with time. This is a simplified form of stochastic differential equations. For the solution of more complex stochastic differential equations, readers are referred to Dixit and Pindyck (1994) .

Let us revisit the isoperimetric problem described earlier but with stochasticity as described in Example 7.5.

Example 7.5: Assume that in the isoperimetric problem, the state variable x_1 that represents the vertical displacement is stochastic but the differential perimeter or the total perimeter ($x_2(t)$ and L are not stochastic) is deterministic. Assume that change in x_1 is normally distributed with a variance parameter $\sigma = 0.5$ and follows a Brownian motion. The perimeter is given to be 16 cm. Solve this isoperimetric problem using stochastic dynamic programming and show the effect of uncertainties on the solution.

Solution: Now the isoperimetric problem formulated earlier can be rewritten as follows.

$$\underset{u_t}{\text{Maximize}} \quad A = \int_0^T x_1(t) \, dt \tag{7.117}$$

subject to

$$dx_1 = u_t dt + \sigma \, dz \quad x_1(0) = 0.0; \tag{7.118}$$

where $dz = \epsilon\sqrt{dt}$ (ϵ is a random number generated from a normal distribution with mean zero and standard deviation of one N(0,1)) and $\sigma = 0.5$.

$$\frac{dx_2}{dt} = \sqrt{1 + u_t^2} \quad x_2(0) = 0.0; \quad x_2(T) = L = 16.0 \tag{7.119}$$

Similar to the deterministic case, introducing a new variable $I(t) = \int_t^T x_1(\tau)d\tau$ leads to

$$\underset{u_t}{\text{Maximize}} \quad A = \int_0^t x_1(t) \, dt + I(t) \tag{7.120}$$

The optimality conditions for this problem derived from Equation (7.116) are

$$\frac{\partial I}{\partial t} + \underset{u_t}{\text{Maximize}} \, x_1(t) + u_t \frac{\partial I}{\partial x_1(t)} + \sqrt{1 + u_t^2} \frac{\partial I}{\partial x_2(t)} + \frac{\sigma^2}{2} \frac{\partial^2 I}{\partial x_1(t)^2} = 0 \tag{7.121}$$

or

$$\text{Maximize}_{u_t} \quad x_1(t) + I_t + I_{x_1}u_t + I_{x_2}\sqrt{1+u_t^2} + I_{x_2x_2}\sigma^2/2.0 = 0 \tag{7.122}$$

From Equation (7.118) and using the definition of I it can be shown that

$$\frac{dI_{x_2x_2}}{dt} = 0.0 \quad I_{x_2x_2}(T) = 0.0 \implies I_{x_2x_2} = 0.0 \tag{7.123}$$

Therefore, maximizing Equation (7.122) with respect to u_t leads to

$$I_{x_1} + \frac{u_t}{\sqrt{1+u_t^2}}I_{x_2} = 0 \tag{7.124}$$

Differentiating Equation (7.122) with respect to x_1 and x_2 leads to

$$\frac{dI_{x_1}}{dt} = -1 \implies I_{x_1} = -t + c_1 \tag{7.125}$$

$$\frac{dI_{x_2}}{dt} = 0 \implies I_{x_2} = c_2 \tag{7.126}$$

Therefore, the velocity parameter u_t follows the path given by

$$u_t = \frac{t - c_1}{\sqrt{c_2 - (t - c_1)^2}} \tag{7.127}$$

This suggests that the deterministic solution and the stochastic solution are the same. However, stochasticity is embedded in the differential equation for x_1 given by Equation (7.118). This is also obvious when one simulates a single instant of stochasticity by choosing a normal random process with a mean of zero and variance σ represented by the parameter ϵ in the following form of Equation (7.128).

$$dx_1 = u_t dt + \sigma \epsilon \sqrt{dt} \quad x_1(0) = 0.0; \tag{7.128}$$

Figure 7.7 shows the deterministic and stochastic path of variable x_1 numerically integrated using the set of equations given above and from Example 7.4.

It can be seen that although the stochastic solution follows a circular path, the expected area obtained in the stochastic case (Figure 7.8) is smaller than the area obtained in the deterministic case (Figure 7.9) for the same perimeter.

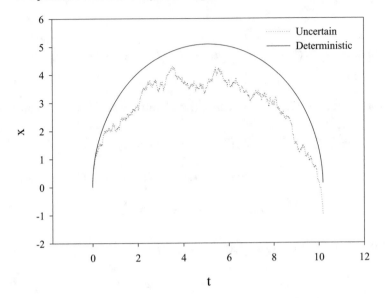

Fig. 7.7. Deterministic and stochastic path of variable x_1

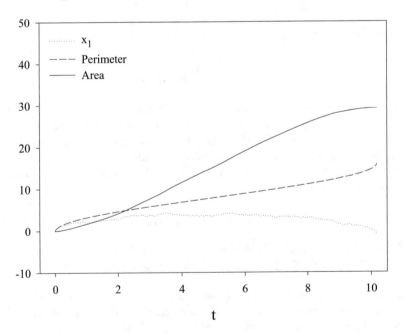

Fig. 7.8. Stochastic solution: x_1, perimeter, and area

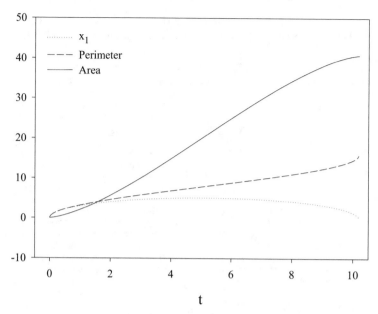

Fig. 7.9. Deterministic solution: x_1, perimeter, and area

7.4.3 Stochastic Maximum Principle

Rico-Ramirez and Diwekar (2004) derived stochastic maximum principle from the dynamic programming optimality condition.

Consider the generalized form of stochastic optimal control problem given below.

$$\text{Maximize } J = j(\overline{x}_t) + \int_0^T k(\overline{x}_t, \theta_t) \, dt \qquad (7.129)$$
$$\theta_t$$

subject to

$$dx_i = f_i(\overline{x}_t, \theta_t) \, dt + g_i(\overline{x}_t, \theta_t) \, dz \qquad (7.130)$$

We can write the optimality condition for stochastic dynamic programming given in Equation 7.116 in the form of two kinds of adjoint variables, namely, $z_i = \frac{\partial J}{\partial x_i}$ and $\omega_{ij} = \frac{\partial^2 J}{\partial x_i \partial x_j}$ as follows.

$$0 = k(\overline{x}_t, \theta_t^*) + \frac{\partial J}{\partial t} + \sum_i z_i f_i(\overline{x}_t, \theta_t^*) + \sum_i \sum_j \frac{g_i g_j}{2} \omega_{ij} \qquad (7.131)$$

Then, the Hamiltonian and adjoint equations from the stochastic maximum principle derived by taking first- and second-order derivative of Equation 7.131 with respect to x_k. For this derivation, we assume that the third-order differential equations like $\frac{\partial^3 J}{\partial x_i \partial x_j \partial x_k}$ can be neglected, Ito's lemma and chain rule are used for the calculation of derivatives, and uncertainties of each state variable can be assumed to be independent of each other $(\frac{\partial g_i}{\partial x_j})_{i \neq j} = 0)$.

$$H(\overline{\omega}_t, \overline{z}_t, \overline{x}_t, \theta_t) = \sum_{i=1}^{nx} z_i f_i(\overline{x}_t, \theta_t) + \sum_{i=1}^{nx} \sum_{j=1}^{nx} \frac{g_i g_j}{2} \omega_{ij} \qquad (7.132)$$

$$\frac{dz_i}{dt} = -\sum_{j=1}^{nx} z_j \frac{\partial f_j}{\partial x_i} - \sum_{j=1}^{nx} \omega_{ji} \frac{g_j}{2} \frac{\partial g_i}{\partial x_i} \qquad (7.133)$$

$$\overline{z}_T = c \qquad (7.134)$$

$$\frac{d\omega_{ik}}{dt} = -\sum_{j=1}^{nx} (\omega_{ji} \frac{\partial f_j}{\partial x_k} + \omega_{jk} \frac{\partial f_j}{\partial x_i} + z_j \frac{\partial^2 f_j}{\partial x_i \partial x_j})$$

$$- \frac{\partial g_i}{\partial x_i} \frac{\partial g_k}{\partial x_k} \omega_{ik} - [\sum_{j=1}^{nx} \frac{g_j}{2} \omega_{ii} \frac{\partial^2 g_i}{(\partial x_i)^2}]_{for \ i=k} \qquad (7.135)$$

$$\overline{\omega}_T = 0 \qquad (7.136)$$

Note that there are additional terms in equations for adjoints (z_is) and there are additional adjoints (ωs) because of uncertainties.

7.5 Reversal of Blending: Optimizing a Separation Process

The hazardous waste case study presented in earlier chapters involved a blending process that essentially mixes several chemical components to form a minimum volume (glass) mixture. The reverse of this blending process is a separation process, where the chemicals are separated using a physical phenomenon such as boiling. Distillation is the oldest separation process commonly used to separate mixtures. Separation of alcohol from water to form desired spirits such as whiskey, brandy, vodka, and the like is a popular example of this process. The basic principle in this separation is that the two components (e.g., alcohol and water) boil at two different temperatures. Simple distillation, also called differential distillation or Rayleigh distillation, is the most elementary example of batch distillation. For a full treatment of the theory and principles of batch distillation, please refer to Diwekar (1995).

In this section we present the deterministic and stochastic optimal control problems in batch distillation. The data, formulation, and computer code for this case study can be found on the Springer website with the book link. Although the deterministic optimal control problems in batch distillation appeared as early as 1963, stochastic optimal control is a recent area of study that draws a parallel between the real option theory in finance and batch distillation optimal control with uncertainties (Rico-Ramirez and Morel, 2003; Ulas and Diwekar, 2004).

As shown in Figure 7.10, in this process, the vapor is removed from the still (reboiler) during each time interval and is condensed in the condenser. This vapor is richer in the more volatile component (lower boiling) than the liquid remaining in the still. Over time, the liquid remaining in the still becomes weaker in its concentration of the more volatile component, and the distillate collected in the condenser gets progressively enriched in the more volatile component. The purity of distillate in this process is governed by the boiling point relations generally characterized by a thermodynamic term called *relative volatility*, defined below.

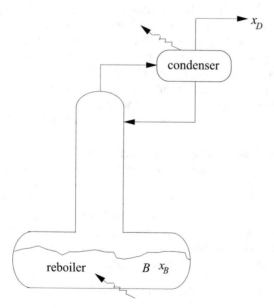

Fig. 7.10. Schematic of a simple distillation operation

Relative volatility, α, of a binary mixture is defined in terms of the ratio of the vapor composition of the more volatile (lower boiling) component (1), y_1 or x_D to the vapor composition of the lower volatile (higher boiling) component (2), y_2 or $(1 - x_D)$, and the ratio of the liquid composition, x_1 or x_B, of the more volatile component and the liquid composition of the lower volatile component, x_2 or $(1 - x_B)$.

$$\alpha = \frac{\alpha_1}{\alpha_2} = \frac{y_1/y_2}{(x_1/x_2)} = \frac{x_D/(1 - x_D)}{x_B/(1 - x_B)} \qquad (7.137)$$

The relative volatility provides the equilibrium relationship between the distillate composition x_D and the liquid composition in the reboiler x_B for the simple distillation process. This is because in the distillation process, it is assumed that the vapor formed within a short period is in thermodynamic equilibrium with the liquid.

One can look at simple distillation as consisting of one equilibrium stage where liquid and vapor are in contact with each other, and the transfer takes place between the two phases, as shown in Figure 7.11a. If N such stages are stacked one above the other, as shown in Figure 7.11b, and are allowed to have successive vaporization and condensation, this multistage process results in a substantially richer vapor and weaker liquid in terms of the more volatile component in the condenser and the reboiler, respectively. This multistage arrangement, shown in Figure 7.11b, is representative of a distillation column, where the vapor from the reboiler rises to the top and the liquid from the condenser is refluxed downward. The contact between the liquid and the vapor phase is established through accessories such as packings or plates. However, it is easier to express the operation of the column in terms of thermodynamic equilibrium stages (in terms of the relative volatile relationships at each stage), which represent the theoretical number of plates in the column, a concept used to design a column. Figure 7.12 shows the schematic of the batch distillation column, where the number of theoretical stages is numbered from top (1) to bottom (N).

In general, as the amount of reflux, expressed in terms of reflux ratio R (defined by the ratio of liquid flow refluxed to the product [distillate] rate withdrawn), increases, the purity of the distillate increases. A similar effect is observed as the number of stages (height) increases in the column. In summary, there is an implicit relation between the top composition x_D and the bottom composition x_B, which is a function of the relative volatility α, reflux ratio R, and number of stages N. The changes in the process can be modeled using differential material balance equations.

$$x_D = f(x_B, R, N) \tag{7.138}$$

Under the assumption of a constant boilup rate V and no holdup (no liquid on plates except in the reboiler) conditions, an overall differential material balance equation over a time dt may be written as

$$\frac{dx_1}{dt} = \frac{dB_t}{dt} = \frac{-V}{R_t + 1}, \quad x_1(0) = B_0 = F, \tag{7.139}$$

where F is the initial feed at time $t = 0$; that is, $x_1(0) = F$. F is also related to the distillate (top product) D by $F = B + D$. Similarly, a material balance for the key component 1 over the differential time dt is

$$\frac{dx_2}{dt} = \frac{V}{R_t + 1} \frac{(x_B^{(1)} - x_D^{(1)})}{B_t}, \quad x_2(0) = x_F^{(1)} \tag{7.140}$$

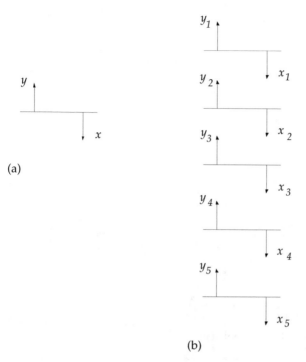

Fig. 7.11. Equilibrium stage processes: **(a)** single stage process, **(b)** multistage process

In the above two equations

B_t = quantity of charge remaining in the reboiler or bottoms (function of time), also represented as B (mol).

D_t = quantity of distillate or product (function of time), also represented as D (mol).

F = initial charge or feed (mol).

R_t = control variable vector, reflux ratio (function of time).

t = batch time (hr).

x_1 = a state variable representing the quantity of charge remaining in the still, B_t (mol).

x_2 = a state variable representing the composition of the key component in the still at time t, $x_B^{(1)}$ (mole fraction).

V = molar vapor boilup rate (mol h^{-1}).

x_B = the bottom or reboiler composition for the key component 1, also represented as $x_B^{(1)}$ (mole fraction).

$x_D^{(1)}$ = the overhead or distillate composition for the key component 1, also represented as x_D (mole fraction).

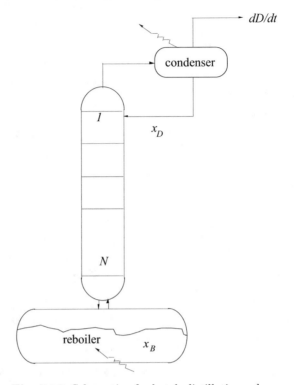

Fig. 7.12. Schematic of a batch distillation column

$x_F^{(1)}$ = the feed composition for the key component 1, also represented as x_F (mole fraction).

Diwekar and co-workers developed a short-cut model for the implicit relation (Equation (7.138)) between x_D and x_B. The additional variables for the short-cut method are given below.

C_1 = constant in the Hengestebeck–Geddes' (HG) equation.
N = number of plates in the column (given).
n = number of components, for binary system $n = 2$.
N_{min} = minimum number of plates.
R_{min} = minimum reflux ratio.

Greek Symbols

α_i = relative volatility of component i.
ϕ = constant in the Underwood equations.

Functional Relationship Between x_D and x_B

At each instant, there is a change in the still composition of the key component 1, resulting in changes in the still composition of all the other components calculated by the differential material balance equations described earlier. The

Hengestebeck–Geddes equation, given below, relates the distillate compositions to the new bottom compositions in terms of a constant C_1.
   The Hengestebeck–Geddes equation:

$$x_D^{(i)} = \left(\frac{\alpha_i}{\alpha_1}\right)^{C_1} \frac{x_D^{(k)}}{x_B^{(k)}} x_B^{(i)}, \quad i = 1, 2, \ldots, n(i \neq k) \tag{7.141}$$

Summation of the distillate composition provides the relation for $x_D^{(1)}$.

$$x_D^{(1)} = \frac{1}{\sum_{i=1}^{n} \left(\frac{\alpha_i}{\alpha_1}\right)^{C_1} \frac{x_B^{(i)}}{x_B^{(1)}}} \tag{7.142}$$

It can be proved that the constant C_1 of the Hengestebeck–Geddes equation is equivalent to the minimum number of plates, N_{min}, in the Fenske
equation.
The Fenske equation is

$$N_{min} = C_1 = \frac{\ln\left[\frac{x_D^{(i)}}{x_D^{(k)}} \frac{x_B^{(k)}}{x_B^{(i)}}\right]}{\ln[\alpha_i]} \tag{7.143}$$

At minimum reflux (R_{min}), an infinite number of equilibrium stages are
required to achieve the given separation.
The Underwood equations for minimum reflux are

$$\sum_{i=1}^{n} \frac{\alpha_i x_B^{(i)}}{\alpha_i - \phi} = 0 \tag{7.144}$$

$$R_{min} + 1 = \sum_{i=1}^{n} \frac{\alpha_i x_D^{(i)}}{\alpha_i - \phi} \tag{7.145}$$

Design variables of the column such as the reflux ratio R and the number of
plates N are related to each other by Gilliland's correlation (Gilliland, 1940)
through the values of R_{min} and N_{min}.
Gilliland's correlation is

$$Y = 1 - \exp\left[\frac{(1 + 54.4X)(X - 1)}{(11 + 117.2X)\sqrt{X}}\right] \tag{7.146}$$

in which

$$X = \frac{R - R_{min}}{R + 1}; \quad Y = \frac{N - N_{min}}{N + 1}$$

From the above equations, it can be seen that the short-cut method has
only two differential equations (Equation (7.139) and (7.140)) and the rest of
the equations are algebraic.

At any instant of time, there is a change in the still composition of the key component (state variable, x_1), the HG equation relates the distillate composition to the new still composition in terms of the constant C_1. This constant C_1 in the HG equation is equivalent to the minimum number of plates N_{min} in the Fenske equation. At this stage, R, C_1, and $x_D^{(1)}$ are the unknowns. Summation of the distillate composition can be used to obtain $x_D^{(1)}$ and the Fenske–Underwood–Gilliland (FUG) equations to obtain C_1. Obtaining the unknown R, referred to as the optimal control variable, is the aim of this case study.

The Maximum Distillate Problem

Where the amount of distillate, D, of a specified concentration for a specified time is maximized.

Converse and Gross (1963) were the first to report the maximum distillate problem for binary batch distillation columns, which was solved using the calculus of variations, Pontryagin's maximum principle, and the dynamic programming scheme.

The maximum distillate problem, described in the literature as early as 1963 (Converse and Gross, 1963), can be represented as follows.

$$\underset{R_t}{\text{Maximize}} \quad J = \int_0^T \frac{dD}{dt} \, dt = \int_0^T \frac{V}{R_t + 1} \, dt, \qquad (7.147)$$

subject to the following purity constraint on the distillate

$$x_{Dave} = \frac{\int_0^T x_D^{(1)} \frac{V}{R_t + 1} \, dt}{\int_0^T \frac{V}{R_t + 1} \, dt} = x_D^* \qquad (7.148)$$

where x_D^* is the specified distillate purity.

Equations (7.139) and (7.140) and the FUG-based short-cut model of the column, which provides correlations between the model parameters and the state variables (Diwekar et al., 1987).

The calculus of variations and the maximum principle formulations for this problem are presented first. This is followed by the dynamic programming formulation, which is then extended further to the stochastic dynamic programming formulation for uncertainty considerations.

7.5.1 Calculus of Variations Formulation

Now we formulate the maximum distillate problem using the calculus of variations. Because this problem contains equality constraints, we need to use the Euler-Lagrangian formulation. First, all three equality constraints (the two differential equations for state variables and the purity constraint) are augmented to the objective function to form a new objective function L given by

$$\text{Maximize} \quad L = \int_0^T \frac{V}{R_t + 1} \left[1 - \lambda(x_D^* - x_D^{(1)})\right] - \mu_1 \left[\frac{dx_1}{dt} - \frac{-V}{R_t + 1}\right]$$
$$x_1,\ x_2,\ R_t$$

$$-\mu_2 \left[\frac{dx_2}{dt} - \frac{V}{R_t + 1} \frac{(x_B^{(1)} - x_D^{(1)})}{x_1}\right] dt$$

$$(7.149)$$

where λ is a scalar Lagrange multiplier and μ_i, $i = 1, 2$ are the Lagrangian multipliers as a function of time. Using $dx/dt = x'$:

$$\text{Maximize} \quad L = \int_0^T \frac{V}{R_t + 1} dt \left[1 - \lambda(x_D^* - x_D^{(1)})\right]$$
$$x_1,\ x_2,\ R_t$$

$$- \mu_1 \left[x_1' - \frac{-V}{R_t + 1}\right] - \mu_2 \left[x_2' - \frac{V}{R_t + 1} \frac{(x_B^{(1)} - x_D^{(1)})}{x_1}\right]$$

$$(7.150)$$

Application of the Euler differential equations leads to the following three Euler-Lagrangian equations.

$$\frac{\partial L}{\partial x_1} - \frac{d(\frac{\partial L}{\partial x_1'})}{dt} = 0 \Longrightarrow \frac{d\mu_1}{dt} = \mu_2 \left[\frac{V}{R_t + 1} \frac{(x_2 - x_D^{(1)})}{x_1^2}\right] \qquad (7.151)$$

$$\frac{\partial L}{\partial x_2} - \frac{d(\frac{\partial L}{\partial x_2'})}{dt} = 0 \Longrightarrow$$

$$\frac{d\mu_2}{dt} = -\frac{V}{R_t + 1} \lambda \left(\frac{\partial x_D^{(1)}}{\partial x_2}\right)_{R_t} - \mu_2 \frac{V}{x_1(R_t + 1)} \left[1 - \left(\frac{\partial x_D^{(1)}}{\partial x_2}\right)_{R_t}\right]$$

$$(7.152)$$

$$\frac{\partial L}{\partial R_t} - \frac{d(\frac{\partial L}{\partial R_t'})}{dt} = 0 \Longrightarrow$$

$$R_t = \frac{\left[\frac{\mu_2}{x_1}(x_2 - x_D^{(1)}) - \mu_1 - \lambda(x_D^* - x_D^{(1)}) + 1\right]}{\frac{\partial x_D^{(1)}}{\partial R_t} \left(\lambda - \frac{\mu_2}{x_1}\right)} - 1 \qquad (7.153)$$

7.5.2 Maximum Principle Formulation

Again, the maximum distillate problem can be written as

$$\text{Maximize} \quad J = \int_0^T \frac{dD}{dt} \, dt = \int_0^T \frac{V}{R_t + 1} \, dt \quad (7.154)$$
$$R_t$$

subject to the following purity constraint on the distillate.

$$x_{Dav} = \frac{\int_0^T x_D^{(1)} \frac{V}{R_t + 1} \, dt}{\int_0^T \frac{V}{R_t + 1} \, dt} = x_D^* \quad (7.155)$$

The constraint on the purity is removed by employing the method of Lagrange multipliers. By combining Equations (7.154) and (7.155):

$$\text{Maximize} \quad L = \int_0^T \frac{V}{R_t + 1} \left[1 - \lambda(x_D^* - x_D^{(1)}) \right] \, dt \quad (7.156)$$
$$R_t$$

where λ is a Lagrange multiplier. Now the objective function is to maximize L, instead of J. To solve this problem, an additional state variable x_3 is introduced, which is given by

$$x_3 = \int_0^t \frac{V}{R_t + 1} \left[1 - \lambda(x_D^* - x_D^{(1)}) \right] \, dt \quad (7.157)$$

The problem can then be rewritten as

$$\text{Maximize} \quad x_3(T) \quad (7.158)$$
$$R_t$$

subject to the following differential equations for the three state variables and the time-implicit model for the rest of the column.

$$\frac{dx_1}{dt} = \frac{-V}{R_t + 1}, \quad x_1(0) = B_0 = F, \quad (7.159)$$

$$\frac{dx_2}{dt} = \frac{V}{R_t + 1} \frac{(x_2 - x_D^{(1)})}{x_1}, \quad x_2(0) = x_F^{(1)} \quad (7.160)$$

$$\frac{dx_3}{dt} = \frac{V}{R_t + 1} \left[1 - \lambda(x_D^* - x_D^{(1)}) \right] \, dt \quad (7.161)$$

The Hamiltonian function, which should be maximized, is

$$H_t = -z_1 \frac{V}{R_t + 1} + z_2 \frac{V(x_2 - x_D^{(1)})}{(R_t + 1)x_1} + z_3 \frac{V}{R_t + 1} \left[1 - \lambda(x_D^* - x_D^{(1)}) \right] \quad (7.162)$$

and the adjoint equations are

$$\frac{dz_1}{dt} = z_2 \frac{V(x_2 - x_D^{(1)})}{(R_t + 1)(x_1)^2}, \ z_1(T) = 0, \tag{7.163}$$

$$\frac{dz_2}{dt} = -z_2 \frac{V(1 - \frac{\partial x_D^{(1)}}{\partial x_2})}{(R_t + 1)x_1} - z_3\lambda \frac{V}{(R_t + 1)}(\frac{\partial x_D^{(1)}}{\partial x_2}), \ z_2(T) = 0 \tag{7.164}$$

and

$$\frac{dz_3}{dt} = 0, \ z_3(T) = 1 \implies z_3(t) = 1 \tag{7.165}$$

The Hamiltonian function in Equation (7.162) can be written as

$$H_t = -z_1 \frac{V}{R_t + 1} + z_2 \frac{V(x_2 - x_D^{(1)})}{(R_t + 1)x_1} + \frac{V}{R_t + 1}\left[1 - \lambda(x_D^* - x_D^{(1)})\right] \tag{7.166}$$

and

$$\frac{dz_2}{dt} = -z_2 \frac{V\left(1 - \frac{\partial x_D^{(1)}}{\partial x_2}\right)}{(R_t + 1)x_1} - \lambda \frac{V}{(R_t + 1)}\left(\frac{\partial x_D^{(1)}}{\partial x_2}\right), \ z_2(T) = 0 \tag{7.167}$$

From the optimality condition $\partial H / \partial R_t = 0$, it follows that

$$R_t = \frac{\left[\frac{z_2}{x_1}(x_2 - x_D^{(1)}) - z_1 - \lambda(x_D^* - x_D^{(1)}) + 1\right]}{\frac{\partial x_D^{(1)}}{\partial R_t}\left(\lambda - \frac{z_2}{x_1}\right)} - 1 \tag{7.168}$$

It can be seen from Equation (7.153) in the calculus of variations formulation and from Equation (7.168) that the two formulations lead to the same results where, in the case of the calculus of variations, the time-dependent Lagrange multipliers μ_i are equivalent to the adjoint variables z_i in the maximum principle formulation.

We have examined the two methods to solve the maximum distillate problem in batch distillation. The two different methods gave the same results. However, in the case of the calculus of variations the problem is in the form of second-order differential equations that may be difficult to solve. However, the maximum principle leads to a two-point boundary value problem. (We know initial boundary conditions for the state variables x_i but not the final boundary conditions for the adjoint variables z_i.) So to obtain the exact solution, one has to solve this problem iteratively using numerical methods. This can be accomplished using several different methods including the shooting method and the method of steepest ascent of the Hamiltonian. Although the details of the methods are beyond the scope of this book, the typical computational intensity involved in solving these problems is illustrated using the method of steepest ascent of Hamiltonian (Diwekar et al., 1987) given in the next section.

7.5.3 Method of Steepest Ascent of Hamiltonian

The solution procedure is described below using the maximum principle formulation and the quasi-steady-state short-cut method to model the batch distillation column.

The maximum principle formulation of the maximum distillate problem involves the solution of the following equations.

State variable differential equations:

$$\frac{dx_1}{dt} = \frac{-V}{R_t + 1}, \quad x_1(0) = B_0 = F, \tag{7.169}$$

$$\frac{dx_2}{dt} = \frac{V}{R_t + 1} \frac{(x_2 - x_D^{(1)})}{x_t^1}, \quad x_2(0) = x_F^{(1)} \tag{7.170}$$

The adjoint equations:

$$\frac{dz_1}{dt} = z_2 \frac{V(x_2 - x_D^{(1)})}{(R_t + 1)(x_1)^2}, \quad z_1(T) = 0, \tag{7.171}$$

$$\frac{dz_2}{dt} = -z_2 \frac{V\left(1 - \frac{\partial x_D^{(1)}}{\partial x_2}\right)}{(R_t + 1)x_1} - \lambda \frac{V}{(R_t + 1)}\left(\frac{\partial x_D^{(1)}}{\partial x_2}\right), \quad z_2(T) = 0 \tag{7.172}$$

Optimality conditions result in the following reflux ratio profile.

$$R_t = \frac{\left[\frac{z_2}{x_1}(x_2 - x_D^{(1)}) - z_1 - \lambda(x_D^* - x_D^{(1)}) + 1\right]}{\frac{\partial x_D^{(1)}}{\partial R_t}\left(\lambda - \frac{z_2}{x_1}\right)} - 1 \tag{7.173}$$

Basically, now the problem is reduced to finding out the solution of Equation (7.173) using Equations (7.169)–(7.172) and the short-cut method equations. The equations also involve the Lagrange multiplier λ, which is a constant for a specific value of final time T. So the above equations must be solved for different values of λ until the purity constraint given below is satisfied.

$$x_{Dav} = \frac{\int_0^T x_D^{(1)} \frac{V}{R_t + 1} dt}{\int_0^T \frac{V}{R_t + 1} dt} = x_D^* \tag{7.174}$$

It can be seen that the solution of these equations involves a two-point boundary value problem, where the initial values of the state variables x_1 and x_2 and the final values of the adjoint variables z_1 and z_2 are known. We seek the maximum value of H by choosing the decision vector R_t. The method of the steepest ascent of the Hamiltonian accomplishes this by using an iterative procedure to find R_t, the optimal decision vector. An initial estimate of R_t is obtained, which is updated during each iteration. If the decision vector R_t is divided into r time intervals, then for the ith component of the decision vector, the following rule is used for proceeding from the jth to $j+1$th approximation.

$$R_i(j+1) \;=\; R_i(j) \;+\; k\frac{\partial H}{\partial R_i}, \quad i = 1, 2, \ldots, r \qquad (7.175)$$

where k is a suitable constant. The iterative method is used until there is no further change in R_t. The value of k should be small enough so that no instability will result, yet large enough for rapid convergence. It should be noted that the sign of k is important, because $\partial H/\partial R_t \to 0$ at and near the total reflux condition, which gives the minimum value of H (i.e., minimum distillate). Also, one has to iterate on the Lagrange multiplier in the outer loop so that the purity constraint is satisfied. Figure 7.13 describes this iterative procedure.

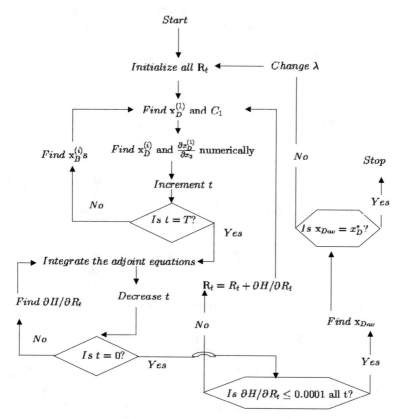

Fig. 7.13. Flowchart for the iterative procedure

7.5.4 Combining Maximum Principle and NLP Techniques

The maximum principle formulation of batch distillation, as also dynamic programming (described later) and the calculus of variations, is widely used in the batch distillation literature. However, solution of the two-point boundary value problem and the additional adjoint equations with the iterative constraint satisfaction can be computationally very expensive. A new approach to optimal control problems in batch distillation, proposed by Diwekar (1992), combines the maximum principle and the NLP techniques. This approach is illustrated in the context of the batch distillation case study.

The Maximum Distillate Problem Revisited

The maximum distillate problem in the original form (without considering the Lagrangian formulation as shown earlier) can be written as

$$\text{Maximize} \quad -x_1(T) \tag{7.176}$$
$$R_t$$

subject to the following differential equations, and the time-implicit FUG model. The Hamiltonian function, which should be maximized, is

$$H_t = -z_1 \frac{V}{R_t + 1} + z_2 \frac{V(x_2 - x_D^{(1)})}{(R_t + 1)x_1} \tag{7.177}$$

The adjoint equations are

$$\frac{dz_1}{dt} = z_2 \frac{V(x_2 - x_D^{(1)})}{(R_t + 1)(x_1)^2}, \quad z_1(T) = -1, \tag{7.178}$$

$$\frac{dz_2}{dt} = -z_2 \frac{V\left(1 - \frac{\partial x_D^{(1)}}{\partial x_2}\right)}{(R_t + 1)x_1}, \quad z_2(T) = 0 \tag{7.179}$$

Combining the two adjoint variables z_1 and z_2 into one using $z_t = z_2/z_1$ results in the following adjoint equation.

$$a\frac{dz_t}{dt} = -z_t \frac{V\left(1 - \frac{\partial x_D^{(1)}}{\partial x_2}\right)}{(R_t + 1)x_1} - (z_t)^2 \frac{V(x_2 - x_D^{(1)})}{(R_t + 1)(x_1)^2} \tag{7.180}$$

The optimality condition on the reflux policy $dH_t/dR_t = 0$ leads to

$$R_t = \frac{B_t - z_t(x_B^{(1)} - x_D^{(1)})}{z_t(\partial x_D^{(1)}/\partial R_t)} - 1 \tag{7.181}$$

It should be remembered that this solution (Equation (7.181)) is obtained by maximizing the Hamiltonian (maximizing the distillate), which does not incorporate the purity constraint. Hence, use of the final boundary condition ($z_T = 0$) provides the limiting solution corresponding to $R = -\infty$ with the lowest overall purity. Because in this formulation the purity constraint is imposed external to the Hamiltonian, the final boundary condition ($z_T = 0$) is no longer valid. We seek the initial value of z_t (z_0) that will satisfy the purity constraint. The iteration variable z_0 can also be interchanged with R_0. As can be seen, this method avoids the multiple iteration loops shown in Figure 7.13.

7.5.5 Uncertainties in Batch Distillation

In order to model a system under uncertainty, a quantitative description of the variations expected must be established. If we consider a mathematical formulation of a dynamic process model as a set of differential algebraic equations (for details, please refer to Naf (1994)):

$$g\left(x, \dot{x}, u, \eta, t\right) = 0 \tag{7.182}$$

with the initial conditions

$$x\left(t = 0\right) = x_0 \tag{7.183}$$

where x are the state variables, u are the input variables, and η are the model parameters, then qualitatively different sources of uncertainty may be located as follows.

1. Uncertainty with respect to the model parameters η. These parameters are a part of the deterministic model and not actually subject to randomness. Theoretically, their value is an exact number. The uncertainty results from the impossibility of modeling the physical behavior of the system exactly.
2. Uncertainty in the input variables u. This kind of uncertainty originates from the random nature and unpredictability of certain process inputs (e.g., feed composition uncertainty).
3. Uncertainty in the initial conditions x_0 (initial charge of a batch, for instance).

The representation of uncertainties for all three categories is usually in terms of distribution functions. Although there are instances of all three sources of uncertainties in batch distillation, in this work we focus on the first category, the case of uncertainty with respect to the model parameters. In general, optimal control problems are considered to be open loop control problems where the optimal reflux profile is generated a priori using a model and then the controller is asked to follow this trajectory. This trajectory would be optimal when the model is an exact replica of the physical phenomena. However, very often this is not the case, and online updating of the profile is necessary. This calls for use of simplified models such as the short-cut model described earlier. The main uncertainty in this model is related to the assumption of constant relative volatility. Therefore, we focus on handling uncertainty in this important thermodynamic parameter here. In practice, this relative volatility parameter α varies with respect to the number of plates in a column as well as with respect to time. We show later that this behavior of the relative volatility can be captured by the geometric Brownian motion representation. Note that with such a representation, we can not only capture the uncertainty in this crucial parameter but also gain all the advantages of using the short-cut model for faster optimal control calculations and efficient online updating.

7.5.6 Relative Volatility: An Ito Process

What is common between the technology stock price example given earlier and the uncertainty in the relative volatility parameter in the batch distillation models?

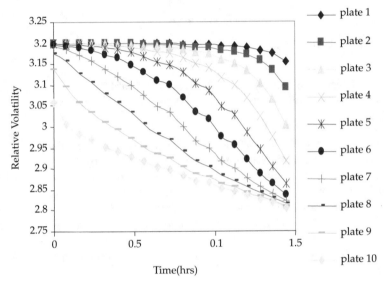

Fig. 7.14. Relative volatility, ideal system as a function of time and number of plate

1. Both have time-dependent variations. The technology stock fluctuates around the mean randomly but over time has a positive expected rate of growth. Relative volatility for an ideal system, however, fluctuates around the geometric mean across the column height, but over a time period the mean decreases (Figure 7.14 shows the relative volatility fluctuations for a pentane–hexane system).
2. Similar to the stock prices, relative volatility can be modeled as a Markov process because, at any time period, the value of relative volatility depends only on the previous value. The changes for both are nonoverlapping.
3. Whether uncertainty in the relative volatility parameter can be represented by a Wiener process can be shown with some simple numerical experiments where the data are generated from a rigorous simulation model (proxy for experiments) for various thermodynamic systems.

In this section, we present the result of a simple numerical experiment to show that the behavior of the relative volatility in a batch column can indeed be represented as an Ito process. We take two examples, the first one is the relative volatility of an ideal system with the pentane–hexane mixture. Figure 7.14 shows the behavior of the relative volatility with respect to time and the plate number for this example. A rigorous simulation with a simulation package $MultiBatchDS^{TM}$ (Diwekar, 1996) was performed in a batch column to obtain the behavior of the relative volatility with respect to time. As we know, the relative volatility is different for each plate of the column at each point in time. This can be captured by a geometric Brownian motion. The equation for geometric Brownian motion (special instance of an Ito process) is

$$da = \alpha\beta dt + \alpha\sigma dz \qquad (7.184)$$

where β and σ are constants.

Equation (7.184) establishes that the changes in relative volatility are log-normally distributed with respect to time. In fact, by using Ito's lemma, it can be shown that Equation 7.184 implies that the change in the logarithm of α is normally distributed (for a finite time interval t) with mean $(\beta - 1/2\sigma^2)\,t$ and variance $\sigma^2 t$, resulting in Equation (7.185) (Dixit and Pindyck, 1994).

$$d(\ln\alpha) = \left(\beta - \frac{1}{2}\sigma^2\right)dt + \sigma dz \qquad (7.185)$$

In Equations (7.184) and (7.185), dz is defined as

$$dz = \epsilon_t\sqrt{dt}$$

where ϵ_t is drawn from a normal distribution with a mean of zero and unit standard deviation. By using the time series data for relative volatility, natural logarithm of relative volatility for a fixed time interval can be used to obtain the mean and variance of the underlying normal distribution. It has been found that the data shown in Figure 7.14 fit well with this representation. Then, $\alpha(t)$ can be calculated by using Equation (7.186) given below.

$$\alpha_t = (1 + \beta\,\Delta t)\,\alpha_{t-1} + \sigma\,\alpha_{t-1}\,\epsilon_t\,\sqrt{\Delta t} \qquad (7.186)$$

Consider now a system of a non-ideal mixture, such as ethanol–water studied by Ulas and Diwekar (2004). This mixture results in a different relative volatility profile from an ideal mixture as shown in Figure 7.15a. It was found that this behavior can be best modeled with a geometric mean reverting process rather than a geometric Brownian motion(Figure 7.15b). The equation for the geometric mean reverting process is

$$da = \eta(\alpha_{avg} - \alpha)dt + \alpha\sigma dz \qquad (7.187)$$

In this equation it is expected that the α value reverts to α_{avg}, but the variance rate grows with α. Here, η is the speed of reversion, and α_{avg} is the "normal" level of α, that is, the level that tends to revert. In order to predict the constants in Equation (7.187), a regression analysis can be performed using the available discrete-time data similar to the ideal system presented earlier. We can write this equation in the discrete form as follows.

$$\alpha_t = \eta\,\alpha_{avg}\,\Delta t + (1 - \eta\,\Delta t)\,\alpha_{t-1} + \sigma\,\alpha_{t-1}\,\epsilon_t\,\sqrt{\Delta t} \qquad (7.188)$$

If we compare the equations for geometric Brownian motion (Equation (7.186)) and geometric mean reverting process (Equation (7.188)), we can see

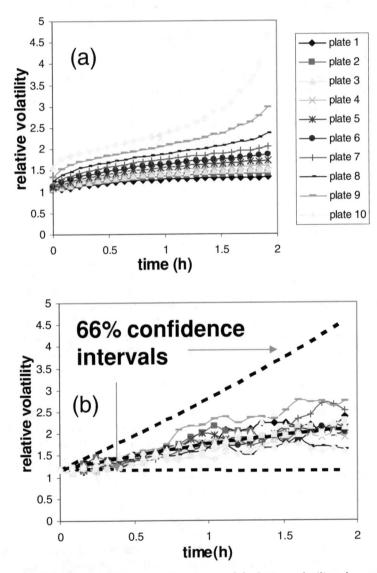

Fig. 7.15. Relative volatility as an Ito process: (**a**)relative volatility changes for a non-ideal mixture, (**b**) Ito process representation

that these equations differ from each other by the constant term $\eta \, \alpha_{avg} \, \Delta t$. This constant term reflects the reversion trend. Using this equation, the sample paths for mean reverting process for a different set of random numbers (ϵ_t in Equation (7.188)) are drawn from a unit normal distribution as shown in Figure 7.15b. Figure 7.15a,b confirms that the relative volatility of this mixture can be represented by the geometric mean reverting process.

7.5.7 Optimal Reflux Profile: Deterministic Case

For the deterministic case, the maximum distillate problem (Problem A) is given by

$$\text{Maximize} \quad L = \int_0^T \frac{V}{R_t + 1} \left[1 - \lambda(x_D^* - x_D^{(1)}) \right] \, dt$$
$$R_t \tag{7.189}$$

subject to

$$\frac{dx_1}{dt} = \frac{-V}{R_t + 1}, \quad x_1(0) = B_0 = F \tag{7.190}$$

$$\frac{dx_2}{dt} = \frac{V}{R_t + 1} \frac{(x_2 - x_D^{(1)})}{x_1}, \quad x_2(0) = x_F^{(1)} \tag{7.191}$$

As mentioned above, for the deterministic case, the optimality conditions (Equation (7.116)) reduce to the H-J-B equation (Equation (7.112)). By applying such conditions to Problem A, we obtain

$$0 = \frac{\partial L}{\partial t} + \text{Maximize} \frac{V}{R_t + 1} \left[1 - \lambda(x_D^* - x_D^{(1)}) \right]$$
$$R_t \qquad - \frac{\partial L}{\partial x_1} \frac{V}{R_t + 1} + \frac{\partial L}{\partial x_2} \left[\frac{V(x_2 - x_D^{(1)})}{(R_t + 1)x_1} \right]$$
$$\tag{7.192}$$

Then, simplifying

$$0 = \left[1 - \lambda \left(x_D^* - x_D^{(1)} \right) - \frac{\partial L}{\partial x_1} + \frac{\partial L}{\partial x_2} \left(\frac{x_2 - x_D^{(1)}}{x_1} \right) \right] \left[-\frac{V}{(R_t + 1)^2} \right]$$
$$+ \left[\frac{V}{R_t + 1} \right] \left[\lambda \frac{\partial x_D^{(1)}}{\partial R_t} - \frac{\partial L}{\partial x_2} \frac{1}{x_1} \frac{\partial x_D^{(1)}}{\partial R_t} \right]$$
$$\tag{7.193}$$

So,

$$R_t = \frac{\frac{\partial L}{\partial x_2}\left(\frac{x_2 - x_D^{(1)}}{x_1}\right) - \frac{\partial L}{\partial x_1} - \lambda(x_D^* - x_D^{(1)}) + 1}{\frac{\partial x_D^{(1)}}{\partial R_t}\left(\lambda - \frac{\frac{\partial L}{\partial x_2}}{x_1}\right)} - 1 \quad (7.194)$$

Equation (7.194) is exactly the same result obtained by solving the maximum distillate problem using the maximum principle, and it is an equivalent solution to the dynamic programming formulation as follows.

$$\frac{\partial L}{\partial x_1} = z_1 \quad (7.195)$$

$$\frac{\partial L}{\partial x_2} = z_2 \quad (7.196)$$

7.5.8 Case in Which Uncertainties Are Present

For this case, the stochastic optimal control problem (Problem B) is expressed as:

$$\text{Maximize } L = E\left[\int_0^T \frac{V}{R_{tU} + 1}\left[1 - \lambda(x_D^* - x_D^{(1)})\right] dt\right]$$

$$R_{tU}$$

subject to

$$dx_1 = \frac{-V}{R_{tU} + 1} dt + x_1\,\sigma_1\,dz, \quad x_1(0) = B_0 = F \quad (7.197)$$

$$dx_2 = \frac{V}{R_{tU} + 1}\frac{(x_2 - x_D^{(1)})}{x_1} dt + x_2\,\sigma_2\,dz, \quad x_2(0) = x_F^{(1)} \quad (7.198)$$

and the optimality conditions developed by Merton and Samuelson (1990) can be stated as

$$\frac{\partial L}{\partial t} + \text{Maximize}\left[k\left(\bar{x}_t, R_{tU}\right) + \frac{1}{dt}E(dL) = 0\right] \quad (7.199)$$

$$R_{tU}$$

$$\frac{\partial L}{\partial t} + \text{Maximize }[k\left(\bar{x}_t, R_{tU}\right) + \frac{\partial L}{\partial x_1}f_1\left(\bar{x}_t, R_{tU}\right) + \frac{\partial L}{\partial x_2}f_2\left(\bar{x}_t, R_{tU}\right)$$

$$R_{tU} \qquad + \frac{\sigma_1^2}{2}\left(x_1\right)^2\frac{\partial^2 L}{\left(\partial x_1\right)^2} + \frac{\sigma_2^2}{2}\left(x_2\right)^2\frac{\partial^2 L}{\left(\partial x_2\right)^2}$$

$$+ \sigma_1\sigma_2 x_1 x_2\frac{\partial^2 L}{\partial x_1\,\partial x_2}] \quad (7.200)$$

Note that if we consider that the uncertainty terms in Equations (7.197) and (7.198) are not correlated, the last term can be eliminated. Hence, by substituting Equations (7.197)–(7.198) into Equation (7.200) we get

$$
0 = \frac{\partial L}{\partial t} + \text{Maximize} \frac{V}{R_{tU}+1}\left[1 - \lambda(x_D^* - x_D^{(1)})\right] - \frac{V}{R_{tU}+1}\frac{\partial L}{\partial x_1} +
$$

$$
R_{tU} \quad \frac{V(x_2 - x_D^{(1)})}{(R_{tU}+1)x_1}\frac{\partial L}{\partial x_2} + \frac{\sigma_1^2}{2}(x_1)^2\frac{\partial^2 L}{(\partial x_1)^2} + \frac{\sigma_2^2}{2}(x_2)^2\frac{\partial^2 L}{(\partial x_2)^2}
$$

$$
0 = \left[1 - \lambda(x_D^* - x_D^{(1)}) - \frac{\partial L}{\partial x_1} + \frac{x_2 - x_D^{(1)}}{x_1}\frac{\partial L}{\partial x_2}\right]\left[-\frac{V}{(R_{tU}+1)^2}\right]
$$

$$
+ \left[\lambda\frac{\partial x_D^{(1)}}{\partial R_{tU}} - \frac{\partial x_D^{(1)}}{\partial R_{tU}}\frac{1}{x_1}\frac{\partial L}{\partial x_2}\right]\left[\frac{V}{R_{tU}+1}\right]
$$

$$
+ \sigma_1\frac{\partial\sigma_1}{\partial R_{tU}}(x_1)^2\frac{\partial^2 L}{(\partial x_1)^2} + \sigma_2\frac{\partial\sigma_2}{\partial R_{tU}}(x_2)^2\frac{\partial^2 L}{(\partial x_2)^2} \tag{7.201}
$$

Simplifying, we get an implicit equation for R_tU:

$$
R_{tU} = \frac{\frac{\partial L}{\partial x_2}\frac{x_2 - x_D^{(1)}}{x_1} - \frac{\partial L}{\partial x_1} - \lambda(x_D^* - x_D^{(1)}) + 1}{\frac{\partial x_D^{(1)}}{\partial R_{tU}}\left[\lambda - \frac{1}{x_1}\frac{\partial L}{\partial x_2}\right]}
$$

$$
- \frac{\left[\sigma_1\frac{\partial\sigma_1}{\partial R_{tU}}(x_1)^2\frac{\partial^2 L}{(\partial x_1)^2} + \sigma_2\frac{\partial\sigma_2}{\partial R_{tU}}(x_2)^2\frac{\partial^2 L}{(\partial x_2)^2}\right]\left[\frac{(R_{tU}+1)^2}{V}\right]}{\frac{\partial x_D^{(1)}}{\partial R_{tU}}\left[\lambda - \frac{1}{x_1}\frac{\partial L}{\partial x_2}\right]} - 1
$$

Note that if we assume $\sigma_1 = 0$ (i.e., uncertainty exists only in x_2 ($x_B^{(1)}$), but does not exist in x_1 (B)), the equation reduces to

$$
R_{tU} = \frac{\frac{\partial L}{\partial x_2}\frac{x_2 - x_D^{(1)}}{x_1} - \frac{\partial L}{\partial x_1} - \lambda(x_D^* - x_D^{(1)}) + 1}{\frac{\partial x_D^{(1)}}{\partial R_{tU}}\left[\lambda - \frac{1}{x_1}\frac{\partial L}{\partial x_2}\right]}
$$

$$
- \frac{\left[\sigma_2\frac{\partial\sigma_2}{\partial R_{tU}}(x_2)^2\frac{\partial^2 L}{(\partial x_2)^2}\right]\left[\frac{(R_{tU}+1)^2}{V}\right]}{\frac{\partial x_D^{(1)}}{\partial R_{tU}}\left[\lambda - \frac{1}{x_1}\frac{\partial L}{\partial x_2}\right]} - 1
$$

$$\tag{7.202}$$

Let us think of what we have accomplished for the uncertain case. By assuming that the state variables of the maximum distillate problem can be represented by Equations (7.197) and (7.198), we have obtained an implicit equation (Equation (7.202)) that allows the calculation of the optimal profile for the reflux ratio. However, we had explained before that this work focused on optimal control problems in which the uncertainty in the calculation is introduced by representing the behavior of the relative volatility as a geometric Brownian motion. If so, then why are we assuming that the state variables are the ones that present such an uncertain behavior? We answer this question in the following section. By using Ito's lemma, we show that the uncertainty in the calculation of the relative volatility affects the calculation of one of the state variables (x_2, which is the same as $x_B^{(1)}$), which can also be represented as an Ito process.

7.5.9 State Variable and Relative Volatility: The Two Ito Processes

Recall that, in the quasi-steady-state method of batch distillation optimal control problems considered in this work, the integration of the state variables leads to the calculation of the rest of the variables assumed to be in quasi-steady-state (Diwekar, 1995). Also, recall that such variables in quasi-steady-state are determined by applying short-cut method calculations.

Let us focus now on the expression for the dynamic behavior of the bottom composition of the key component, Equation (7.198):

$$dx_2 = \frac{V}{R_{tU} + 1} \frac{(x_2 - x_D^{(1)})}{x_1} dt + x_2 \, \sigma_2 \, dz \qquad (7.203)$$

The question here is how to calculate the term corresponding to uncertainty in α. To relate the relative volatility to the state variable x_2 ($x_B^{(1)}$), we have to consider the HG equation, which relates the relative volatility to the bottom composition $x_B^{(1)}$ through the constant C_1:

$$1 = \sum_{i=1}^{n} \left(\frac{\alpha_i}{\alpha_1} \right)^{C_1} \frac{x_D^{(1)}}{x_B^{(1)}} x_B^{(i)} \qquad (7.204)$$

Note that the equation contains the relative volatility to the power of C_1. Rearranging,

$$1 = \frac{x_D^{(1)}}{x_B^{(1)}} \alpha_1^{-C_1} \sum_{i=1}^{n} \alpha_i^{C_1} x_B^{(i)} \qquad (7.205)$$

Taking the derivatives of this expression implicitly with respect to $x_B^{(i)}$ and $\alpha_i^{C_1}$,

$$x_B^{(i)} d\alpha_i^{C_1} + dx_B^{(i)} \alpha_i^{C_1} = 0$$

$$\frac{d\alpha_i^{C_1}}{\alpha_i^{C_1}} = -\frac{dx_B^{(i)}}{x_B^{(i)}} \qquad (7.206)$$

If we express the behavior of relative volatility by the general equation for an Ito process:

$$d\alpha = f_1(\alpha, t)dt + f_2(\alpha, t)dz \qquad (7.207)$$

For the geometric Brownian motion and the geometric mean reverting process, $f_2(\alpha, t)$ is the same for both of these processes. Therefore we can write Equation (7.207) in the following form.

$$d\alpha = f_1(\alpha, t)dt + \sigma\alpha dz \qquad (7.208)$$

Then, by using Ito's lemma (Equation (7.108)):

$$dF = \frac{\partial F}{\partial t} dt + \frac{\partial F}{\partial x} dx + \frac{1}{2}\frac{\partial^2 F}{\partial x^2}(\sigma)^2(x)^2 dt$$

We can obtain an expression for the relative volatility to the power of C_1, α^{C_1},

$$d\alpha^{C_1} = \frac{\partial \alpha^{C_1}}{\partial \alpha}d\alpha + \frac{1}{2}\sigma^2\alpha^2\frac{\partial^2 \alpha^{C_1}}{\partial \alpha^2}dt \qquad (7.209)$$

Simplifying:

$$d\alpha^{C_1} = C_1\alpha^{C_1-1}d\alpha + \frac{1}{2}\sigma^2\alpha^{C_1}C_1(C_1-1)dt$$

$$\frac{d\alpha^{C_1}}{\alpha^{C_1}} = C_1\left[\frac{f_1(\alpha,t)}{\alpha}dt + \sigma dz\right]dt + \frac{1}{2}\sigma^2 C_1(C_1-1)dt \qquad (7.210)$$

$$= f_{new}(\alpha, t)dt + \sigma_{new}dz \qquad (7.211)$$

where

$$f_{new}(\alpha, t) = C_1\frac{f_1(\alpha,t)}{\alpha} + \frac{1}{2}\sigma^2 C_1(C_1-1)$$

and

$$\sigma_{new} = C_1\sigma$$

.

Substituting Equation (7.211) in Equation (7.206) implies that

$$\frac{dx_2}{x_2} = \frac{dx_B^{(1)}}{x_B^{(1)}} = -f_{new}(\alpha, t)dt + \sigma_{new}dz \qquad (7.212)$$

Note that Equation (7.212) establishes that the uncertain behavior for the relative volatility results in a similar behavior for the dynamics of x_2. That is, if α is an Ito process, then x_2 is represented by a similar Ito process. For an ideal system, this process is shown to be a geometric Brownian motion, whereas for a non-ideal system such as ethanol–water it is found to be a geometric mean reverting process .

7.5.10 Coupled Maximum Principle and NLP Approach for the Uncertain Case

Although Ito's lemma and dynamic programming helped us to provide an analytical expression for the reflux ratio profile, these equations are cumbersome and computationally inefficient to solve. One of the fastest and simplest methods to solve optimal control problems in batch distillation with no uncertainty is the coupled maximum principle and NLP approach described earlier. Such an approach can also be used in this work for the solution of the optimal control problem in the uncertain case, but in order to do that, the derivation of the appropriate adjoint equations is required. In this section, we show the maximum principle formulation that results from the analysis of the uncertain case (similar to the formulation presented earlier for the deterministic case; we are not considering the Lagrangian expression of the objective function).

The problem is expressed as

$$\text{Maximize} \; -x_1(T) \qquad (7.213)$$
$$R_{tU}$$

subject to

$$\frac{dx_1}{dt} = \frac{-V}{R_t + 1}, \; x_1(0) = B_0 = F \qquad (7.214)$$

$$dx_2 = \frac{V}{R_{tU} + 1} \frac{(x_2 - x_D^{(1)})}{x_1} \, dt + x_2 \, \sigma_2 \, dz, \; x_2(0) = x_F^{(1)} \qquad (7.215)$$

The Hamiltonian, which should be maximized, is

$$H = \frac{V}{R_{tU} + 1} \frac{\partial L}{\partial x_1} + \frac{V}{R_{tU} + 1} \frac{\left(x_2 - x_D^{(1)}\right)}{x_1} \frac{\partial L}{\partial x_2} + \frac{\sigma_2^2}{2} (x_2)^2 \frac{\partial^2 L}{(\partial x_2)^2} \qquad (7.216)$$

The adjoint equations are

$$\frac{dz_1}{dt} = z_2 \frac{V \left(x_2 - x_D^{(1)}\right)}{(R_{tU} + 1)(x_1)^2}, \; z_1(T) = -1 \qquad (7.217)$$

$$\frac{dz_2}{dt} = -z_2 \frac{V \left(1 - \frac{\partial x_D^{(1)}}{\partial x_2}\right)}{(R_{tU} + 1)x_1} - \sigma_2^2 x_2 \frac{\partial^2 L}{(\partial x_2)^2}, \; z_2(T) = 0 \qquad (7.218)$$

Recall that

$$\frac{\partial L}{\partial x_1} = z_1$$

$$\frac{\partial L}{\partial x_2} = z_2$$

Also, if we define

$$\frac{\partial^2 L}{(\partial x_2)^2} = \omega_t$$

Then it can be shown that

$$\frac{d\omega_t}{dt} = -\omega_t \frac{V\left(1 - \frac{\partial x_D^{(1)}}{\partial x_2}\right)}{(R_{tU} + 1)x_1} + z_2 \frac{V\left(1 - \frac{\partial^2 x_D^{(1)}}{(\partial x_2)^2}\right)}{(R_{tU} + 1)x_1}$$

$$-\omega_t \sigma_2^2 - 2\sigma_2^2 x_2 \frac{\partial^3 L}{(\partial x_2)^3}, \quad \omega_T = 0$$

$$(7.219)$$

The optimality conditions on the reflux ratio results in

$$R_{tU} = \frac{-\frac{\partial L}{\partial x_2} \frac{x_2 - x_D^{(1)}}{x_1} + \frac{\partial L}{\partial x_1}}{\frac{\partial x_D^{(1)}}{\partial R_{tU}} \frac{1}{x_1} \frac{\partial L}{\partial x_2}} + \frac{\left[\sigma_2 \frac{\partial \sigma_2}{\partial R_{tU}} (x_2)^2 \frac{\partial^2 L}{(\partial x_2)^2}\right]}{\frac{\partial x_D^{(1)}}{\partial R_{tU}} \frac{1}{x_1} \frac{\partial L}{\partial x_2}} \left[\frac{(R_{tU}+1)^2}{V}\right] - 1 \quad (7.220)$$

Now, if we define

$$\xi = \frac{\frac{\partial^2 L}{(\partial x_2)^2}}{\frac{\partial L}{\partial x_1}} = \frac{\omega_t}{z_1} \qquad (7.221)$$

$$z = \frac{\frac{\partial L}{\partial x_2}}{\frac{\partial L}{\partial x_1}} = \frac{z_2}{z_1} \qquad (7.222)$$

and consider negligible third partial derivatives, then, without loss of information, Equations (7.217), (7.218), (7.219), and (7.220) can be reformulated as

$$\frac{dz}{dt} = -z_2 \frac{V\left(x_2 - x_D^{(1)}\right)}{(R_{tU} + 1)(x_1)^2} - z \frac{V\left(1 - \frac{\partial x_D^{(1)}}{\partial x_2}\right)}{(R_{tU} + 1)x_1} - \sigma_2^2 x_2 \xi, \quad z_2(T) = 0 \quad (7.223)$$

$$\frac{d\xi}{dt} = -\xi \frac{V\left(1 - \frac{\partial x_D^{(1)}}{\partial x_2}\right)}{(R_{tU} + 1)x_1} + z \frac{V \frac{\partial^2 x_D^{(1)}}{(\partial x_2)^2}}{(R_{tU} + 1)x_1}$$

$$-\sigma_2^2 \xi - \xi z \frac{V\left(x_2 - x_D^{(1)}\right)}{(R_{tU} + 1)(x_1)^2}, \quad \xi_T = 0$$

$$(7.224)$$

$$R_{tU} = \frac{x_1 - z(x_2 - x_D^{(1)})}{\frac{\partial x_D^{(1)}}{\partial R_{tU}} z} + \frac{x_1 \left[\sigma_2 \frac{\partial \sigma_2}{\partial R_{tU}} (x_2)^2 \xi \right] \left[\frac{(R_{tU}+1)^2}{V} \right]}{\frac{\partial x_D^{(1)}}{\partial R_{tU}} z} - 1 \quad (7.225)$$

This representation allowed us to use the coupled maximum principle–NLP solution algorithm. In such an approach, the Lagrangian formulation of the objective function is not used in the solution. Most important of all, the algorithm avoids the solution of the two-point boundary value problem for the pure maximum principle formulation, or the solution of partial differential equations for the pure dynamic programming formulation. Note that Equation (7.225) is obtained by maximizing the Hamiltonian (maximizing the distillate) and does not incorporate the purity constraint. Hence, the use of the final boundary condition ($\mu_T = 0$, $\xi_T = 0$) provides the limiting solution resulting in all the reboiler charge instantaneously going to the distillate pot ($R = -\infty$) with the lowest overall purity. Because in this approach the purity constraint is imposed external to the Hamiltonian, the final boundary condition is no longer valid. Instead, the final boundary condition is automatically imposed when the purity constraint is satisfied. The algorithm involves the solution of the NLP optimization problem for the scalar variable R_0, the initial reflux ratio, subject to

1. The dynamics of the state variables given by Equations (7.214) and (7.215).
2. The adjoint equations (Equations (7.223) and (7.224)), and the initial conditions for these adjoint equations, derived in terms of the decision variable R_0.
3. The optimality conditions for the control variable (reflux ratio, Equation (7.225)).

Earlier, we established by numerical experiments that the uncertainties in relative volatility can be represented as an Ito process. For the optimal control problem, the system considered is 100 kmol of ethanol–water being processed in a batch column with 1 atm pressure, 13 theoretical stages, 33 kmol/h vapor rate, and the batch time of 2 h. For this problem, the purity constraint on the distillate is specified as 90%. The optimal reflux profile (a) and optimal distillate flow rates (b) for the stochastic case and the deterministic case are shown in Figure 7.16. There is a significant difference between the two profiles. These two profiles for the reflux ratio are given to a rigorous simulator ($MultiBatchDS^{TM}$, Diwekar, 1996) to compare the process performances. The average purity is found to be almost the same at about 90% for both of these cases. However, for the deterministic case the distillate amount is 69% lower than the stochastic case. This case study shows that representing uncertainties in relative volatility with Ito processes can significantly improve the system performance in terms of product yield.

7.6 Sustainable Mercury Management: An Optimal Control Problem

As explained in Chapter 2, the complicated cycling of mercury necessitates mitigation strategies at various stages of the mercury cycle. The last four chapters proposed mercury trading to achieve greater discharge reductions at the reduced overall cost, thereby aiding ecological as well as economic sustainability. This will reduce the amount of mercury that is released into the earth's atmosphere, and consequently, the quantity of mercury entering various water bodies. However, the presence of mercury in water bodies cannot be totally eliminated. This is not only because it is economically infeasible to eliminate the anthropogenic mercury contribution to the atmosphere but also because the atmosphere exhibits background mercury cycling independent of the anthropogenic contribution that affects water body concentrations.

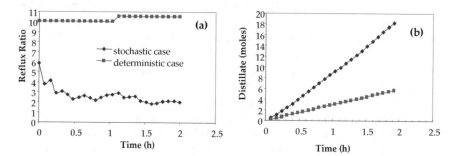

Fig. 7.16. Optimal profiles for deterministic and stochastic cases

Once present in water, even in small amounts, mercury is highly dangerous not only to the aquatic communities but also to humans through direct and indirect effects. Methylation of inorganic mercury leads to the formation of methyl mercury that accumulates up the aquatic food chains, so that organisms in higher trophic levels have higher mercury concentrations. Since methyl mercury is selectively bioaccumulated, one mitigation option at the ecosystem level is to restrict mercury methylation in water bodies. Although the exact mechanism of mercury methylation is not well understood, the literature based on experimental studies show that acidic lakes (low pH lakes) have high mercury bioaccumulation rates. Hence, controlling lake/river pH is an option to mitigate harmful effects due to bioaccumulation.

This section explores the idea of lake liming to control water pH for bioaccumulation control. Since natural systems such as lakes and rivers are dynamic in nature, static decisions will lead to suboptimal results, and time-dependent liming is expected to result in more accurate pH control. This section is based on the paper by Shastri and Diwekar (2008) and uses optimal control theory to derive the time-dependent liming strategy.

It is well known that natural systems are not very well understood and consequently are associated with considerable uncertainties that need to be incorporated for realistic analysis. Effective uncertainty modeling techniques from real options theory are implemented for the same. The resulting stochastic optimal control problem is solved using the stochastic maximum principle.

7.6.1 Mercury Bioaccumulation

Figure 7.17 presents the overall mercury cycling in water bodies such as lakes and rivers. Mercury can exist in the water bodies in various forms, the important forms being: elemental mercury (Hg), inorganic mercury (Hg(II)), organic methylmercury (CH_3Hg), and complexes of these with dissolved organic carbon or suspended particulate matter. A number of pathways exist by which mercury and its compounds can enter the freshwater environment: Hg(II) and methylmercury from atmospheric deposition (wet and dry) can enter the water bodies directly; Hg(II) and methylmercury can be transported to the water bodies in runoff (bound to suspended soil/humus or attached to dissolved organic carbon), or Hg(II) and methylmercury can leach into the water body from groundwater flow in the upper soil layers. Mercury and its compounds exist in different segments of the water body, such as the water column, sediment (active and passive), and the biota (fish).

Once in the water bodies, mercury can exist in dissolved as well as particulate form and can undergo the following different yet simultaneous transformations: Elemental Hg can be oxidized to Hg(II) or volatilized to the atmosphere (evasion); Hg(II) can be methylated in sediments and water column to form methylmercury or can undergo reduction to form elemental Hg; methylmercury can be alkylated to form dimethylmercury, and Hg(II) and methylmercury can form organic and inorganic complexes with sediment and suspended particulate matter. The concentration of each chemical form depends on the extent of these various transformations and can be different for each water body. Of the various chemical forms of mercury, methylmercury (MeHg) is considered to be the most dangerous.

To summarize, mercury methylation to MeHg is a key step in the bioaccumulation of mercury in aquatic food chains. The concentration of MeHg in water depends on the equilibrium between the methylation and demethylation reactions, which occur in the water column as well as sediments. It has been shown that a strong correlation between acidic conditions, i.e., low pH values, and high mercury bioaccumulation in fish. Therefore, the liming of lakes and rivers to control pH can control the harmful effects of mercury in water.

7.6.2 Mercury pH Control Model

The parameters of the model that are used in the dynamic model are Distribution Coefficient (Dc), Internal Loading Rate (ILR), Dynamic Ratio (Dr), Lake Water Retention Time (Rt), Sedimentation Rate (Sr), and Active

Sediment Age (ASA). The control variable is Lime Input (u).

These parameters are computed from the various other lake-related basic parameters. The various state variables for the model are

y_1: Lime in water

y_2: Lime in active sediment

y_3: Lime in passive sediment

y_4:Lake/river pH

The set of ODEs for the complete lake liming problem are presented below.

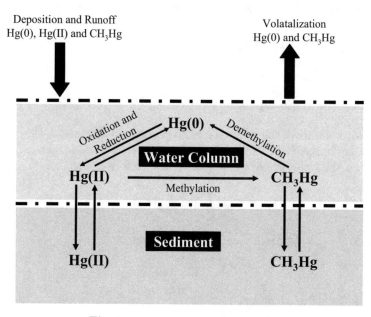

Fig. 7.17. Mercury cycling in water

$$f_1 = \frac{dy_1}{dt} = u.Dc.(0.422 * 0.712)$$

$$+ y_2.ILR.Dr - \frac{(y_1 - k(4))}{Rt.52} - (y_1 - k(4)).Sr \qquad (7.226)$$

$$f_2 = \frac{dy_2}{dt} = (y_1 - k(4)).Sr + u.(1 - Dc).(0.422 * 0.712) - y_2.ILR.Dr - \frac{y_2}{ASA}$$
$$(7.227)$$

$$f_3 = \frac{dy_3}{dt} = \frac{y_2}{ASA} \qquad (7.228)$$

$$f_4 = \frac{dy_4}{dt} = \frac{-10}{\left[1 + 10^{[k_1.(\frac{k_4}{y(1)})^4.(k_2 - 20.\frac{y(1)}{k_5})]}\right]^2}.k_3.10^{[k_1.(\frac{k_4}{y(1)})^4.(k_2 - 20.\frac{y(1)}{k_5})]}.$$

$$\left[k_1.k_4^4.\left(\frac{-4}{y(1)^5}\frac{dy(1)}{dt}\right)k_2 - \frac{k_1.20.k_4^4}{k_5}\left(\frac{-3}{y(1)^4}\frac{dy(1)}{dt}\right)\right] \qquad (7.229)$$

where

- $k_1 = 0.62$
- $k_2 = log_{10}(5) + 10$
- $k_3 = \ln(10)$
- $k_4 = 6825$
- $k_5 = 12000$

Further simplification of the equations can be made using symbols given below.

- $P_1 = 10^{[k_1.(\frac{k_4}{y(1)})^4.(k_2-20.\frac{y(1)}{k_5})]}$
- $P_2 = 1 + P_1$

7.6.3 Deterministic Optimal Control

For the formulation of the optimal control problem in this model, the objective function is expressed in terms of pH. If $\bar{y}$ represents the targeted pH value (=7), then the time-dependent objective is therefore defined as:

$$J = \int_0^T (y_4 - \bar{y})^2 \, dt \qquad (7.230)$$

Representing the objective function equation as:

$$F = (y_4 - \bar{y})^2 \qquad (7.231)$$

The Hamiltonian is given by

$$H = F + \sum_{i=1}^4 z_i f_i \qquad (7.232)$$

Optimality Condition

The general optimality condition for any control variable u is given by

$$\frac{\partial H}{\partial u} = 0 \qquad (7.233)$$

$$\sum_{i=1}^4 z_i \frac{\partial f_i}{\partial u} + \frac{\partial F}{\partial u} = 0 \qquad (7.234)$$

$$0 = z_1.Dc.(0.422 * 0.712) + z_2.(1 - Dc).(0.422 * 0.712)$$

$$z_4.\left\{ \frac{-10}{P_2^2}.k_3.P_1. \right.$$

$$\left[k_1(k_4)^4.\left(\frac{-4}{y(1)^5}(Dc.0.422*0.712) \right)k_2 \right.$$

$$\left. \left. - \frac{k_1(20)k_4^4}{k_5}\left(\frac{-3}{y(1)^4}(Dc.0.422*0.712) \right) \right] \right\} \tag{7.235}$$

Adjoint Equations

The general adjoint equation is given as:

$$-\frac{dz_i}{dt} = \frac{\partial H}{\partial y_i} \quad i = 1,\ldots,4 \tag{7.236}$$

The adjoint equations are accordingly given as:

$$-\frac{dz_1}{dt} = \frac{\partial F}{\partial y_1} + \sum_{i=1}^{4} z_i.\frac{\partial f_i}{\partial y_1} \tag{7.237}$$

$$\frac{\partial f_4}{\partial y_1} = \left\{ \frac{-10}{P_2^2}.k_3.P_1\left[k_1.k_4^4.k_2.\left(\frac{-16}{y(1)^5} \right)\left(\frac{1}{Rt.52} + Sr \right) \right.\right.$$

$$\left.\left. - \frac{k_1.20.k_4^4}{k_5}.\left(\frac{-9}{y(1)^4} \right)\left(\frac{1}{Rt.52} + Sr \right) \right] \right\} +$$

$$\left[k_1.k_4^4.\left(\frac{-4}{y(1)^5}\frac{dy(1)}{dt} \right)k_2 - \frac{k_1.20.k_4^4}{k_5}\left(\frac{-3}{y(1)^4}\frac{dy(1)}{dt} \right) \right].\left\{ -10.k_3. \right.$$

$$\left.\left[\frac{P_2^2.k_3.P_1.\left[\frac{-4.k_1.k_4^4.k_2}{y(1)^5} + \frac{20.k_1.k_4^4.3}{k_5.y(1)^4} \right] - P_1.2.P_2.k_3.P_1.\left[\frac{-4.k_1.k_4^4.k_2}{y(1)^5} + \frac{20.k_1.k_4^4.3}{k_5.y(1)^4} \right]}{P_2^4} \right] \right\} \tag{7.238}$$

$$-\frac{dz_1}{dt} = z_1.\left(\frac{-1}{Rt.52} - Sr \right) + z_2.Sr + z_4.\frac{\partial f_4}{\partial y_1} \tag{7.239}$$

$$-\frac{dz_2}{dt} = \frac{\partial F}{\partial y_2} + \sum_{i=1}^{4} z_i.\frac{\partial f_i}{\partial y_2} \tag{7.240}$$

$$\frac{\partial f_4}{\partial y_2} = \frac{-10}{P_2^2}.k_3.P_1\left[k_1.k_4^4.k_2.\left(\frac{-4}{y(1)^5} \right)\left(ILR \right) \right.$$

$$\left. - \frac{k_1.20.k_4^4}{k_5}.\left(\frac{-3}{y(1)^4} \right)\left(ILR \right) \right] \tag{7.241}$$

$$-\frac{z_1}{dt} = z_1.ILR.Dr + z_2.\left(-ILR.Dr - \frac{1}{ASA} \right) + z_3.\frac{1}{ASA} + z_4.\frac{\partial f_4}{\partial y_2} \tag{7.242}$$

$$-\frac{dz_3}{dt} = \frac{\partial F}{\partial y_3} + \sum_{i=1}^{4} z_i \cdot \frac{\partial f_i}{\partial y_3} \qquad (7.243)$$

$$-\frac{dz_3}{dt} = 0 \qquad (7.244)$$

$$-\frac{dz_4}{dt} = \frac{\partial F}{\partial y_4} + \sum_{i=1}^{4} z_i \cdot \frac{\partial f_i}{\partial y_4} \qquad (7.245)$$

$$-\frac{dz_4}{dt} = 2.(y_4 - \bar{y}) \qquad (7.246)$$

7.6.4 Stochastic Optimal Control

As stated earlier, there is uncertainty in predicting the natural pH of water. Shastri and Diwekar (2008) have shown that this uncertainty can be captured as a mean reverting Ito process. The time-dependent parameter, which represents the natural variation in pH, now becomes an additional state variable in the model and hence the stochastic lake liming model is given as:

$$f_1 = \frac{dy_1}{dt} = u.Dc.(0.422 * 0.712)$$

$$+ y_2.ILR.Dr - \frac{(y_1 - k(4))}{Rt.52} - (y_1 - k(4)).Sr \qquad (7.247)$$

$$f_2 = \frac{dy_2}{dt} = (y_1 - k(4)).Sr + u.(1 - Dc).(0.422 * 0.712) - y_2.ILR.Dr$$

$$- \frac{y_2}{ASA} \qquad (7.248)$$

$$f_3 = \frac{dy_3}{dt} = \frac{y_2}{ASA} \qquad (7.249)$$

$$f_4 = \frac{dy_4}{dt} = \frac{-10}{P_2^2}.k_3.P_1.\left[k_1.k_4^4.\left(\frac{-4}{y(1)^5} \frac{dy(1)}{dt} \right) k_2 - \frac{k_1.20.k_4^4}{k_5} \right.$$

$$\left. \times \left(\frac{-3}{y(1)^4} \frac{dy(1)}{dt} \right) \right] + y(5)$$

$$\qquad (7.250)$$

$$f_{ito} = \frac{dy_5}{dt} = \eta(\bar{dpH} - y_5) + \frac{\sigma \ \epsilon}{\sqrt{\Delta t}} \qquad (7.251)$$

Here, $\bar{dpH}$ is the mean value of fractional natural pH variation.

$\bar{y}$ represents the targeted pH value. The time-dependent objective is, therefore, defined as:

$$J = \int_0^T (y_4 - \bar{y})^2 \ dt \qquad (7.252)$$

Representing the objective function equation as:

$$F = (y_4 - \bar{y})^2 \qquad (7.253)$$

The Hamiltonian is given by

$$H = F + \sum_{i=1}^{5} z_i f_i + \frac{1}{2} w \sigma^2 \tag{7.254}$$

Optimality Condition

The general optimality condition for any control variable u is given by

$$\frac{\partial H}{\partial u} = 0 \tag{7.255}$$

$$\sum_{i=1}^{5} z_i \frac{\partial f_i}{\partial u} + \frac{\partial F}{\partial u} + \frac{\partial}{\partial u} \frac{1}{2} w \sigma^2 = 0 \tag{7.256}$$

$$0 = z_1.Dc.(0.422 * 0.712) + z_2.(1 - Dc).(0.422 * 0.712)$$
$$z_4.\left\{ \frac{-10}{P_2^2}.k_3.P_1. \right.$$
$$\left[k_1(k_4)^4.\left(\frac{-4}{y(1)^5}(Dc.0.422 * 0.712) \right) k_2 \right.$$
$$\left. \left. - \frac{k_1(20)k_4^4}{k_5}\left(\frac{-3}{y(1)^4}(Dc.0.422 * 0.712) \right) \right] \right\} \tag{7.257}$$

Adjoint Equations

The general adjoint equation is given as:

$$-\frac{dz_i}{dt} = \frac{\partial H}{\partial y_i} \qquad i = 1, \dots, 5 \tag{7.258}$$

$$\frac{dw}{dt} = -2 w \frac{\partial}{\partial y_5} f_5 - \frac{1}{2} w \frac{\partial^2}{\partial y_5^2}(\sigma^2) - z_5 \frac{\partial^2}{\partial y_5^2} f_5 \tag{7.259}$$

The adjoint equations are accordingly given as:

$$-\frac{dz_1}{dt} = \frac{\partial F}{\partial y_1} + \sum_{i=1}^{5} z_i.\frac{\partial f_i}{\partial y_1} \tag{7.260}$$

$$\frac{\partial f_4}{\partial y_1} = \left\{ \frac{-10}{P_2^2}.k_3.P_1 \left[k_1.k_4^4.k_2.\left(\frac{-16}{y(1)^5} \right)\left(\frac{1}{Rt.52} + Sr \right) \right.\right.$$
$$\left. - \frac{k_1.20.k_4^4}{k_5}.\left(\frac{-9}{y(1)^4} \right)\left(\frac{1}{Rt.52} + Sr \right) \right] \right\} +$$
$$\left[k_1.k_4^4.\left(\frac{-4}{y(1)^5}\frac{dy(1)}{dt} \right) k_2 - \frac{k_1.20.k_4^4}{k_5}\left(\frac{-3}{y(1)^4}\frac{dy(1)}{dt} \right) \right].\left\{ -10.k_3. \right.$$

$$\left[\frac{P_2^2.k_3.P_1.\left[\frac{-4.k_1.k_4^4.k_2}{y(1)^5} + \frac{20.k_1.k_4^4.3}{k_5.y(1)^4}\right] - P_1.2.P_2.k_3.P_1.\left[\frac{-4.k_1.k_4^4.k_2}{y(1)^5} + \frac{20.k_1.k_4^4.3}{k_5.y(1)^4}\right]}{P_2^4}\right]\right\}$$

$$(7.261)$$

$$-\frac{dz_1}{dt} = z_1.\left(\frac{-1}{Rt.52} - Sr\right) + z_2.Sr + z_4.\frac{\partial f_4}{\partial y_1} \tag{7.262}$$

$$-\frac{dz_2}{dt} = \frac{\partial F}{\partial y_2} + \sum_{i=1}^{5} z_i.\frac{\partial f_i}{\partial y_2} \tag{7.263}$$

$$\frac{\partial f_4}{\partial y_2} = \frac{-10}{P_2^2}.k_3.P_1\left[k_1.k_4^4.k_2.\left(\frac{-4}{y(1)^5}\right)\left(ILR\right)\right.$$
$$\left. - \frac{k_1.20.k_4^4}{k_5}.\left(\frac{-3}{y(1)^4}\right)\left(ILR\right)\right] \tag{7.264}$$

$$-\frac{dz_2}{dt} = z_1.ILR.Dr + z_2.\left(-ILR.Dr - \frac{1}{ASA}\right) + z_3.\frac{1}{ASA} + z_4.\frac{\partial f_4}{\partial y_2} \tag{7.265}$$

$$-\frac{dz_3}{dt} = \frac{\partial F}{\partial y_3} + \sum_{i=1}^{5} z_i.\frac{\partial f_i}{\partial y_3} \tag{7.266}$$

$$-\frac{dz_3}{dt} = 0 \tag{7.267}$$

$$-\frac{dz_4}{dt} = \frac{\partial F}{\partial y_4} + \sum_{i=1}^{5} z_i.\frac{\partial f_i}{\partial y_4}d \tag{7.268}$$

$$-\frac{dz_4}{dt} = 2.(y_4 - \bar{y}) \tag{7.269}$$

$$-\frac{dz_5}{dt} = \frac{\partial F}{\partial y_5} + \sum_{i=1}^{5} z_i.\frac{\partial f_i}{\partial y_5} \tag{7.270}$$

$$-\frac{dz_5}{dt} = 1 + \eta \tag{7.271}$$

$$\frac{dw}{dt} = -2\,w\,\frac{\partial}{\partial y_5}f_5 - \frac{1}{2}w\frac{\partial^2}{\partial y_5^2}(\sigma^2) - z_5\frac{\partial^2}{\partial y_5^2}f_5 \tag{7.272}$$

$$\frac{dw}{dt} = -2\,w\,\eta - z_5\,\eta \tag{7.273}$$

7.6.5 Results and Discussions

Lake A

The parameter values for lake A are
Initial lake pH = 6.15
Lake area = 1.26 km^2
Lake mean depth = 8.5 m
Lake maximum depth = 26.2 m
Drainage area = 51.5 km^2
Mean annual precipitation = 602 mm/year
Active sediment age = 519.6 weeks
Internal loading rate = 0.001 (1/month)
Distribution coefficient = 0.5
Settling velocity = 0.074 m/week
Additive constant = 2.375

The other parameters are computed using these basic parameters as per the following relationships.
Lake volume = Lake area * Lake mean depth
Water discharge = 0.01*DrainageArea*Precipitation/600
Lake water retention time = Lake volume/(Water discharge*60*60*24*365/7) (weeks)
Sedimentation rate = Settling velocity/Lake mean depth (1/week)
Dynamic ratio = Lake area$^{0.5}$/Lake mean depth

It should be noted that the lake has a relatively low lime internal loading rate and a high lime setting velocity. Due to these characteristics, the effect of lime addition is not prolonged. Any lime addition results in a fast rise in the lake pH. However, the effect of liming is not maintained for a longer duration since the lime particles quickly settle in the sediment compartment, and the re-suspension rate of lime from active sediment to the water compartment is low.

In Figure 7.18, the top profile presents the result of the stochastic control problem solution, indicating that the targeted lake pH is effectively achieved. The lake pH rises very quickly within about 1 year and then fluctuates around the targeted pH for the remaining time horizon. The control variable profile for this result is shown in Figure 7.19. The liming rate is initially high to raise the lake pH value to the targeted value as quickly as possible. After the targeted range has been achieved, the lime addition rate settles at a nonzero value for the remaining time horizon. The value fluctuates continuously to account for the natural variations in the lake pH. The plots also show that use of the stochastic optimal control (top plot) leads to better lake pH control than deterministic control (middle plot). This emphasizes that taking liming decisions while ignoring the natural pH variations will lead to suboptimal results, thereby highlighting the importance of uncertainty incorporation in decision making.

7.7 Summary

An optimal control problem involves vector decision variables. These problems are a subset of differential algebraic optimization problems. If the underlying differential equations can be discretized into a set of algebraic equations, then these problems can be solved using traditional NLP techniques. Otherwise, one has to resort to either the calculus of variations, the maximum principle, or the dynamic programming approach. The calculus of variations represents the first systematic theory for optimization that was derived to solve optimal control problems. The name optimal control comes from the solution method proposed for control problems, popularly known as Pontryagin's maximum principle. Dynamic programming presents another alternative to solve these problems. These different methods follow different paths to arrive at the same solution (for convex problems). In the presence of uncertainties, stochastic calculus enters in optimal control theory. Ito's lemma and dynamic programming can together provide a way to handle stochastic optimal control problems. Recent advances provided the stochastic maximum principle as a better alternative to dynamic programming for solution of these problems.

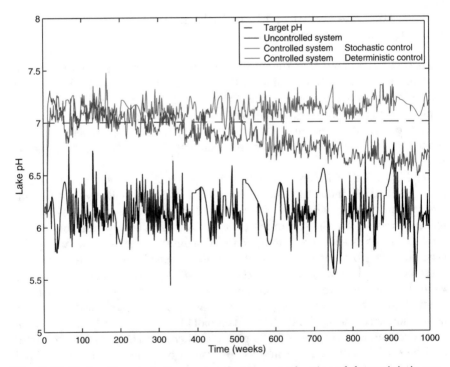

Fig. 7.18. Basic pH control: Comparison between stochastic and deterministic control for Lake A

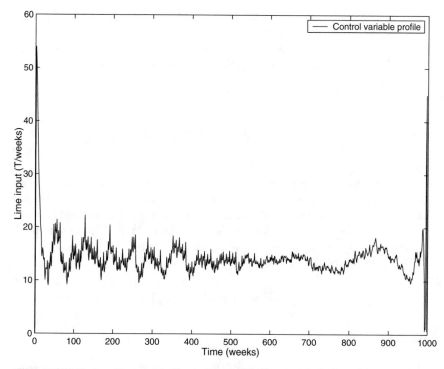

Fig. 7.19. Basic pH control: Control variable (lime addition) profile for Lake A

Exercises

7.1 A performance equation for a simple process is given by

$$\frac{dx_1}{dt} = -ax_1 + \theta_t, x_1(o) = \alpha, 0 \le t \le T$$

The objective is to maximize the following index of performance.

$$J = \frac{1}{s} \int_0^T \left[(x_1)^2 + (\theta_t)^2 \right] dt$$

Solve the above problem using the calculus of variations and the maximum principle.

7.2 Solve Problem 7.1 above using the dynamic programming method and compare the results.

7.3 A man is considering his lifetime plan of investment and expenditure. He has an initial level of savings S and no income other than which he obtains from investment at a fixed interest rate. His total capital x is therefore governed by the equation

$$\frac{dx(t)}{dt} = \alpha x(t) - r(t)$$

where $\alpha \geq 0$ and r denotes his rate of expenditure. His immediate enjoyment due to expenditure is $U(r)$, where U is his utility function. In his case $U(r) = r^{0.5}$. Future enjoyment at time t is counted less today, at time 0, by incorporation of a discount term $e^{-\beta t}$. Thus, he wishes to maximize

$$J = \int_0^T e^{-\beta t} U(r) dt$$

subject to the terminal constraint $x(T) = 0$.
Solve this problem (find $r(t)$) using the maximum principle and dynamic programming.

7.4 Zermelo's problem. We consider the problem of navigating a boat across a river, in which there is strong current, so as to get to a specified point on the other side in minimum time. We assume that the magnitude of the boat's velocity with respect to water is a constant V. The downstream current at any point depends only on the distance from the bank.
The equations of the boat's motion are

$$\frac{dx}{dt} = V \cos \theta + u(y)$$

$$\frac{dy}{dt} = V \sin \theta$$

where x is the downstream position along the river, y is the distance from the origin bank, $u(y)$ is the downstream current, and θ is the heading angle of the boat. The heading angle is the control, which may vary along the path.
The problem is to minimize the time of crossing. Frame the problem as an optimal control problem and solve it.

7.5 At what point is it optimal to pay a sunk cost I in return for a project whose value is P, given that P evolves according to the following Geometric Brownian motion

$$dP = \alpha P \, dt + \alpha P \, dz$$

where dz is the increment of a Wiener process? This equation implies that the current value of the project is known but the future values are lognormally distributed.
The value of the investment opportunity to be maximized is given by the function $F(P)$

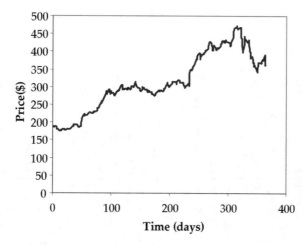

Fig. 7.20. One-year historical prices of a technology stock (http://www.finance.yahoo.com)

$$F(P) = \max E[(P_T - I)\exp(-\gamma t)]$$

where E denotes the expectation, T is the future time that the investment is made, and γ is the discount rate.

- Find the deterministic solution.
- Find the stochastic solution for different values of α.
- Compare the two solutions.

7.6 Identify and model the following time-dependent uncertainties.

(a) Figure 7.20 shows the 1-year performance for a stock price. (http://www.finance.yahoo.com)

(b) Tables 7.1 and 7.2 show relative volatility variations of two binary systems distilled in a batch column (Ulas and Diwekar, 2004).

(c) Table 7.3 shows the data for insulin dynamics obtained from 4 patients. The subjects were studied while lying down during continuous enteral nutrition (90 Cal/h), and blood samples were taken at 10-min intervals (Simon et al. (1987), Ulas and Diwekar, 2010) over a 24-h time period.

Table 7.1. Relative volatility change for a binary system distilled in a batch column

Time	Relative Volatility in the Column, α									
0	3.238945	3.238752	3.238339	3.237547	3.2355995	3.2322937	3.223829	3.20298	3.153673	3.058115
0.04	3.238887	3.23862	3.238099	3.237056	3.2348889	3.2300876	3.218605	3.190123	3.125928	3.02251
0.08	3.238874	3.238583	3.238009	3.236823	3.2342099	3.2279723	3.212138	3.173219	3.096759	2.999475
0.12	3.238864	3.238555	3.237937	3.236618	3.2335865	3.2260524	3.206723	3.161346	3.080729	2.988859
0.16	3.238844	3.238496	3.237769	3.236124	3.2320901	3.2217099	3.195793	3.140916	3.057358	2.97442
0.2	3.238822	3.238429	3.237573	3.235537	3.2303807	3.2171416	3.185604	3.124498	3.040747	2.96437
0.24	3.238808	3.238388	3.237448	3.235171	3.2293529	3.2145405	3.180147	3.116436	3.03108	2.959767
0.28	3.238756	3.238225	3.23697	3.233814	3.225736	3.2059594	3.163726	3.093926	3.012905	2.947534
0.32	3.23872	3.238115	3.236653	3.232947	3.2235476	3.2010864	3.155056	3.08295	3.003553	2.941822
0.36	3.238676	3.237981	3.236276	3.231943	3.2210887	3.1958344	3.146121	3.072159	2.994596	2.93629
0.4	3.238604	3.237768	3.235692	3.230425	3.2174783	3.1884426	3.134161	3.058334	2.983393	2.929319
0.44	3.238512	3.237502	3.234983	3.228653	3.2133841	3.1802903	3.121734	3.044681	2.972699	2.92259
0.48	3.2384	3.237179	3.234133	3.226579	3.208778	3.1715407	3.108874	3.031184	2.962315	2.915969
0.52	3.23826	3.236791	3.233138	3.224185	3.2035869	3.1620584	3.095588	3.017819	2.952232	2.90946
0.56	3.238068	3.236264	3.231823	3.221101	3.1970549	3.1505491	3.08024	3.003059	2.941299	2.902311
0.6	3.237837	3.235631	3.230257	3.217509	3.1896544	3.138009	3.064347	2.988441	2.93067	2.895263
0.64	3.237556	3.234877	3.228404	3.213315	3.1812119	3.1243762	3.047942	2.973911	2.920315	2.8883
0.68	3.237216	3.233984	3.226231	3.208441	3.1717114	3.1095666	3.031065	2.959594	2.910207	2.881407
0.72	3.237216	3.233984	3.226231	3.208441	3.1717114	3.1095666	3.031065	2.959594	2.910207	2.881407
0.76	3.236727	3.232699	3.223167	3.201677	3.1589352	3.0907886	3.010882	2.943205	2.898771	2.873451
0.8	3.236132	3.23111	3.219405	3.193591	3.1441961	3.0703349	2.990333	2.927202	2.887687	2.865583
0.84	3.235398	3.229166	3.214767	3.18381	3.1272096	3.0482852	2.969538	2.911683	2.876973	2.857808
0.88	3.235398	3.229166	3.214767	3.18381	3.1272096	3.0482852	2.969538	2.911683	2.876973	2.857808
0.92	3.234465	3.226742	3.20905	3.171937	3.1077058	3.0248275	2.948872	2.896739	2.866636	2.85017
0.96	3.233293	3.223661	3.201908	3.157659	3.0855408	3.0002305	2.928577	2.88247	2.856691	2.842598

Table 7.2. Relative volatility change for a binary system distilled in a batch column

Time	Relative Volatility in the Column, α								
0	1.0522762	1.0621427	1.0745572	1.0904452	1.1111968	1.1391496	1.1786381	1.2376235	1.3345339	1.522994
0.08	1.060233	1.0732714	1.0889895	1.1090644	1.135699	1.1725089	1.2249458	1.3031469	1.4277388	1.6444457
0.16	1.0681077	1.0838086	1.1027857	1.1269087	1.1586705	1.2019888	1.2617935	1.3478542	1.4782737	1.6942083
0.24	1.0820835	1.1018312	1.1254601	1.1548379	1.1927939	1.2420634	1.3077886	1.3991966	1.5321229	1.7480417
0.32	1.091015	1.1129383	1.138999	1.1710048	1.2106975	1.2628269	1.3307525	1.4228786	1.55291	1.7749849
0.4	1.1023375	1.1268915	1.155204	1.1896269	1.2322514	1.286208	1.3553962	1.4499282	1.5859735	1.806484
0.48	1.1134975	1.1402204	1.1708671	1.2072346	1.2511778	1.3065384	1.3777133	1.473424	1.6119249	1.8370812
0.56	1.1255244	1.1548791	1.1873393	1.2253319	1.2715166	1.3275467	1.4004528	1.4982183	1.639484	1.8720006
0.64	1.1367036	1.1680233	1.2028205	1.2421587	1.2891118	1.3473645	1.4211588	1.5207846	1.6660685	1.9077169
0.72	1.1485878	1.1825652	1.2185648	1.2599255	1.3085956	1.3677541	1.436443	1.5460012	1.6965012	1.9515886
0.8	1.1538502	1.1891414	1.2262039	1.2679951	1.3174898	1.3775018	1.4542029	1.5581317	1.7117269	1.9746554
0.88	1.1602837	1.1966916	1.2353232	1.2780467	1.3281551	1.3893661	1.4673454	1.5733476	1.7313144	2.0054996
0.96	1.17215	1.2107858	1.2511007	1.2960565	1.3481717	1.4113422	1.4922928	1.6032873	1.7715012	2.0728553
1.04	1.1777324	1.2181165	1.2597229	1.3053921	1.3585867	1.4231897	1.5060243	1.6202903	1.7954477	2.1156877
1.12	1.1828447	1.2243916	1.2675844	1.3144537	1.3685725	1.4345327	1.5194686	1.6373722	1.8203254	2.1624705
1.2	1.1879406	1.2302236	1.2744059	1.3227913	1.3782276	1.4456004	1.5327661	1.654706	1.8464017	2.2140512
1.28	1.1925078	1.2362538	1.2811364	1.3303596	1.3871823	1.4563016	1.5459651	1.6724313	1.8741497	2.2718873
1.36	1.1963485	1.2415793	1.2877814	1.3378799	1.3958197	1.466916	1.5591169	1.6906525	1.9038574	2.3371383
1.44	1.200165	1.2460939	1.2935418	1.3451192	1.4044408	1.4770525	1.5723648	1.7094756	1.9360342	2.4136622
1.52	1.2038805	1.2508519	1.2990539	1.3517309	1.4126043	1.4873238	1.5859922	1.7295383	1.9715452	2.5048239
1.6	1.2062229	1.2541476	1.3031609	1.356611	1.4185523	1.494909	1.5963518	1.7454316	2.0010414	2.5853757
1.68	1.2100478	1.2589818	1.3094049	1.3644644	1.4283321	1.507563	1.6140622	1.7737113	2.0575691	2.7602649
1.76	1.2130304	1.2629693	1.3143709	1.3706866	1.4363377	1.5183192	1.6298072	1.8005826	2.1170006	2.9735475
1.84	1.2160225	1.2669797	1.3195688	1.3773472	1.4450846	1.5305337	1.6487687	1.8356362	2.2043635	3.3605658
1.92	1.218953	1.2709608	1.3247993	1.3842399	1.4545065	1.5444523	1.6722319	1.8846106	2.3528533	4.2365735

Table 7.3. Insulin data for four patients

Patient 1		Patient 2		Patient 3		Patient 4	
Time	Insulin	Time	Insulin	Time	Insulin	Time	Insulin
4.25	9.73	6.34	11.2	6.32	10.4	4.23	11.2
14.8	11	19	12.2	12.6	11.7	12.7	10.5
25.5	10	27.4	14.4	21	12.3	27.5	12.3
36.1	11	29.6	17	31.6	12.3	44.4	13
42.5	12.8	38	20.2	40	11.5	59.2	15.3
46.7	14.2	44.4	23.7	48.4	10.9	63.5	17.9
59.5	15	59.2	24.8	59	10.7	72	20.2
70.1	13.1	67.6	21.6	69.5	10.9	86.8	21.7
76.5	12.1	71.8	19.2	73.7	11.7	101	22
85	11	78.2	15.4	82.2	11.7	110	20.2
93.5	12.3	86.6	12.2	86.4	10.9	116	18.4
99.9	13.9	101	10.1	94.8	9.94	120	16.6
106	15	112	7.73	94.8	9.16	131	14.3
114	16.5	118	6.93	101	9.94	146	17.4
121	18.1	131	9.86	111	9.42	150	19.2
123	19.4	139	12.8	122	8.9	163	20.7
127	20.5	148	15.7	132	9.42	182	21
131	21.5	160	17.3	137	10.9	196	20
134	23.1	171	18.9	141	11.7	213	20.7
140	24.2	186	19.2	143	12.8	230	21.5
140	25.2	198	16.8	143	13.8	262	17.1
142	26.3	213	14.4	151	14.1	249	20
144	27.1	219	12.5	158	14.9	268	14.6
155	28.4	228	9.33	168	14.3	273	12.5
161	29.4	238	11.4	174	13.6	285	11.2
170	30	245	13.8	177	12.8	298	14.8
180	30	249	15.2	181	11.7	307	19.2
182	28.9	262	15.4	185	10.9	313	24.1
185	27.8	266	13.3	189	10.4	317	28.4
191	26.8	270	11.7	193	9.42	324	32.8
197	26	277	9.86	200	8.63	330	37.6
202	25.2	281	7.46	208	8.11	336	40.2
206	24.2	289	9.06	219	7.32	349	38.2
210	23.6	296	12.2	225	8.11	351	34.8
216	22.8	300	14.4	234	8.63	360	30.2
223	23.4	302	16.2	242	9.16	364	24.8
225	24.2	308	18.4	246	10.2	372	21.2
227	25.2	312	20.5	253	10.4	376	18.9

Table 7.3. (Continued)

Patient 1		Patient 2		Patient 3		Patient 4	
Time	Insulin	Time	Insulin	Time	Insulin	Time	Insulin
231	26	323	22.1	261	10.7	385	16.4
233	27.1	334	20.2	269	9.68	400	20
246	27.1	338	17.8	278	9.16	406	24.1
255	26	342	15.2	284	8.11	410	29.2
255	24.7	348	12.8	290	7.59	415	33
259	23.9	357	10.6	293	6.8	425	36.4
265	23.1	365	8.53	301	6.02	429	33.5
272	22.6	372	6.93	309	6.28	432	31
278	22.1	382	5.06	318	7.59	436	27.6
285	21.8	401	8	322	9.68	440	23.5
291	21.5	408	11.4	322	12	444	18.9
302	21.5	416	14.9	326	14.6	451	15.8
306	22.8	422	17.8	328	18	459	13.8
310	23.6	429	20	333	21.4	470	12.3
312	24.7	433	22.9	335	24.6	478	14.3
319	25.5	444	24	337	27.4	489	17.6
325	25	452	22.1	343	31.1	495	22
336	25.7	454	20	345	33.7	499	24.3
340	26	460	16.8	347	37.1	508	25.8
344	27.1	465	14.9	358	35.3	520	22.8
346	28.1	471	13	364	32.9	527	18.9
353	28.4	475	11.4	368	29.3	548	20.7
361	28.4	488	10.4	373	27.2	563	20.2
363	27.6	496	10.4	375	24.6	582	21.7
363	26.5	511	12	381	20.9	599	20.2
367	25.5	524	14.4	385	18.8	609	19.2
367	24.4	526	16.5	394	16.7	616	21.7
374	25.5	532	17.8	404	18.3	626	22.5
378	26.5	537	19.4	411	19.6	645	20
384	25.7	547	21.3	415	20.9	654	17.6
391	25	556	18.9	419	22.7	660	15.3
395	24.2	562	16	425	21.9	675	18.4
397	23.6	568	13	430	20.6	686	20.2
399	22.8	581	14.4	432	19.3	690	22.8
402	22.1	589	16.8	434	17.8	703	21.7
406	21	600	18.4	436	16.2	707	20.2
406	19.7	613	16.2	438	14.3	724	20.7
406	18.9	621	14.1	442	13	732	18.7
406	17.8	634	15.2	444	11.5	739	16.4
406	17.1	636	17	446	10.2	751	16.9
408	16	644	18.4	453	9.42	758	18.9

Table 7.3. (Continued)

Patient 1		Patient 2		Patient 3		Patient 4	
Time	Insulin	Time	Insulin	Time	Insulin	Time	Insulin
408	15.2	647	20	459	7.85	764	21.2
414	14.2	651	21.3	468	9.68	772	22.8
412	13.1	657	18.9	472	11.5	792	23.5
416	12.6	659	17	472	13.3	806	23.5
421	12.1	663	15.2	474	14.9	828	22
425	11.3	670	13	480	17	828	20.2
431	10.5	674	15.7	480	19.1	834	18.2
440	10.2	680	18.1	482	21.9	844	15.8
448	11	685	20.2	484	25.3	849	14.1
455	11.5	689	23.2	489	29	864	14.1
463	12.6	689	26.6	491	31.9	874	16.9
467	13.4	695	28.5	495	35	880	18.9
472	14.2	697	30.6	497	38.4	883	21
478	15	699	33	499	41.3	887	23
480	16.3	712	31.4	503	39	902	22.5
482	17.3	716	30.1	510	36.3	912	20.7
487	18.1	716	28.8	510	35	925	19.4
487	19.4	718	25.8	512	32.9	933	16.9
491	20.5	723	22.9	512	31.9	940	16.4
495	21.5	725	21	516	28.7	950	19.4
495	22.6	729	18.9	520	24	955	22
499	23.4	729	17	522	18.5	961	24.8
499	24.7	742	16	529	13.6	963	27.1
510	24.4	752	16.8	537	10.2	972	28.4
512	23.6	761	15.7	550	12.8	982	29.7
516	22.6	771	14.1	558	14.1	999	28.4
518	21.5	778	12.5	565	16.2	1010	26.6
523	20.7	782	11.4	565	17.5	1010	24.6
523	19.7	795	11.4	569	20.6	1020	22.8
525	18.9	814	10.9	573	22.7	1030	22.3
525	18.1	828	11.7	575	25.1	1040	21.5
527	17.1	833	14.4	586	25.9	1050	20.2
529	16.3	837	17.3	592	24	1050	17.9
531	15.2	841	19.7	596	22.7	1060	16.1
536	14.4	845	21.8	596	20.6	1070	14.3
538	13.1	847	24.5	600	18.5	1080	14.3
542	12.3	847	26.9	605	16.7	1090	16.1
546	11.5	856	28.2	613	14.6	1100	18.7
550	11	864	28.8	628	13.3	1100	21.2
559	11.5	875	27.2	634	11.5	1110	24.1

Table 7.3. (Continued)

Patient 1		Patient 2		Patient 3		Patient 4	
Time	Insulin	Time	Insulin	Time	Insulin	Time	Insulin
565	12.3	881	25.8	640	9.94	1110	26.6
565	13.4	890	25	649	8.9	1130	27.9
574	14.2	898	25.8	659	8.11	1140	26.6
578	15.5	907	26.9	672	8.37	1150	25.3
582	16.3	917	28.2	683	9.94	1150	24.1
584	17.1	928	28.8	689	12.3	1160	22.8
589	17.6	940	26.1	693	14.3	1180	25.3
597	17.8	955	23.2	697	15.9	1180	27.1
604	18.4	962	20	702	17.5	1190	28.7
606	17.1	970	18.1	706	19.1	1210	28.4
610	16	987	17.3	712	20.6	1220	28.9
614	14.7	1000	16.5	718	20.4	1230	27.1
616	13.6	1020	16.5	725	19.3	1230	24.6
621	12.6	1030	18.6	729	18	1240	23
631	11.8	1040	20.8	731	16.2	1240	21
642	11.8	1050	22.6	737	13.3	1250	18.4
652	11	1060	20.8	750	15.4	1260	21
661	11.5	1070	18.4	763	16.7	1260	23.3
667	12.6	1070	16.5	771	18.3	1260	25.6
674	13.6	1080	15.2	777	19.8	1270	27.9
682	13.9	1080	14.1	790	22.5	1270	29.2
691	14.4	1090	16.2	803	22.7	1280	29.2
695	15.7	1100	17.8	815	22.7	1280	27.4
699	17.1	1110	18.1	818	24.3	1290	25.1
701	18.6	1120	16.8	822	25.9	1290	22
704	19.7	1130	15.2	824	27.2	1300	20
706	20.7	1140	13.8	828	28.5	1300	20.5
706	22.3	1150	12.8	834	28.5	1320	19.2
708	23.9	1160	11.7	845	27.7	1330	20.7
712	25	1170	10.4	847	26.7	1330	22.8
712	26	1170	9.33	853	26.1	1330	24.6
714	27.6	1190	8.26	862	25.6	1340	26.9
714	28.9	1210	7.73	872	25.9	1350	26.6
716	30.2	1220	9.6	883	25.9	1360	24.6
721	31.8	1230	10.9	887	26.9	1370	25.1
723	32.8	1240	13	891	27.7	1390	23.5
725	34.2	1250	15.4	893	28.7	1400	22.8
731	35.7	1260	17.6	900	29.8	1400	20.7
738	35.2	1270	19.7	906	28.7	1410	19.2
742	34.4	1280	21.6	910	26.9	1420	17.6

Table 7.3. (Continued)

Patient 1		Patient 2		Patient 3		Patient 4	
Time	Insulin	Time	Insulin	Time	Insulin	Time	Insulin
748	35.2	1290	19.4	917	24.6	1430	19.4
757	35.7	1290	17.8	921	22.5		
763	35.5	1300	15.4	933	20.9		
767	35	1310	13.8	942	18.8		
767	33.9	1320	12	944	17.8		
769	33.1	1330	10.4	950	15.4		
774	32.3	1340	13	957	13.6		
774	31.8	1350	15.2	959	12.3		
776	30.7	1360	17.3	967	10.7		
776	29.7	1360	19.4	974	10.7		
780	28.6	1360	21.6	976	12		
782	27.1	1380	21.6	980	13.3		
784	25.5	1380	20	984	14.3		
789	24.2	1390	18.4	988	15.4		
789	23.1	1410	17.3	993	16.4		
799	22.1	1410	19.4	1000	15.4		
812	25.7	1420	21.3	1010	14.6		
812	26.5	1430	23.7	1010	13.3		
816	27.3	1430	25.8	1020	13.8		
806	22.8			1030	14.3		
808	23.6			1040	14.1		
808	24.7			1040	12.5		
816	28.1			1050	13.6		
823	29.2			1060	15.1		
829	28.4			1060	16.4		
831	27.6			1070	18.5		
833	27.1			1070	20.4		
838	25.7			1080	21.9		
838	25			1080	23.5		
838	24.2			1080	25.1		
840	23.4			1090	26.4		
844	22.8			1090	25.6		
848	22.6			1100	24.6		
855	22.3			1100	23.8		
861	23.1			1110	22.5		
867	23.1			1120	20.6		
869	22.1			1120	19.6		
872	21			1130	17.5		
872	20			1130	15.1		
874	18.9			1140	13.6		

Table 7.3. (Continued)

| Patient 1 | | Patient 2 | | Patient 3 | | Patient 4 | |
Time	Insulin	Time	Insulin	Time	Insulin	Time	Insulin
876	18.1			1140	12		
878	17.3			1150	14.3		
878	16			1160	16.7		
878	15			1160	19.1		
884	14.4			1160	21.7		
891	14.2			1160	25.6		
893	15			1170	31.1		
901	16.5			1170	36.1		
906	17.8			1180	38.2		
910	19.2			1190	35.8		
912	20.5			1190	34.2		
916	21.5			1200	32.7		
925	22.6			1200	31.4		
933	21.8			1200	26.9		
938	20.5			1210	23.8		
940	19.4			1210	20.6		
942	17.8			1220	17		
944	16.5			1220	14.9		
948	15.2			1230	13.3		
952	13.9			1240	16.2		
957	12.3			1250	18.8		
961	11			1250	21.2		
965	10.2			1260	23		
969	9.47			1260	25.1		
974	8.42			1270	26.4		
984	7.63			1280	24.6		
993	9.47			1290	22.7		
1010	11			1290	20.6		
1010	12.8			1300	18.5		
1020	14.2			1300	16.2		
1020	15.5			1310	14.3		
1030	16.8			1320	13		
1030	18.6			1330	11.2		
1040	20.2			1330	9.42		
1040	21.3			1350	8.37		
1050	22.3			1360	9.42		
1060	24.2			1370	10.9		
1070	24.2			1370	12.5		
1080	22.6			1380	14.6		

Table 7.3. (Continued)

Patient 1		Patient 2		Patient 3		Patient 4	
Time	Insulin	Time	Insulin	Time	Insulin	Time	Insulin
1080	21.3			1390	16.7		
1080	20			1390	18.8		
1090	21.5			1390	21.2		
1100	22.6			1390	23.5		
1100	23.6			1400	26.4		
1110	25			1400	28.7		
1110	26.3			1400	29.8		
1110	28.1			1410	32.1		
1120	29.7			1420	30.3		
1130	31.3			1420	29		
1130	29.2			1420	27.2		
1140	27.8			1430	25.6		
1140	26			1430	24.8		
1150	24.7						
1150	23.1						
1150	21.8						
1170	23.4						
1180	25.2						
1190	23.9						
1200	23.6						
1210	23.4						
1220	24.4						
1230	25.7						
1230	27.6						
1240	28.6						
1240	30						
1240	31						
1240	32.3						
1240	33.9						
1240	36.3						
1240	38.1						
1250	39.4						
1250	41						
1250	42.6						
1250	44.4						
1250	46						
1260	47.6						
1270	48.4						
1280	48.9						
1280	47.3						
1280	46						
1290	44.4						

Table 7.3. (Continued)

Patient 1		Patient 2		Patient 3		Patient 4	
Time	Insulin	Time	Insulin	Time	Insulin	Time	Insulin
1300	41.5						
1300	38.9						
1310	36.3						
1320	34.2						
1330	32.3						
1340	34.2						
1340	35.5						
1350	36.8						
1360	35.2						
1370	34.4						
1370	32.8						
1380	31						
1380	30						
1390	28.4						
1400	26.8						
1410	25.7						
1410	24.4						
1420	22.3						
1420	20.5						
1430	20.5						

Bibliography

Aris R. (1961), *The Optimal Design of Chemical Reactors*, Academic Press, London.

Bellman R. (1957), *Dynamic Programming*, Princeton University Press, Princeton, NJ.

Betts J. T. (2001), *Practical Method for Optimal Control using Nonlinear Programming*, SIAM, Philadelphia.

Boltyanskii V. G., R. V. Gamkrelidze, and L. S. Pontryagin (1956), On the theory of optimum processes (in Russian), *Doklady Akad. Nauk SSSR*, **110**, no. 1.

Converse A. O. and G. D. Gross (1963), Optimal distillate policy in batch distillation, *Industrial Engineering Chemistry Fundamentals*, **2** , 217.

Diwekar U. M. (1995), *Batch Distillation: Simulation, Optimal Design and Control*, Taylor & Francis Publishers, Washington DC.

Diwekar U. M. (1992), Unified approach to solving optimal design-control problems in batch distillation, *AIChE Journal*, **38**, 1551.

Diwekar U. M. (1996), *User's Manual for MultiBatch$\underline{DS}^{TM}$* , BPRC, Pittsburgh.

Diwekar U. M., R. K. Malik, and K. P. Madhavan (1987), Optimal reflux rate policy determination for multicomponent batch distillation columns, *Computers and chemical Engineering*, **11**, 629.

Dixit A. K. and R. S. Pindyck (1994), *Investment Under Uncertainty*, Princeton University Press, Princeton, NJ.

Fan L. T. (1966), *The Continuous Maximum Principle*, John Wiley & Sons, New York.

Gilliland E. R. (1940), Multicomponent rectification. Estimation of the number of theoretical plates as a function of reflux, *Industrial Engineering Chemistry*, **32**, 1220.

Kirk D. E. (1970), *Optimal Control Theory An Introduction*, Prentice Hall, Englewood Cliffs, NJ.

Merton R. C., and P. A. Samuelson (1990), *Continuous-Time Finance*, B. Blackwell, Cambridge, MA.

Naf U. G. (1994) Stochastic simulation using gPROMS, *Computers and chemical Engineering*, **18**, S743.

Pontryagin L. S. (1957), Basic problems of automatic regulation and control (in Russian), *Izd-vo Akad Nauk SSSR*.

Pontryagin L. S. (1956), Some mathematical problems arising in connection with the theory of automatic control system (in Russian), Session of the Academic Sciences of the USSR on Scientific Problems of Automatic Industry, October 15.

Rico-Ramirez V. and U. Diwekar (2004), Stochastic maximum principle for optimal control under uncertainty,*Computers and Chemical Engineering*, **28**, 2845.

Rico-Ramirez V., U. Diwekar, and B. Morel (2003), Real option theory from finance to batch distillation, *Computers and Chemical Engineering*, **27**, 1867.

Shastri Y. and U. Diwekar. Optimal control of lake pH for mercury bioaccumulation control, (2008), *Ecological Modelling*, **216**,1.

Shastri Y. and U. Diwekar. (2009) Freshwater ecosystem conservation and management: A control theory approach. Chapter in *Freshwater Ecosystems: Biodiversity, Management and Conservation*, Nova Science Publishers.

Simon C., G. Brandenberger, and M. Follenius (1987), Ultradian oscillations of plasma glucose, insulin and c-peptide in man during continuous enteral nutrition, *Journal of Clinical Endocrinology and Metabolism*, **64 (4)**, 669.

Thompson G. L. and S. P. Sethi (1994), *Optimal Control Theory*, Martinus Nijhoff, Boston.

Troutman J. L. (1995), *Variational Calculus and Optimal Control*, Second Edition, Springer, New York.

Ulas S. and U. Diwekar (2004),Thermodynamic uncertainties in batch processing and optimal control, *Computers and Chemical Engineering*, **28**, 2245.

Appendix A

Details of Glass Property Constraints

Notation

C_1	Bound for Crystal 1 -3.0
C_2	Bound for Crystal 2 -0.08
C_3	Bound for Crystal 3 -0.225
C_4	Bound for Crystal 4 -0.18
C_5	Bound for Crystal 5 -0.18
k_{min}	Lower limit for conductivity -18
k_{max}	Upper limit for conductivity -50
μ_{min}	Lower limit for viscosity (PaS) -2.0
μ_{max}	Upper limit for viscosity (PaS) -10.0
D^{PCT}_{max}	Max release rate (product consistency test) (g per m_2) -10.0
D^{MCC}_{max}	Max release rate (materials characterization center) (g per m^2) -28.0
μ^i_a	Linear coefficients of viscosity model
μ^{ij}_b	Cross term coefficients of viscosity model
k^i_a	Linear coefficients of electrical conductivity model
k^{ij}_b	Cross term coefficients of electrical conductivity model
Dp^i_a	Linear coefficients of durability (PCT) model (for Boron)
Dp^{ij}_b	Cross term coefficients of durability (PCT) model for Boron
Dm^i_a	Linear coefficients of durability (MCC) model (for Boron)
Dm^{ij}_b	Cross term coefficients of durability (MCC) model (for Boron)

© Springer Nature Switzerland AG 2020

U. M. Diwekar, *Introduction to Applied Optimization*, Springer
Optimization and Its Applications 22,
https://doi.org/10.1007/978-3-030-55404-0

1. Component Bounds:
   (a) $0.42 \leq p^{(SiO_2)} \leq 0.57$
   (b) $0.05 \leq p^{(B_2O_3)} \leq 0.20$
   (c) $0.05 \leq p^{(Na_2O)} \leq 0.20$
   (d) $0.01 \leq p^{(Li_2O)} \leq 0.07$
   (e) $0.0 \leq p^{(CaO)} \leq 0.10$
   (f) $0.0 \leq p^{(MgO)} \leq 0.08$
   (g) $0.02 \leq p^{(Fe_2O_3)} \leq 0.15$
   (h) $0.0 \leq p^{(Al_2O_3)} \leq 0.15$
   (i) $0.0 \leq p^{(ZrO_2)} \leq 0.13$
   (j) $0.01 \leq p^{(other)} \leq 0.10$
2. Five glass crystallinity constraints:
   (a) $p^{(SiO_2)} > p^{(Al_2O_3)} * C_1$
   (b) $p^{(MgO)} + p^{(CaO)} < C_2$
   (c) $p^{(Fe_2O_3)} + p^{(Al_2O_3)} + p^{(ZrO_2)} + p^{('Other')} < C_3$
   (d) $p^{(Al_2O_3)} + p^{(ZrO_2)} < C_4$
   (e) $p^{(MgO)} + p^{(CaO)} + p^{(ZrO_2)} < C_5$
3. Solubility Constraints:
   (a) $p^{(Cr_2O_3)} < 0.005$
   (b) $p^{(F)} < 0.017$
   (c) $p^{(P_2O_5)} < 0.01$
   (d) $p^{(SO_3)} < 0.005$
   (e) $p^{(Rh_2O_3 + PdO + Ru_2O_3)} < 0.025$
4. Viscosity Constraints:
   (a) $\sum_{i=1}^{n} \mu_a^i * p^{(i)} + \sum_{j=1}^{n} \sum_{i=1}^{n} \mu_b^{ij} * p^{(i)} * p^{(j)} > \log(\mu_{min})$
   (b) $\sum_{i=1}^{n} \mu_a^i * p^{(i)} + \sum_{j=1}^{n} \sum_{i=1}^{n} \mu_b^{ij} * p^{(i)} * p^{(j)} < \log(\mu_{max})$
5. Conductivity Constraints:
   (a) $\sum_{i=1}^{n} k_a^i * p^{(i)} + \sum_{j=1}^{n} \sum_{i=1}^{n} k_b^{ij} * p^{(i)} * p^{(j)} > \log(k_{min})$
   (b) $\sum_{i=1}^{n} k_a^i * p^{(i)} + \sum_{j=1}^{n} \sum_{i=1}^{n} k_b^{ij} * p^{(i)} * p^{(j)} < \log(k_{max})$
6. Dissolution rate for boron by PCT test (DissPCTbor):
   $\sum_{i=1}^{n} Dp_a^i * p^i + \sum_{j=1}^{n} \sum_{i=1}^{n} Dp_b^{ij} * p^{(i)} * p^{(j)} < \log(D_{max}^{PCT})$
7. Dissolution rate for boron by MCC test (DissMCCbor):
   $\sum_{i=1}^{n} Dm_a^i * p^i + \sum_{j=1}^{n} \sum_{i=1}^{n} Dm_b^{ij} * p^{(i)} * p^{(j)} < \log(D_{max}^{MCC})$

Waste Composition Data

	Fractional Composition of Wastes						
Comp.	AY-102	AZ-101	AZ-102	SY-102	SY-101	SY-103	B-103
SiO_2	0.072	0.092	0.022	0.020	0.000	0.019	0.011
B_2O_3	0.026	0.000	0.006	0.003	0.000	0.000	0.000
Na_2O	0.105	0.264	0.120	0.154	0.300	0.230	0.100
Li_2O	0.000	0.000	0.000	0.000	0.000	0.000	0.000
CaO	0.061	0.012	0.010	0.030	0.007	0.006	0.000
MgO	0.040	0.000	0.003	0.012	0.000	0.001	0.000
Fe_2O_3	0.328	0.323	0.392	0.133	0.000	0.039	0.155
Al_2O_3	0.148	0.157	0.212	0.318	0.659	0.546	0.214
ZrO_2	0.002	0.057	0.063	0.002	0.000	0.001	0.000
Other	0.217	0.096	0.173	0.328	0.034	0.159	0.520
Total	1.000	1.000	1.000	1.000	1.000	1.000	1.000
Cr_2O_3	0.016	0.007	0.005	0.089	0.002	0.116	0.000
F	0.006	0.001	0.001	0.005	0.002	0.001	0.000
P_2O_5	0.042	0.001	0.021	0.088	0.013	0.005	0.037
SO_3	0.001	0.018	0.009	0.027	0.005	0.002	0.007
NobMet	0.000	0.000	0.000	0.000	0.000	0.000	0.000
Mass	59772	40409	143747	359609	167510	185990	6170

	Fractional Composition of Wastes $w^{(i)}/g^{(i)}$						
Comp.	BY-104	BY-110	C-103	C-105	C-106	C-108	C-109
SiO_2	0.030	0.040	0.412	0.359	0.437	0.001	0.001
B_2O_3	0.000	0.000	0.000	0.000	0.000	0.000	0.000
Na_2O	0.082	0.089	0.006	0.012	0.014	0.010	0.007
Li_2O	0.000	0.000	0.000	0.000	0.000	0.000	0.000
CaO	0.141	0.046	0.041	0.044	0.046	0.000	0.737
MgO	0.000	0.000	0.028	0.026	0.031	0.000	0.000
Fe_2O_3	0.067	0.051	0.338	0.064	0.214	0.206	0.003
Al_2O_3	0.344	0.462	0.057	0.372	0.168	0.693	0.013
ZrO_2	0.007	0.003	0.043	0.004	0.008	0.032	0.000
Other	0.330	0.309	0.075	0.119	0.082	0.058	0.238
Total	1.000	1.000	1.000	1.000	1.000	1.000	1.000
Cr_2O_3	0.000	0.000	0.002	0.005	0.004	0.002	0.000
F	0.001	0.001	0.000	0.000	0.000	0.000	0.000
P_2O_5 0.016	0.022	0.013	0.012	0.031	0.047	0.003	0.000
SO_3	0.002	0.003	0.000	0.002	0.000	0.000	0.000
NobMet	0.000	0.000	0.000	0.000	0.000	0.000	0.000
Mass	155473	103492	85211	207127	367165	46919	53271

	Fractional Composition of Wastes						
	C-111	C-112	S-102	SX-106	TX-105	TX-118	U-107
SiO_2	0.002	0.001	0.000	0.033	0.010	0.060	0.008
B_2O_3	0.000	0.000	0.000	0.000	0.000	0.000	0.000
Na_2O	0.011	0.005	0.337	0.280	0.168	0.425	0.038
Li_2O	0.000	0.000	0.000	0.000	0.000	0.000	0.000
CaO	0.426	0.593	0.000	0.000	0.000	0.000	0.000
MgO	0.000	0.000	0.000	0.000	0.000	0.000	0.000
Fe_2O_3	0.042	0.002	0.023	0.102	0.167	0.026	0.077
Al_2O_3	0.256	0.097	0.582	0.388	0.595	0.240	0.650
ZrO_2	0.007	0.000	0.000	0.000	0.000	0.000	0.000
Other	0.256	0.302	0.058	0.197	0.060	0.250	0.228
Total	1.000	1.000	1.000	1.000	1.000	1.000	1.000
Cr_2O3	0.000	0.000	0.024	0.020	0.000	0.000	0.000
F	0.000	0.000	0.000	0.001	0.000	0.004	0.001
P_2O_5	0.012	0.005	0.006	0.038	0.002	0.159	0.020
SO_3	0.000	0.000	0.000	0.003	0.001	0.009	0.001
NobMet	0.000	0.000	0.000	0.000	0.000	0.000	0.000
Mass	24485	65673	36537	45273	42200	412495	11504

Appendix B

Nonlinear Models for the Mercury Treatment

A real-world case study of mercury trading presented starting, with the Linear Programming . In Chapters 3, 4, and 5, we have used nonlinear models for the cost of different technologies. Details of these models are presented here. The parameters that are considered to be stochastic are identified. Chapter 2 identified three technologies to treat water waste containing mercury, namely: coagulation and filtration, granular activated carbon adsorption, and ion exchange. However, the cost estimates through these models are only approximations. The following sections present the model details for the three nonlinear process models.

Coagulation and Filtration

Industrial experience shows that filtration costs can be comparable to coagulation. Since the cost models for the coagulation process could not be found, the model incorporated the filtration costs in this work.

The input parameters required to compute the cost of microfiltration (MF) are:

- Design product flow rate (gallons per day)
- Plant availability (%)
- Microfilters system equipment cost ($)
- Cost per MF membrane ($)
- MF modular system flow rate (gallons per minute)
- Number of membranes per microfilter
- Pump efficiency (%)

© Springer Nature Switzerland AG 2020
U. M. Diwekar, *Introduction to Applied Optimization*, Springer
Optimization and Its Applications 22,
https://doi.org/10.1007/978-3-030-55404-0

- Motor efficiency (%)
- Design feed pressure (psi)
- Backflush pressure (psi)
- Backwash intervals (minutes)
- Backwash and backflush duration (minutes)

The operation and maintenance cost inputs are:

- Electricity rate ($/kwh)
- Chemical costs (sodium hypochlorite, $/L)
- Design dosage (mg/L)
- Specific gravity of sodium hypochlorite
- Solution concentration (%)
- Membrane life (year)
- Staff days/day
- Labor rate (salary and benefits, $/hr)
- Amortization time (year)
- Interest rate (%)

Process Flow Calculations

MF feed flow is the total feed flow to the microfiltration plant. It is calculated by:

$$MFF = \frac{MFP}{Y}$$

where

- MFF = Microfiltration feed flow (L/sec)
- MFP = Microfiltration product flow (L/sec)
- Y = Recovery rate

MF reject flow (MFR (L/sec)) is the amount of water used for the backwash and cleaning of the membranes. It is calculated by:

$$MFR = \frac{BBD * BBF}{BI}$$

where

- BBD = backwash and backflush duration (sec)
- BBF = backwash flow rate (L/sec)
- BI = backwash interval (sec)

Recovery rate (R) is calculated by:

$$R = \frac{MFP}{MFF}$$

Feed pump brake horsepower (HP) is calculated by:

$$HP = \frac{MFF * DFP * 2.31}{PP\%.3960}$$

where

- DFP = design feed pressure (psi)
- PP% = pump efficiency (%)
- 2.31 = conversion factor for feet of vertical head of water per lb/in^2
- 3960 = another English-Metric conversion factor.

Feed pump kilowatt-hour (kWh) is calculated by:

$$kWh = \frac{MFF * DFP * 2.31 * 0.00315}{PP\% * M\% * 1000}$$

where

- M% = motor efficiency (%)
- 0.00315 = conversion factor for consumption of electrical energy

The building area in the square meter is estimated to be 1.23% of the design product flow rate in cubic meters per day.

Capital Cost Estimation

Direct capital costs are the sum of microfilters, building, MF installation, miscellaneous, plant interconnecting piping, engineering. These cost elements are discussed below:

- Microfilters: The actual price for microfilters is obtained from membrane manufacturers. The price will vary upon the type of microfilters and quantities involved. The total microfilters cost is estimated as the cost per skid unit times the number of units.
- Building: The building cost is estimated at $1076 per square meter times the total building area in square meters.
- MF installation: The microfilter installation cost is estimated at $70,000 per unit for a large system (at 37.85 L/s flow rate).
- Miscellaneous: This cost includes any miscellaneous items needed to complete the project. It is estimated 5% of the total microfilter cost.
- Plant interconnecting piping: This cost is estimated at 5% of the sum of total microfilter and miscellaneous costs.
- Engineering: Engineering cost is estimated at 10% of the sum of total microfilter and miscellaneous costs.

Indirect Capital Cost

The indirect capital costs are the sum of:

- Interest during construction (6% of the total direct capital cost)
- Contingencies (20% of the total direct capital cost)
- A&E fees and project management (10% of the total direct capital cost)
- Working capital (4% of the total direct capital cost)

Operation and Maintenance Cost Estimation

Operation and maintenance costs include:

- Electricity
- Labor
- Chemicals (sodium hypochlorite)
- Membrane replacement
- Cleaning chemicals
- Repairs and replacement and miscellaneous

Total annual cost equals the capital recovery cost plus the total operation and maintenance costs. These major O&M cost elements are discussed below:

- Electricity: Electricity cost is the total kilowatt-hour for the feed pump and backflush pump times the electricity cost ($/kWh).
- Labor: This cost is estimated by the number of staff days times the going rate per day.
- Chemicals: The cost of Sodium hypochlorite for disinfection is estimated based on the correlated formula from the Microfiltration membrane quotation data:

$$SHC * (0.0025 * MFP - 333.33)$$

where SHC = sodium hypochlorite cost ($/L)
- Membrane replacement: The cost is estimated by:

$$\frac{Elements * \$/Element}{Membrane\ life}$$

- Cleaning chemicals: Sodium hypochlorite cost is estimated based on the correlated formula from the Microfiltration membrane quotation data:

$$(0.00005 * MFP + 66.67) * SHC$$

- Repairs and replacement and misc.: The cost for repairs and replacements assumed to be 0.5% of the total direct capital cost.

For the modeling of the stochastic cost model, uncertain parameters in the coagulation and filtration model are cost per membrane, electricity rate, sodium hypochlorite, and membrane life.

Granular Activated Carbon Adsorption

The capital, as well as the operating costs, are based entirely on flow rate and bed life. Costs are estimated using relationships derived from cost data in the 1979 EPA report. It is apparent from this data that there is a change in size versus cost relationship at $4000\,\mathrm{m}^3/\mathrm{day}$. Capital costs for GAC are fairly constant with respect to capacity until a production level of $4000\,\mathrm{m}^3/\mathrm{day}$. Above this level, there are different cost curves for a bed life of 3, 6, and 12 months. Regeneration costs are not included. The cost parameter used in these equations, x, is the volumetric flow rate in $\mathrm{m}^3/\mathrm{day}$. The composition of the water is not considered. The cost equations are given below. Here, CC represents the capital cost, and OM represents the operation and maintenance cost.

- 3, 6, or 12 months of bed life, capacity $\leq 4000\,\mathrm{m}^3/\mathrm{day}$

$$CC = 9875 * x^{(1-0.4596)}$$

- 12 months of bed life:

$$OM_{\leq 4000} = 2631.18 * x^{(1-0.4706)}$$
$$CC_{>4000} = 1948.8 * x^{(1-0.2569)}$$
$$OM_{>4000} = 225.42 * x^{(1-0.1692)}$$

- 6 months of bed life:

$$OM_{\leq 4000} = 2089.46 * x^{(1-0.4187)}$$
$$CC_{>4000} = 150 * x$$
$$OM_{>4000} = 235.91 * x^{(1-0.15)}$$

- 3 months of bed life:

$$OM_{\leq 4000} = 1563.45 * x^{(1-0.3463)}$$
$$CC_{>4000} = 200 * x$$
$$OM_{>4000} = 515.91 * x^{(1-0.203)}$$

Ion Exchange

The model provides a cost estimation for an ion exchange unit based on available design parameters. Data required from the model calculations include:

- Desired flow rate (L/sec)
- Equivalents/L of Cation $> +1$ in water (Equiv/L)
- Equivalents/L of Anion > -1 in water (Equiv/L)

Parameters with default values for the process are given below.

- Desired run cycle: 7 days
- Resin expansion coefficient: 200%
- Cost factor for pressure: 1
- Aspect ratio: 2 (height/diameter)
- Cost of NaCl: 0.02 \$/Kg

Various resin parameters with default values are given below.

- Required service flow rate: Range 16–40 L/(hr*L resin)
- Cation equivalents/L of Resin: 1.9 Equiv/L
- Anion equivalents/L of Resin: 1.4 Equiv/L
- Resin price: 6700 \$/m^3
- Volume NaCl/volume resin for regeneration: 483 Kg/m^3
- Regeneration fluid concentration: 10%

The various cost calculations are given below.

Resin Medium

The minimum resin volume(m^3) is calculated by:

$$\text{Min resin Volume}(m^3) = \frac{\text{Desired flow rate (L/s)}}{\text{Service flow rate (L/hr*Lresin)}}$$

Time until resin exhaustion (days) is calculated by:

$$\text{Time until resin exhaustion (days)} = \frac{\text{MRV * (EQC + EQA)}}{\text{FR * (ECR + EAR)}}$$

where

- MRV = minimum resin volume, m^3
- EQC = Equivalents/L of Cation ¿ +1 in water, Equiv/L
- EQA = Equivalents/L of Anion ¿ -1 in water, Equiv/L
- ERC = Cation Equivalents/L of Resin, Equiv/L
- EAR = Anion Equivalents/L of Resin, Equiv/L
- FR = Desired flow rate (L/s)

An 'if' statement is built-in for the resin volume required to meet exhaustion time. It states that if time until resin exhaustion is greater than the desired run cycle, then the resin volume required to meet exhaustion time is equal to the minimum resin volume. Otherwise, the resin volume required to meet exhaustion time is calculated by:

$$RVET = \frac{RC*FR*(EQC+EQA)}{(ECR + EAR)}$$

where

- RVET = resin volume required to meet exhaustion time, days
- RC = desired run cycle, days

Resin manufacturers recommend an expansion coefficient of two to provide ample room for the resin to expand during upflow regeneration.

Total Vessel Volume (TVV) is calculated by:

$$TW = RVET * \text{Resin expansion coefficient}$$

Resin Cost (RC) is calculated by:

$$RC=MRV*RP$$

where RP = nominal resin price, $/m^3

Vessel Cost

The fiber glass pressure vessel cost is calculated by the following formula:

$$\text{Log}(\$) = 3.44609 + 0.561757 * \text{Log}(TVV)$$

Regeneration

NaCl is used resin regeneration. Amount of NaCl required is calculated by the following equation:

$$\text{NaCl required} = \rho_{NaCl} * RVET$$

where ρ_{NaCl} = density of NaCl, kg/m^3

The total chemical cost per year is calculated by:

$$NaCl_{required} * NaCl_{cost} * \frac{365}{DRC + I}$$

where

- $NaCl_{cost}$= sodium chloride cost, $/kg$
- DRC = desired run cycle, days
- 365 = days per year

Storage tank cost is calculated by:

$$\text{Tank Cost} = 0.1427X^3 - 5.6691X^2 + 257.56X - 467.45$$

where X is the tank volume in m3. This formula is developed from the Snyder cone bottom tank, HDLPE model tank prices.

Regeneration, and Backwashing Pump

Construction cost and O&M cost formulas for regeneration and backwashing pump are developed from the 1979 EPA report.
    Construction cost (CC):

$$CC = 36000 + 1254.21X - 0.1212X^2$$

Operating and Maintenance cost (O&M):

$$\text{O+M} = 73.3X^{0.75} + 2200$$

where X is the filter area in m^2.

    Total construction costs include resin cost, resin operating tank cost, storage tank cost, and regeneration and backwashing pump cost.

Index

© Springer Nature Switzerland AG 2020
U. M. Diwekar, *Introduction to Applied Optimization*, Springer
Optimization and Its Applications 22,
https://doi.org/10.1007/978-3-030-55404-0

Printed in the United States
by Baker & Taylor Publisher Services